Sport Mechanics for Coaches

Second Edition

Vanessa
Brooks

Sport Mechanics for Coaches

Second Edition

Gerry Carr
University of Victoria

Human Kinetics

Library of Congress Cataloging-in-Publication Data

Carr, Gerald A., 1936-
Sport mechanics for coaches / Gerry Carr.--2nd ed.
p. cm.
"This book is a revised edition of Mechanics of Sports, published in 1997 by Human Kinetics"--T.p. verso.
Includes bibliographical references and index.
ISBN 0-7360-3972-4 (Soft Cover)
1. Kinesiology. 2. Human mechanics. 3. Sports--Physiological aspects. I. Carr, Gerald A., 1936- Mechanics of sport. II. Title.
QP303.C375 2004
612'.044--dc21
2004007917

ISBN-10: 0-7360-3972-4

ISBN-13: 978-0-7360-3972-7

This book is a revised edition of *Mechanics of Sport,* published in 1997 by Human Kinetics.

Acquisitions Editor: Scott Parker; **Developmental Editors:** Anne Rogers and Marni Basic; **Assistant Editors:** Amanda S. Ewing and Jennifer L. Davis; **Copyeditor:** Patricia MacDonald; **Proofreader:** Pam Johnson; **Permission Manager:** Dalene Reeder; **Graphic Designer:** Robert Reuther; **Graphic Artist:** Kathleen Boudreau-Fuoss; **Photo Manager:** Dan Wendt; **Cover Designer:** Keith Blomberg; **Photographer (cover):** © Sport The Library/Robb Cox; **Art Manager:** Kelly Hendren; **Illustrators:** Argosy and Paul To; **Printer:** United Graphics

Copies of this book are available at special discounts for bulk purchase for sales promotions, premiums, fund-raising, or educational use. Special editions or book excerpts can also be created to specifications. For details, contact the Special Sales Manager at Human Kinetics.

Printed in the United States of America 10 9 8 7 6 5 4

Human Kinetics
Web site: www.HumanKinetics.com

United States: Human Kinetics
P.O. Box 5076, Champaign, IL 61825-5076
800-747-4457
e-mail: humank@hkusa.com

Canada: Human Kinetics
475 Devonshire Road Unit 100, Windsor, ON N8Y 2L5
800-465-7301 (in Canada only)
e-mail: orders@hkcanada.com

Europe: Human Kinetics
107 Bradford Road, Stanningley, Leeds LS28 6AT, United Kingdom
+44 (0) 113 255 5665
e-mail: hk@hkeurope.com

Australia: Human Kinetics
57A Price Avenue, Lower Mitcham, South Australia 5062
08 8372 0999
e-mail: liaw@hkaustralia.com

New Zealand: Human Kinetics
Division of Sports Distributors NZ Ltd.
P.O. Box 300 226 Albany, North Shore City, Auckland
0064 9 448 1207
e-mail: info@humankinetics.co.nz

To all
who want to understand
how superior sport techniques
are based on the best use
of scientific concepts and natural laws.

CONTENTS

PREFACE

I am delighted to have been given the opportunity by Human Kinetics to write a second edition of *Sport Mechanics for Coaches.* The excellent response to the first edition that I have received from coaches, physical educators, and students in my biomechanics classes at the University of Victoria has been very encouraging. At the same time, suggestions for improvements have been a stimulus for writing a second edition.

You will notice that many aspects of the second edition remain unchanged from the first edition. As with the first edition, the primary purpose of this edition is to assist readers in understanding how the laws that control movement on earth apply to techniques used in sport. Terminology like *inertia, drag, velocity,* and *impulse* are all explained in a sports context and in a manner that is easy to understand. There are no difficult calculations and very few formulas. And throughout the chapters, readers will find boxed, sports-related, and interesting facts about athletes, world records, and equipment advances—all updated from the first edition and increased in number.

In addition, there are more interesting explanations and improved illustrations and photographs to highlight the discussions of sport skills and sport techniques. Where appropriate, explanations have been updated to reflect the changes that have occurred both in sport technique and equipment in recent years. Important terms and concepts are highlighted where they are discussed in the text and listed at the end of each chapter; concise definitions are supplied in a glossary provided at the back of the book.

New to the second edition is a series of practical activities that challenge the reader beyond the review questions found at the end of chapters 2 through 8. These chapters provide the tools to help readers develop a good, basic understanding of sport mechanics. Chapters 7 and 8 show how to make use of these tools in the analysis, identification, and correction of errors in an athlete's performance, and chapter 9 discusses the mechanics of selected sport skills. The descriptions and mechanics of these sport skills have been improved and brought up to date.

Human Kinetics and I feel that *Sport Mechanics for Coaches, Second Edition,* will be of great use to everyone involved in teaching, coaching, or working in sport, dance, or rehabilitation—indeed, in all areas of human movement.

ASEP SILVER LEVEL SERIES PREFACE

The American Sport Education Program (ASEP) Silver Level curriculum is a series of practical texts that provide coaches and students with an applied approach to sport performance. The curriculum is designed for coaches and for college undergraduates pursuing professions as coaches, physical education teachers, and sport fitness practitioners.

For instructors of undergraduate courses, the ASEP Silver Level curriculum provides an excellent alternative to other formal texts. In most undergraduate programs today, students complete basic courses in exercise physiology, mechanics, motor learning, and sport psychology—courses that are focused on research and theory. Many undergraduate students are looking for ways to directly apply what they learn in the classroom to what they can teach or coach on the court or playing field. ASEP's Silver Level series addresses this need by making the fundamentals of sport science easy to understand and apply to enhance sport performance. The Silver Level series is specifically designed to introduce these sport science topics to students in an applied manner. Students will find the information and examples user friendly and easy to apply in the sports setting.

Books available in the ASEP Silver Level curriculum include the following:

Sport Mechanics for Coaches—an explanation of the mechanical concepts underlying performance techniques; designed to enable coaches and students to observe, analyze, develop, and correct the mechanics of sport technique for better athletic performance.

Sport Physiology for Coaches—an applied approach to exercise physiology; designed to enable coaches and students to assess, initiate, enhance, and refine human performance in sport participation and to improve sport performance.

Sport Psychology for Coaches—a practical discussion of motivation, communication, stress management, mental imagery, and other cutting-edge topics; this text is designed to enhance the coach–athlete relationship and to stimulate improved sport performance.

Teaching Sport Skills for Coaches—a practical approach for learning to teach sport skills, guided by a practical understanding of the stages of learning and performance, individual differences and their impact on skill acquisition, and the critical elements required to create a learning environment that enhances optimal sport skill development and performance.

A variety of educational elements make these texts student- and instructor-friendly:

- Learning objectives introduce each chapter.
- Boxes illustrate sport-specific applications of key concepts and principles.
- Chapter summaries review the key points covered in the chapter and are linked to the chapter objectives by content and sequence.
- Key terms at the end of most chapters list the terms introduced in that chapter and remind coaches and students that "These are words you should know." The first occurrence of the word in the chapter is boldfaced.
- Chapter review questions at the end of each chapter allow coaches and students to check their comprehension of the chapter's contents. Answers to questions appear in the back of the book.
- Real-world application scenarios called practical activities follow the review questions. These scenarios provide problem situations for the students to solve. The solutions require the students to describe how the concepts discussed in the chapter can be applied in real-world scenarios. Sample solutions appear in the back of the book.
- A glossary defines all of the terms covered in the book.
- A bibliography section included at the end of the book serves as a resource for additional reading and research.
- A general index lists subjects covered in the book.

These texts are also the basis for a series of Silver Level online courses being developed by Human Kinetics. These courses will be offered through ASEP's Online Education Center for coaches and students wanting to increase their knowledge through practical and applied study of the sport sciences.

ACKNOWLEDGMENTS

I very much appreciate the assistance from Larry Yore, Jim Haddow, Keith Russell, Alex Van Netten, and James Hay, and the encouragement and continued support from the members of the faculty of the School of Physical Education, University of Victoria, British Columbia, Canada. I also want to thank the enthusiastic and supportive students in the sport mechanics classes at the University of Victoria—the guinea pigs for testing my explanations of the principles discussed in this text. Lastly, I want to thank my wife, Catherine, who by now knows this second edition as well as she did the first!

PART I

Understanding Mechanics and Sport Technique

1

Making a Smart Move

Sport Mechanics for Coaches is written for coaches, teachers, athletes, and sport fans. It tells you how a knowledge of sport mechanics helps produce better performances. Those of you who coach will find that this book helps you become a better coach. Athletes will discover information that helps improve their performances. *Sport Mechanics for Coaches* is valuable also to those who neither coach nor compete but who are enthusiastic sport fans. You'll find that *Sport Mechanics for Coaches* helps you become a more critical and appreciative observer of the sports you love.

When you read *Sport Mechanics for Coaches* you'll find that it doesn't concentrate on any particular sport—such as football, rowing, or basketball—or on any specific sport skill—such as punting, passing, serving, or spiking. Instead, *Sport Mechanics for Coaches* explains how and why a basic understanding of mechanical principles helps produce an improved performance. You'll be able to look at an athlete's performance and say to yourself, "Some of the actions in the athlete's technique are good. But I can see actions that are inefficient and need correcting. What I know of mechanics tells me that they're wrong, and I know what kind of movements should replace them. When the corrections are made, the athlete will have a more efficient technique and produce a better performance."

Sport Mechanics

Scientists who work in the fields of **mechanics** and **biomechanics** ("bio" means living) study the effects of forces (such as gravity and air resistance) on living and nonliving objects. They use their knowledge of mechanics to design objects that we use in everyday life, such as buildings, bridges, automobiles, boats, and planes. They also assess the effects of forces on humans and, vice versa, the effects of forces that humans apply. It's obvious that gravity and air resistance or even the forces that occur during collisions make no distinction between nonsporting and sporting activities. A high jumper fights against gravity just as a person climbing stairs or a plane taking off does. Likewise, air resistance opposes both an automobile and an Olympic sprint cyclist. This tells us that the same mechanical principles used in our everyday world can also be applied to sport.

Mechanical Principles

In sport, mechanical principles are nothing more than the basic rules of mechanics and physics that govern an athlete's actions. For example, if a coach and an athlete understand the characteristics of the earth's gravitational force, they know what must be done to best counteract the effect of this force and, conversely, what actions must be performed to make use of it. A springboard diver who is aware that gravity acts perpendicularly to the earth's surface has a better understanding of what trajectory gives an optimal flight path for a dive. Similarly, wrestlers quickly learn that gravity is their friend when they've got their opponents off balance. On the other hand, if they don't maintain their own stability, gravity can team up with their opponents! Ski jumpers understand that if they flex their legs and bend forward as they accelerate down the in-run, they reduce air resistance. This body position allows them to accelerate to an optimal speed in preparation for the takeoff. Once in flight the ski jumpers counteract the force of gravity by making use of air resistance. They extend their legs and lean forward to deflect air downward. In response, the air pushes them upward. It is this variation in the use of gravity and air resistance that helps Olympic ski jumpers fly to distances in excess of 130 m **(meters)**.

There are many more forces on earth besides gravity and air resistance. These forces act in different ways, and if you're in a contact sport you must also consider the forces produced by your opponents. If you are a coach and you understand how all these forces interrelate, you'll be better able to analyze an athlete's technique and improve the athlete's performance. If you are an athlete and you have this knowledge, you'll understand why it's better to apply muscular force at one instant than at another and why certain movements in your technique are best performed in a particular manner. Even as a spectator and sport fan, you'll find that an understanding of basic mechanical principles helps you become more knowledgeable and appreciative of what it takes to produce an excellent performance.

In sport, the laws of mechanics don't apply to the athlete alone. Mechanical principles are used to improve the efficiency of sport equipment and playing surfaces. Modern track shoes, speedskates, skis, slick bodysuits for swimming and cycling, and safety equipment (such as pole-vault landing pads) are all designed with an understanding of the external forces that exist on earth and the forces that an athlete produces. This knowledge has been instrumental in raising the standard of performance in every sport.

Technique

When we compare the performances of two athletes, we often say that one of the two athletes has better form, or more precisely, that one athlete has better technique than the other. By **technique** we mean the pattern and sequence of movements that the athletes use to perform a sport skill, such as a forearm pass in volleyball, a hip throw in judo, or a somersaulting dive from the tower.

Sport skills vary in number and type from one sport to the next. In some sports (e.g., discus and javelin) there is only one skill to perform. A discus thrower spins and throws the discus—nothing more. But in tennis, players perform forehands, backhands, volleys, and serves. Each skill, whether it's a tennis serve or a discus throw, has a specific objective determined by the rules of the sport. In a serve, a tennis player wants to hit the ball over the net and into the service area in such a way that the opponent cannot return it. A discus thrower aims to throw the discus as far as possible, making sure that it lands in the designated area. Both athletes try to use good technique so that the objectives of each skill are achieved with the highest degree of efficiency and success.

Good Technique

An athlete can perform a skill with good or poor technique. Poor technique is ineffective and fails to produce the very best results. You can see plenty of poor technique at any public driving range—and, along with poor technique, inferior results! Hooks and slices are mixed in with wild swings that miss the ball entirely. Even if you know little about golf, you'll be amazed by the variation you see in the performances of a single stroke. Now compare these recreational golfers with elite professional players. Although elite players differ in height, strength, and weight, the basic technique they use in their strokes is very much the same. From backswing to follow-through, you're likely to see a smooth application of force that appears graceful and fluid. This efficiency of motion tells you that elite golfers use good technique. They practice for hours to hone this technique so that their actions become highly effective and get the job done.

Apart from minor differences, all top-class athletes, no matter what their sport, use superior technique based on the best use of the mechanical principles that control human movement. But it's important to remember that the refined, polished movements you see in the technique of an elite athlete seldom occur by chance. They usually occur as a result of hours of practice. Likewise, it's virtually impossible nowadays for an athlete to reach world-class status without the assistance of coaches and analysts who know why it's better to perform the actions in a sport skill in one way rather than in another. Today's top athletes get help from knowledgeable coaches who critically observe their performances and tell them what is efficient movement and what is not. The coaches' knowledge, coupled with the athletes' talent and discipline, helps produce first-rate performances.

Teaching Good Technique

What must you know in order to teach good technique? As an example, let's look at what's necessary when you teach a novice to drive a golf ball. When you introduce this skill, it's an asset if you can demonstrate good technique, although this ability is certainly not essential. What's more important is that you are able to analyze and correct faults in the novice's performance and that you use good teaching progressions to lead the novice to a more refined performance. To do this you need a basic understanding of the mechanics of the golf drive, which means that you

The Human Body

Our bodies have a framework of over 200 bones. These bones form the skeleton, which supports and protects our inner organs and provides attachments for our muscles, tendons, and ligaments. Our bones articulate with other bones at the joints. We have a variety of joints, and depending on their design, they determine the type of movement that occurs. As far as sport mechanics is concerned, the most important joints are those in our hips, legs, shoulders, arms, and spinal column—all of which allow considerable movement. These joints form axes of rotation around which major parts of our bodies (called body segments) rotate.

must know why certain actions in a golf drive are best performed one way and not another. It's the same when you coach or instruct classes in volleyball. A volleyball coach needs to know the mechanical reasons why certain movements get a player up in the air for a spike and why other movements do not. In baseball, a pitching coach aims to teach the most efficient actions for the windup, delivery, and follow-through. Similarly, a batting coach works to make the batter a more effective hitter. The golf, baseball, and volleyball coaches are using their knowledge of mechanical principles when they eliminate poor movements and replace them with actions that are more efficient.

Failure of Traditional Training Methods

Many coaches and athletes still follow traditional methods during training. They reason that "this is how it was done in the past and it worked well, so this is how we should do it now." They have no idea why some movements may be good and others bad. Then there are those coaches who are happy using a trial-and-error method. Occasionally they get good results, but more often they don't. Many coaches teach their athletes a technique based on a world champion's technique without taking into account differences in physique, training, and maturity. Similarly, young athletes often copy every action of a world-class performer, including idiosyncrasies that are mechanically ineffectual. Al Oerter, four-time Olympic discus champion from 1956 to 1968, frequently inverted the discus as he swung his arm back during his windup. This action was simply a personal trait that added nothing to the mechanical efficiency of Oerter's throwing technique, yet many young athletes copied it, believing that it would add distance to their throws. A more comical example of copying idiosyncrasies was seen in the number of kids who attempted slam dunks with their mouths open and tongues hanging out. Why? Because this was a common characteristic of Michael Jordan!

Being able to distinguish between mechanically correct movements and those that serve no purpose is essential for skill development. Coaches and athletes who blindly mimic the methods and techniques of others progress only so far. *Sport Mechanics for Coaches* will help you eliminate this haphazard approach. By developing a basic understanding of mechanics you'll be able to analyze performances and teach movement patterns that produce efficient technique. This will lead to better performances.

Getting the Most From *Sport Mechanics for Coaches*

Most people involved in coaching are reluctant to study sport mechanics; from past experience they know it has meant tackling dry, boring texts loaded with formulas, calculations, and scientific terminology. These texts are frequently written by academics who fail to explain the relationship of good technique to the principles of mechanics in a manner that is meaningful to coaches and sport enthusiasts. You'll be happy to find that *Sport Mechanics for Coaches* is a very different type of book. This book contains few formulas or calculations; and it uses familiar measurements, like pounds, feet, and inches, while giving you metric equivalents as necessary. Whether you coach, teach, perform as an athlete, or watch as a fan, *Sport Mechanics for Coaches* is a book that you can learn from immediately.

How *Sport Mechanics for Coaches* Will Help You

To get the most from *Sport Mechanics for Coaches* all you need is a desire to know how and why things work in the world of sport. In other words, if you have curiosity and a desire to improve, you'll get a lot of useful information from this text. Here's how:

- **You will learn to observe, analyze, and correct errors in performance.** This is the most important benefit you'll get from reading *Sport Mechanics for Coaches*. This text will help you develop a basic understanding of mechanics, and by using this knowledge you will be able to distinguish between efficient and inefficient movements in an athlete's technique. The information in this text will help you pick out unproductive movements and follow up with precise instructions that help optimize performance. You won't waste time with vague advice like "Throw harder" or "Try to be more dynamic." Obscure tips like these only confuse and frustrate the athletes you're trying to help. On the other hand, if you're the athlete and your coach is not present, a basic knowledge of sport mechanics will help you understand why you should eliminate certain movements in your technique and instead emphasize other actions.

- **You'll be better able to assess the effectiveness of innovations in sport equipment.** When Greg LeMond of the United States won the Tour de France by a few seconds over Laurent Fignon of France, Greg certainly illustrated the value of fitness and determination. But equally important, Greg and the technicians that assisted him knew the importance of reducing wind resistance to a minimum, particularly during the Tour's final time trial. They realized that if Greg could maintain a low dartlike body position and have the air flow smoothly over and past his body, then he would spend less energy pushing air aside and more energy could be spent propelling himself at high speed. This knowledge of mechanics paid off! It's no different in other sports. You need to know what is gained from design changes in such items as golf clubs, tennis rackets, skis, speedskates, mountain bikes, and swimsuits. *Sport Mechanics for Coaches* cannot teach you all there is to know in the world of sport equipment design because changes and modifications will continue to occur at an ever increasing pace. But this text will certainly give you a foundation of knowledge on which you can build.

- **You'll be better prepared to assess training methods for potential safety problems.** Think of an athlete squatting with a barbell on

Advances in Equipment Require Changes in Technique

In the Nagano Olympics, world records in speedskating were continually broken by athletes using clap skates. The blade on a clap skate is hinged at the front of the shoe but not at the heel. The hinge allows the blade to stay in contact with the ice longer and the skater is able to thrust at the ice for a longer time period. The characteristic clapping noise occurs at the end of each stroke when the blade "claps" back into contact with the heel of the shoe. The clap skate requires a technique in which the skater must push more directly backward and from the toe rather than out to the side and from the heel. Athletes who were accustomed to the older skates (where the blade was permanently joined to the boot at heel and toe) have had to adapt to the new equipment and change their technique.

Adapted from "The Athletic Arms Race" by Mike May in *Scientific American* special issue: "Building the Elite Athlete" Nov 27, 2000. page 74.

his shoulders. Where should your athlete position the bar? Should it be placed high on the shoulders or lower down? And what about the angle of the athlete's back during the squatting action? What are the mechanical implications of a full squat compared with half and three-quarter squats, and how fast should the athlete lower into the squat position? If you know about levers and torque, you'll understand why it's dangerous to bend forward when you squat. Likewise, if you are familiar with the characteristics of momentum and understand how every action has an equal and opposite reaction, you'll know that dropping quickly into a full squat puts tremendous stress on the lower back, knees, and hips. It's possible that you have been teaching good technique but don't fully understand why one way of performing the technique is potentially dangerous and another is not. *Sport Mechanics for Coaches* will give you the reasons.

In gymnastics you will frequently see spotting techniques that provide a high level of safety, and then you'll see other techniques that endanger both the gymnast and the spotter. By reading *Sport Mechanics for Coaches* you'll discover why efficient spotting requires an understanding of balance, levers, torque, and the momentum generated by the gymnast performing the skill. This information will help you teach safe spotting techniques in gymnastics and good technique in weight training. Of course, *Sport Mechanics for Coaches* is not limited to these two sports. You can apply the mechanical principles that you read about to every sport.

- **You'll be better able to assess the value of innovations in the ways sport skills are performed.** In sport, our capacity for reasoning and creativity has been responsible for the advances in talent selection, technique, training, and equipment design. We all possess a tremendous capacity for creativity, and to be a good coach you must use this creativity to search for better ways to improve your athlete's performance. All athletes differ in physique, temperament, and physical ability; what works for one athlete will not necessarily work for another. Similarly, a young athlete will differ dramatically from a mature athlete. To help your athletes achieve top-flight performances, it's good to learn why sport techniques are performed as they are and to be prepared to modify certain aspects of these techniques in order to fit the age, maturity, and experience of your athletes.

There are many examples of the willingness of coaches and athletes to try out new ideas. In team games, think of how coaches modify attack and defense formations relative to the team they will face in an upcoming contest. Among athletes, think of how the creativity and experimentation of Dick Fosbury revolutionized the high jump and how the glide and rotary techniques have increased the distances thrown in the shot put. In gymnastics, consider the number of skills named after their creative inventors (such as the "Thomas Flair," named after former U.S. gymnast Kurt Thomas). So be curious and learn the how and why of technique. At the same time, be creative and willing to experiment, and be sure to encourage your athletes to use their own creative capacities as well. Always look for ways to improve your understanding of the sport you are coaching. Be a coach, an analyst, and an innovator all at the same time!

- **You will know what to expect from different body types and different levels of maturity.** If you understand the mechanical principles governing the techniques of your sport, you'll understand why young athletes who are growing fast have a tough time maneuvering, changing direction, and coordinating their movements in a manner similar to more mature athletes. You'll realize that you cannot and should not expect young athletes to follow the same training regimens that you would demand of a more mature athlete. You'll also understand why tall athletes with long arms and legs have an edge in some sports but are at a disadvantage in others. Similarly, you will realize why smaller athletes tend to have a good strength-to-weight ratio and can cut, turn, and shift more quickly than athletes who are taller and heavier.

How *Sport Mechanics for Coaches* Is Organized

Sport Mechanics for Coaches is divided into two parts, each with a different focus. Part I contains chapters 1 through 6, which get into the meat of sport mechanics. They have purposely been

Never Discount Individual Creativity and Inventiveness

Just as Dick Fosbury revolutionized high jumping, so did Graeme O'Bree from Scotland, who set the cycling world talking with his self-designed bike. O'Bree removed the bike's crossbar, shortened the length of the bike, and made its profile as narrow as possible. With no crossbar, O'Bree cycled with his legs brushing against each other. His chest lay flat on top of the handlebars. O'Bree's body position reduced air resistance to a minimum. Although lying with the chest on the handlebars was subsequently declared illegal by the Cycling Federation, O'Bree was not to be beaten by the rules. He modified his bike even further so that he could cycle without his chest contacting the handlebars but with his upper body still stretched forward horizontally like an arrow in what was called the "Superman Position." O'Bree became a world champion in the velodrome and knew the importance of reducing air resistance to a minimum. Although the "Superman Position" is illegal in the Tour de France, athletes attempt to cycle with their upper bodies in a horizontal position during the Tour's time trials. Reducing air resistance to a minimum in these high-speed time trials is absolutely essential.

given informal titles: "Making a Smart Move," "Starting With Basics," "Getting a Move On," "Rocking and Rolling," "Don't Be a Pushover," and "Going With the Flow." You'll understand why these titles were chosen when you read the content of each chapter and read about the various forces at work as athletes perform in their sports. These chapters explain the interaction between an athlete, the equipment, and the ever-present external forces that assist or oppose the athlete during the performance of a sport skill. You'll learn what forces are at work when sprinters accelerate, gymnasts spin in the air, and pitchers hurl curveballs. You'll understand the mechanics of good technique and what advantages and disadvantages exist when athletes compete at high altitudes.

In part II, chapters 7 and 8 explain how you can put to work the information you learned in part I. These chapters discuss why athletes must make their muscles work as a team and why it's so important to synchronize and coordinate muscle actions. This synchronization and coordination produce superior technique—technique that has resulted in a 20-ft pole vault, an 8-ft high jump, triple back somersaults in gymnastics floor exercises, quintuple-twisting triple somersaults in ski aerials, and a long jump close to 30 ft.

Chapters 7 and 8 are particularly useful to coaches and physical educators because they explain how to observe and analyze an athlete's technique and how to set about correcting errors that are found. Each chapter gives you a series of steps to follow. You'll learn how to break a skill down into phases and what to look for as you analyze each phase. Then you'll read a series of important mechanical principles that you can refer to when you start correcting errors.

In chapter 9 you'll learn about techniques and mechanics in a wide range of sport skills, including sprinting, jumping, swimming, lifting, throwing, and kicking. First you'll read descriptions of the most prominent features in the performance of these skills. Then you'll read the mechanical reasons that the technique of each skill is best performed in a certain way. The goal in this chapter is to show you how technique and mechanics are inseparable, no matter what the sport.

Sport Mechanics for Coaches finishes with a glossary and a list of references that will help you expand your knowledge of sport mechanics. The glossary avoids dull, scientific explanations. Instead it relates scientific principles to athletes and to the movement of sport implements such as bats, balls, and javelins.

After you've read *Sport Mechanics for Coaches,* you should be able to watch an athlete perform and immediately know what external forces the athlete must contend with. You'll be able to analyze an athlete's movements and immediately recognize how they can be improved. You'll spot poor actions and replace them with efficient movements that produce quality technique—technique based on sound mechanical principles.

Understanding how physical laws influence sport performances will help you, whether you are coaching, competing, or simply watching as a spectator. If you coach, remember that sport mechanics is just one tool that you'll use. You'll also need to improve your knowledge in such areas as sport psychology, physiology, nutrition, sport injuries, and the teaching of sport skills. The American Sport Education Program can provide you with all the necessary information in these areas.

2

Starting With Basics

When you finish reading this chapter, you should be able to explain

- how gravity affects athletic performance;
- the relationship between body weight, mass, and inertia;
- how the center of gravity of an athlete or an object can vary in position;
- the difference between speed, velocity, and acceleration;
- how athletes make use of the earth's reaction force;
- how athletes make use of force vectors;
- the difference between linear, angular, and general motion; and
- the factors that influence the flight paths of athletes or objects (such as baseballs and javelins).

In this chapter we'll look at some basic mechanical concepts that affect athletic performance. We'll also spend some time discussing the effect of the earth's gravitational pull. Even though the force of gravity can help an athlete, there are many sports (e.g., jumping and throwing) where gravity is the toughest opposition that the athlete has to face. So it's worthwhile knowing as much as you can about the characteristics of this ever-present force. You'll learn in this chapter how an athlete's body mass and body weight are related and what is meant by inertia. You'll see that a massive athlete has more inertia than one who has less body mass. Like gravity, inertia can be helpful in some sport skills, but in other sport skills it presents real problems for an athlete. You'll also learn how every action has an equal and opposite reaction, and you'll read some examples of how this law influences sport skills.

Chapter 2 concludes with a discussion of the factors that affect the trajectories (i.e., the flight paths) of athletes (e.g., divers, gymnasts, and high jumpers) and the trajectories of objects (e.g., balls, discuses, and javelins). The discussion of trajectories will bring you back again to earth's gravitational force and the part it plays in determining the shape of flight paths that athletes and objects follow. Check how all the mechanical concepts discussed in this chapter interrelate, and watch for this interrelationship to continue in later chapters.

Mass

Mass simply means substance, or **matter.** If an object has substance and occupies space it has mass. We frequently talk of NFL linemen as being massive or having tremendous body mass, indicating that the athletes are enormous and have plenty of muscle, bones, fat, tissue, fluids, and other substances that make up their bodies. An

Getting Weighed at the Earth's Core

Standing on the surface of the earth, your body mass pulls on the earth, and the mass of the earth pulls on you, and the direction of the earth's pull is toward the core of the earth. This gives you your body weight. So how much would you weigh at the core of the earth—more or less than on the surface? The answer is a lot less! In fact, you would be weightless. At the core of the earth there would be equal amounts of the earth's mass above you as below you. So the mass of the earth above your head would pull you up by the same amount as the mass of the earth below pulls you down.

important characteristic of all objects that have mass is that they are attracted to other objects that also have mass. Like the earth's gravitational pull on our bodies, the attractive force between two objects that have mass (such as the two NFL linemen) is also called a gravitational attraction. The actual amount of **gravitational attraction** between any two objects (human or otherwise) changes according to how much mass they have and how far apart they are. So two athletes standing together in conversation exert a gravitational attraction on each other. This attractive force is so minuscule as to be hardly worth considering. A different situation exists when you consider gigantic objects with tremendous mass, like the planets in our solar system. Gravity holds the moon in its orbit around the earth and the earth in its orbit around the sun. We stay in contact with the surface of the earth because the mass of the earth pulls on us, and we, with what little mass we have, pull on the earth.

Weight

Athletes who want to perform well in their chosen events carefully monitor their body weight. They know that too much or too little weight can seriously affect their performance. For all of us, checking our body weight is a means of assessing our general health and fitness.

When we get on a scale, the dial gives us a reading that we associate with the amount of body mass that we carry around. A common assumption is that an athlete's body weight compresses the springs in the scale, and the readout on the dial represents the amount that the springs are squeezed together. This is true, but what actually happens is a little more complex.

In mechanical terms an athlete's **weight** represents the earth's gravity pulling on the athlete's body and, vice versa, the pull of the athlete's body on the earth. The readout on the scale represents how much pull or attraction exists between the two. The earth pulls the athlete downward and, in reverse, the athlete pulls the earth upward. So, an athlete with more body mass compresses the springs to a greater extent than an athlete who has less body mass. As a result the needle on the scale moves farther around the dial.

How Weight and Mass Are Related

When you stand on the surface of the earth, you travel around in a circle as the earth rotates daily around its axis. The size of the circle that you follow depends on the latitude where you are standing. You rotate on the spot at the North or South Pole and speed around an enormous circle when you are at the equator. The size of the circular pathway that you follow when you stand at the equator depends on the distance of the equator to the earth's core and how quickly the earth spins (i.e., once every 24 hours). The spin of the earth also makes the equator bulge outward approximately 13 miles farther from the earth's core than the North or the South Pole. Standing at the equator (or even better, at the top of a mountain at the equator!) puts a considerable distance between you and the earth's core. This large distance reduces the gravitational attraction that exists between you and the earth. Now let's add one other factor. When you stand at the equator, the earth's spin will cause you (and everything else at that latitude) to travel around a circular pathway at approximately 1,000 mph. At this high speed your body mass tries to fly away from the surface of the earth like water flying away from wet clothing during the spin cycle in a washing machine. Earth's gravitational attraction is strong enough to stop you flying off into space, but the outward tug of your body mass causes it to "fight" against the inward pull of earth's gravity. These factors cause you and everything else that has mass to weigh just a little less at the equator and progressively more as you move toward the poles.

Consider the impact of the characteristics just discussed: An athlete who weighs 200 lb (90.72 kg) at the poles will weigh about 198.94 lb (90.24 kg) at sea level at the equator, and an athlete who weighs 200 lb at sea level will weigh approximately 199.77 lb (90.61 kg) at an altitude of 12,000 ft (3,657.61 m). This means that at the top of Mount Everest, which is over 29,000 ft (8,839.22 m) in height, you will weigh slightly less than at sea level. These small changes in weight tell you that an athlete's body weight is a function of variations in the earth's gravitational pull on the athlete's mass.

An athlete's body mass can remain constant, yet the same person's body weight can fluctuate, depending on where the athlete is on the earth. The same principle applies to the weight of a shot, a javelin, or any kind of equipment used in a sporting contest.

So although weight and mass are different, they are related. The difference between the two is not tremendously important in sport because sport skills take place on or close to the surface of the earth. Under these conditions, weight and mass experience the same proportions (i.e., an athlete who weighs more than another athlete will also have more mass).

How Mass Is Related to Inertia

Mass is directly related to inertia. The more mass an object has, the more inertia it has too. **Inertia** means resistance to change. We use the word *inertia* in everyday life to characterize people who are slow to commit themselves to action. So you could say that there's a relationship between inertia and laziness. In mechanical terms, inertia means more than just laziness because it describes the desire of an object (or an athlete) to continue doing whatever it's doing. If it's motionless it will want to remain motionless. If it's moving slowly it will want to continue moving slowly, and if it's moving fast it will want to continue moving fast. We must also consider one more important characteristic of inertia. Once on the move, objects always want to move in a straight line. They will not willingly travel around circular pathways; it's necessary to pull or push on them to produce a curved pathway. A ball thrown by an outfielder will travel in a straight line following its release trajectory were it not for air resistance slowing it down and gravity curving its flight path toward the earth's surface.

The more massive an athlete, the more the athlete's body mass resists change. A giant 300-lb (136-kg) athlete needs to exert great muscular force to get his body mass moving. Once moving in a particular direction, the athlete must again produce an immense amount of muscular force to stop or to change direction. Athletes with less body mass have less inertia and therefore need to apply less force to get themselves going. Likewise, they need less force than a more massive athlete to maneuver or stop themselves once they're on the move.

There are many examples in everyday life of inertia at work. Oil tankers that cross our oceans have tremendous mass and inertia. They need powerful engines to get them going and huge distances to stop and to turn around. The same can be said of giant trucks that carry provisions across the country. A high number of gears in the transmission (many more than a family car) help the loaded truck get under way and overcome its inertia at rest. Once the truck is rolling, the driver is well aware that it is dangerous to attempt a tight turn at high speed. The truck would roll over because mass (i.e., inertia) on the move wants to travel in a straight line. To stop or to slow down and get around tight turns, the driver makes use of superpowerful air brakes that are designed to battle mass and inertia on the move. The same principles apply in sport. Consider Japanese sumo wrestlers or defensive and offensive linemen in football. Just like the oil tanker and the truck, these athletes must apply tremendous force to get their body mass moving and then apply a huge amount of force to change direction or to maneuver the great mass of their opponents.

In sports like squash or badminton, it's possible for the immense mass and inertia of huge athletes to work against them. It's no good being massive when sudden and varied movement changes are required unless you have the power to move your mass quickly and to control it once it's moving. Massive athletes tend to have a poorer strength-to-weight ratio than do smaller, less massive athletes; so they have a tougher time stopping, starting, and changing direction. That's why badminton and squash players are lean, lightweight, and anything but massive. If you're a small, lightweight squash player, you can get a lot of pleasure from making your massive opponent crash into the side walls. You have a friend helping you in the court—your opponent's inertia!

An interesting example of inertia at work occurs when athletes are in flight. Consider two athletes who decide to bungee jump from a bridge. One athlete is twice as massive as the other. They step off the bridge at the same instant. Surprisingly, both accelerate toward the earth at approximately the same rate. Because the earth attracts the more massive bungee jumper twice as much, you might think that this athlete would accelerate downward twice as fast. But this same athlete has twice the inertia of the other thrill seeker and so resists being accelerated by gravity twice as much. In this situation, air resistance plays a negligible role and both athletes accelerate downward at approximately the same rate.

Think of inertia as an enemy when an athlete wants to get moving. To defeat this enemy, it's good if the athlete's mass is made up of powerful muscles that are able to generate the required amount of force. Once on the move, inertia can become an athlete's friend because the second characteristic of inertia is that it wants to keep the athlete going. The difference between resting inertia and moving inertia causes athletes to expend much more energy at the start of a 100-m dash than when sprinting in the middle of the race. The two characteristics of inertia, to resist motion and then to persist in motion, occur not only in linear situations where objects and athletes move in a straight line but also in rotary situations where objects such as bats and clubs are made to follow a circular pathway. As long as the athlete makes a baseball bat travel around in an arc, the bat will try to continue moving along this circular pathway. If the bat slips out of the hands of the athlete, then it will immediately go back to its initial preference, which is to move at a constant speed along a straight line.

In linear movement, mass is synonymous with inertia. The more mass, the more inertia. The characteristics of inertia are described in the first of Isaac Newton's three famous laws of motion. We commonly call it Newton's first law, Newton's law of inertia, or simply Newton I. Newton's first law also applies to rotary situations. But rotary inertia (also called rotary resistance or moment of inertia) involves more than just the mass of the object. We also need to know how the mass is distributed (i.e., spread out or compressed) relative to the axis around which the object is spinning. It's not difficult to understand, and we will read about rotational movement in chapter 4.

Speed, Acceleration, and Velocity

Earlier we talked about two bungee jumpers and their acceleration as they fell toward the earth. Let's check out the difference between speed and acceleration and then introduce velocity—a term that you'll find frequently throughout this text.

Speed is a **scalar** measure indicating how fast an object is traveling at a particular instant in time. If an elite sprinter runs 100 m in 10 sec, we know that the athlete has run a certain distance (100 m, or 109.36 yd) in a certain time (10 sec). From this information you can work out the sprinter's average speed, which is 22.36 mph, or 36 km/h (22.36 mph = 10.9 yd/sec).

The speed that the sprinter averaged over a distance of 100 m is 22.36 mph—nothing more. These numbers don't tell you the sprinter's top speed (which could be as high as 26 mph), and they don't tell you anything about the sprinter's **acceleration** or **deceleration,** which is the rate that speed changes. A sprinter who averages 22.36 mph over 100 m runs faster and slower than 22.36 mph during different phases of the race. Why? Because immediately after the starter's gun goes off, the athlete is gaining speed and for a while runs much slower than 22.36 mph. The athlete then has to run faster somewhere else in the race to average 22.36 mph over the whole distance.

Modern radar equipment is often used in sport research, and it can give you the speed of a baseball as it leaves the pitcher's hand or as it leaves the bat at the instant it is hit. This equipment is now so precise that it can determine over what distance a sprinter was accelerating. It can also give you the distance that the athlete held a particular speed during a race and by how much the athlete decelerated at the end of a race.

Rates of acceleration vary dramatically from one athlete to another. Some athletes rocket out of the blocks and have tremendous acceleration over the first 40 m of a 100-m race. Thereafter their rate of acceleration drops off, and close to the tape they may even decelerate. Athletes who raced against multiple Olympic champion Carl Lewis were well aware that he could still be accelerating at the 70-m mark in the 100-m dash. His rate of acceleration may have been less than that of his opponents at the start of the race, but his acceleration continued longer. Over the last 30 m, Carl frequently caught and passed athletes who were "tying up" (i.e., breaking proper form because of fatigue) and decelerating.

It's possible for athletes to reduce their rate of acceleration and still increase speed. As long as acceleration exists, even if it's minimal, speed will increase. If deceleration occurs, speed will be reduced. How much an athlete's speed increases or decreases depends on the athlete's rate of acceleration and deceleration.

Uniform acceleration and **uniform deceleration** mean that an athlete or an object speeds up or slows down at a regular rate. An example of uniform acceleration occurs when a four-man bobsled slides down the track in the Winter Olympics and accelerates to a speed of 15 ft/sec by the first second, 30 ft/sec by the second, and 45 ft/sec by the third. For every second that the bobsled is moving, it is increasing speed at a uniform rate of 15 ft/sec. You write this acceleration as 15 ft/sec/sec, or 15 ft/sec^2. Notice that there is one distance unit (i.e., 15 ft) and two time units (i.e., sec/sec) whenever you refer to acceleration. This indicates the rate of change of speed, or the amount of speed added (i.e., 15 ft/sec), with each successive time unit (i.e., 1 sec) that passes. If the bobsled decelerates at a uniform rate, then the reverse occurs. In this case it is slowing, or losing speed, at a uniform rate.

Uniform acceleration and deceleration does not happen that often in sport. When athletes (or objects like balls or javelins) are on the move, varying oppositional forces ranging from opponents to air resistance cause their acceleration (or deceleration) to be varied (or nonuniform). However, one of the best examples of uniform acceleration and deceleration occurs in flights of short duration such as in high jump, long jump, diving, trampoline, and gymnastics. In these situations air resistance is so minimal as to be considered negligible. Gravity uniformly slows, or decelerates, the athletes as they rise in flight by a speed of 32 ft/sec for every 1 sec of flight (i.e., 32 ft/sec^2) and then accelerates them at a uniform rate of 32 ft/sec^2 on the way down. (In the metric system, 32 ft/sec^2 = 9.8 m/sec^2 approximately.) Sometimes you'll see deceleration described as **negative acceleration** and acceleration as **positive acceleration.** A minus sign in front of 32 ft/sec^2 (i.e., –32 ft/sec^2) indicates that the diver is decelerating at a rate of 32 ft/sec for each second that he is rising in the air.

How does velocity fit into this description of speed, acceleration, and deceleration? **Velocity** is nothing more than a more precise description of speed. It means both speed and direction. For example, 20 mph simply indicates speed; 20 mph due south indicates velocity. Speed tells you how fast. Velocity tells you how fast and in what direction.

How the Earth's Gravity Affects Athletic Performance

Earlier in this chapter, we saw how the earth's gravitational pull varies. How do these differences affect performances in sport? As an example, let's look at some of the venues where the Olympic Games were held. Athletes experienced slightly less gravitational pull at the 1968 Olympics in Mexico City, which is at higher altitude and closer to the equator, than at the 1952 Olympics in Helsinki or the 1980 Games in Moscow, both of which are in northern latitudes and closer to sea level. Considering gravity by itself and not air resistance, Peter Brancazio, an avid sport fan and physics professor at Brooklyn College, calculated that a 70-ft shot put in Oslo, Norway (latitude 60 degrees N), would travel 1 in. (2.54 cm) farther in Montreal, Canada (45 degrees N), 2 in. (5.08 cm) farther in Cairo, Egypt (30 degrees N), and 3 in. (7.62 cm) farther in Caracas, Venezuela (10 degrees N). A 300-ft (91.44-m) javelin throw in Moscow (56 degrees N) would travel 301 ft (91.74 m) in Lima, Peru (12 degrees S).

Of greater importance than the slight reduction in gravity's pull is the so-called thin air that occurs at high altitudes. Although air contains the same proportions of oxygen (21%), nitrogen (78%), and other gases (1%) at high altitudes as at sea level, in a similar volume of air there is less of each the more you go up in altitude. This characteristic greatly affected athletes who competed in the 1968 Olympic Games at Mexico City, which is 7,350 ft (2,240 m) above sea level. At Mexico City athletes had to breathe more vigorously and more often to get the oxygen they needed. This caused a serious problem for athletes in endurance events, but it assisted those athletes in short sprints because they run on stored energy supplies in their bodies. When Bob Beamon set his world record in the long jump in Mexico City, he benefited (as did all the other long jumpers) from a slight reduction in gravity, reduced air resistance from less dense air (as he sprinted down the runway), and the fact that his approach was a short sprint and not a distance run.

After standing for many years, Bob Beamon's record was beaten by Mike Powell in the 1991 World Track and Field Championships in Tokyo. Tokyo is close to sea level and at a much lower altitude than Mexico City. So if you considered only the differences in atmospheric conditions, you can assume that Mike Powell's jump in Tokyo would have produced a greater distance had he performed it in Mexico City. (Note: The track and the approach in Tokyo used an ultra-springy artificial surface that subsequently was banned from further use. This surface assisted the sprinters and was likely of assistance to all long jumpers as well.)

Acceleration Due to Gravity

When a pole-vaulter drops from above the bar, gravity accelerates the athlete toward the pit. If the athlete clears the bar at 20 ft (6.10 m) rather than at 15 ft (4.57 m), the extra distance gives the earth more time to accelerate the athlete on the way down. Dropping from 20 ft, a pole-vaulter hits the pit at a greater velocity than when clearing 15 ft.

The pole-vaulter's acceleration toward the earth is similarly experienced by a tower diver heading toward the water. Because of earth's gravitational pull, a diver continuously accelerates during the fall to the water. Figure 2.1 discounts air resistance and shows an athlete stepping off and dropping from a height of 256 ft (78 m) in height. After 1 sec of fall, the athlete is traveling at a velocity of 32 ft/sec (9.8 m/sec), or 21.8 mph (35 km/h). After 2 sec, the athlete's velocity has reached 64 ft/sec (19.6 m/sec), or 43.6 mph (70 km/h). At the 3-sec mark, the athlete has reached 96 ft/sec (29.4 m/sec), or 65.4 mph (105 km/h). Finally at the 4-sec mark, the athlete's velocity has increased to an incredible 128 ft/sec (39.2 m/sec), or 87.2 mph (140 km/h). As noted earlier, because of the regular addition of a speed of 32 ft/sec (9.8 m/sec) for every second, and discounting air resistance, we say that the force of gravity uniformly accelerates a falling athlete, such as a diver, trampolinist, or pole-vaulter, by a velocity of 32 ft/sec (9.8 m/sec) for every second of fall, or 32 ft/sec^2 (9.8 m/sec^2).

How can we get some idea of the effect of **gravitational acceleration?** Look again at figure 2.1 and check the distance the athlete covers with each second of fall. After 1 sec the athlete has fallen 16 ft (or approximately 4.8 m). This is not very far, but gravity has only had 1 sec to accelerate the athlete downward from board level. By the 2-sec mark, the athlete has fallen a distance of 64 ft (19.6 m). At the 3-sec mark the athlete has reached 144 ft (43.9 m), and finally

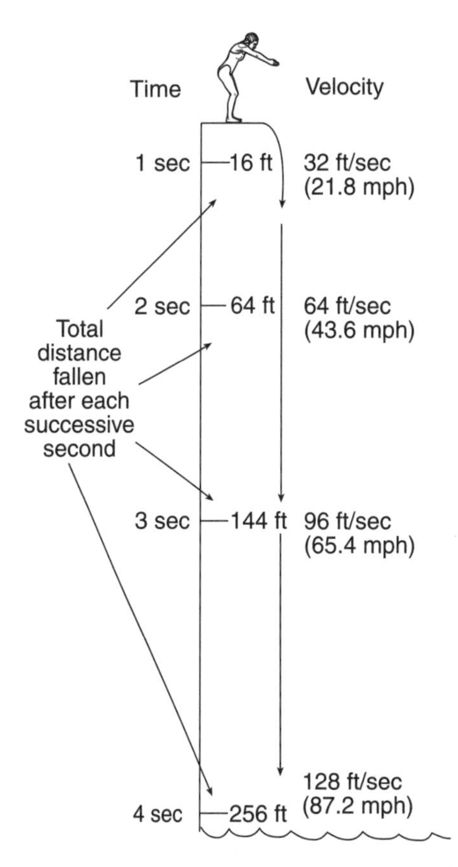

FIGURE 2.1 Distance, velocity, and acceleration due to gravity. (Air resistance has been discounted.)

Breaking the Sound Barrier in a Free Fall

During the Cold War, the United States experimented with different techniques for saving the lives of pilots who had to bail out of high-altitude spy planes. In 1960, as part of Operation Excelsior, Joe Kittinger was lifted in a balloon above the New Mexico desert to 102,000 ft—more than 19 miles up. Wearing a special spacesuit to protect himself from sub-zero temperatures and to provide oxygen in the near vacuum of the upper stratosphere, Joe stepped out of the cabin under the balloon and performed a free fall that lasted for 4 1/2 minutes prior to deploying his parachute. Joe's free fall was the longest recorded in history and during this fall he reached a velocity of 714 mph, breaking the sound barrier and causing a sonic boom that was heard on earth.

by 4 sec the distance covered is an amazing 256 ft (78.0 m). The athlete accelerates downward at a uniform rate of 32 ft/sec^2 (9.8 m/sec^2), and so with each passing second, the athlete adds on an additional 21.8 mph (35 km/h). Because of the constant increase in velocity, the athlete covers increasingly large stretches of distance with each second that passes.

The phenomenal acceleration caused by gravity's pull makes tower diving a risky sport. A standard tower is 10 m (just under 33 ft) from the surface of the water. Tower divers take between 1.50 to 1.75 sec to reach the surface of the water, and they are traveling at close to 38 mph when they enter! Water is hard when you hit it at that speed. Divers often wear wraps to provide additional support to their wrists, because they use a palm-first entry rather than one with the fingertips leading. A palm-first position with the hands bent back at the wrists helps produce a bubbling, splashless entry (i.e., a "rip entry").

Does the uniform acceleration of springboard and tower divers mean that a parachutist in a free fall from several thousand feet up accelerates continuously toward the earth? No, because air resistance in the denser atmosphere close to the earth's surface increases to a point where a parachutist in a free fall reaches a constant (or terminal) velocity of about 125 mph (201.16 km/h). In this situation, the pull of gravity is counterbalanced by the resistance generated by the atmosphere. Variations in terminal velocity can occur depending on the body position used by the parachutist during the free fall. The commonly used stomach-down spread-eagle position produces a much slower terminal velocity than when a parachutist falls headfirst with legs together and arms by the sides. Differences in the types of parachutes also cause variations in the speed with which the parachutist returns to earth. Modern wing-shaped ram-air parachutes are designed to provide greater maneuverability and a softer landing than the older half-sphere parachutes that were commonly used during World War II.

An Athlete's Center of Gravity

The attraction of earth's gravity is one of the most powerful forces an athlete must face. To get up in the air as high as possible, to maintain balance, and to throw far all require an understanding of how this ever-present force operates.

The earth's gravitational pull on an athlete is concentrated at the athlete's center of gravity. A similar situation occurs with all other living and nonliving objects. To understand what is meant by center of gravity, let's first consider an iron shot that is perfectly spherical and made up of iron particles throughout. We'll assume that each particle of the ball has the same mass. The earth's gravitational attraction pulls on each of these particles with the same force. If you join all these "pulls" together and combine them into a single force, the place where this force is concentrated would be the shot's **center of gravity.** In a perfectly spherical iron shot, the center of gravity is dead center (i.e., in the middle of the shot and equidistant in any direction from its surface). In this case you'd find as much mass directly above

Cliff Divers Respect Gravity's Acceleration

Cliff divers in Acapulco, Mexico, dive from 118 ft and hit the water at close to 60 mph—more than double the top speed reached by an Olympic sprinter in the 100-m dash. The extra distance that the divers fall (compared with just over 32 ft for a tower diver) gives gravity more time to accelerate them. To clear the cliff face on the way down, divers must thrust outward in excess of 20 ft. They time their entries into the water so that incoming waves make the water deep enough. Failure can mean hitting the rocks a mere 16 ft below the surface!

the shot's center as directly below, and likewise to the left as compared to the right.

Here's a different example. Imagine a wooden ruler used for measuring short distances. You can balance this ruler as a seesaw on your fingertip at a point halfway in from the ends and halfway in from the sides. Your fingertip supports the ruler directly below its center of gravity, and there are equal amounts of the ruler's mass lengthwise and equal amounts widthwise from the ruler's center of gravity. What would happen if we attached a small piece of lead to one end of the ruler? To balance the ruler, you need to move your supporting finger toward the end of the ruler where the lead is attached. The lead plus a smaller section of the ruler now balances a longer section on the opposing side of your supporting finger. This occurs because the lead has a considerable amount of mass concentrated in the space that it occupies. The amount of mass in the lead plus a small section of the wooden ruler on one side of your finger balances the mass existing in the longer section of the ruler on the opposing side. The same situation occurs when you balance a metal hammer with a wooden handle horizontally on your finger. The balance point would be very close to the head of the hammer.

An athlete's body is similar to the example of the hammer because it is not made of the same material throughout and an athlete's body mass is not uniformly distributed from his head to his feet. Instead, an athlete's body is made up of different substances, such as bone, muscle, fat, and tissue, all of which have different densities and different shapes. **Density** refers to the amount of substance (or mass) contained in a particular space (i.e., quantity of mass per unit volume). Bones and muscles are more dense and have more mass concentrated in the space they occupy than an equal volume of body fat. This means that there is more attraction between the earth and a cubic inch (or a cubic centimeter) of bone or muscle than between the earth and the same volume of body fat. The result of these variations in density is that an athlete's center of gravity is seldom equal in distance from the athlete's head as from the athlete's feet. But even though it is not equidistant from head to feet, an athlete's center of gravity "positions itself" so that there will be as much mass directly above his center of gravity as directly below, and as much mass to the left of his center of gravity as there is to the right.

Shifting the Center of Gravity

Living beings can shift their center of gravity from one position to another, and training teaches athletes to position their center of gravity in certain ways so that they can produce an optimal performance. During a golf drive, when athletes shift their body weight from the rear foot to the forward foot, they are in essence shifting their center of gravity so that they have optimal stability for the application of force to the ball. You will see the same repositioning of an athlete's center of gravity in all sport skills, irrespective of whether they are performed in the air, on the land, or in the water.

The center of gravity for most adult male athletes, standing still with their arms by their sides, is approximately at belt level, or about an inch

G Forces

Designers of roller coasters give riders the opportunity to safely experience accelerations of up to 3 to 4 gs. The rider's normal body weight is 1 g, so at 3 to 4 gs the rider experiences an acceleration that makes him or her feel three to four times heavier. There are limits to the g forces that a healthy individual can tolerate, and 3 to 4 gs is considered to be the limit for most roller coasters. How many g forces have humans experienced? In 1954, Col. John Stapp of the USAF rode a rocket sled that reached a speed of 632 mph in 5 sec. The sled was then brought to a halt in 1.25 sec. Such phenomenal deceleration subjected Col. Stapp to 40 gs and momentarily raised his body weight to 6,800 lb! Col. Stapp survived this ordeal and died in 1999 at the ripe old age of 89. Since Col. Stapp's exploits, using special survival suits humans have experienced over 80 gs.

above the navel. The center of gravity of female athletes is usually a little lower. The reason for the difference is that males tend to have more body mass in the shoulders and less in the hips, whereas for females the reverse occurs. Several factors cause the center of gravity of athletes to shift from the average positions that we've just indicated. An athlete who naturally has long and heavily muscled legs and a lighter build in the upper body will have a center of gravity that is positioned lower on his body than average. Through training, athletes can also change the position of their center of gravity. For example, a bodybuilder who works out for years on his upper body and neglects to develop his legs will shift his center of gravity toward the more massive upper parts of his body. But by far the most important factor is that all of us can maneuver our center of gravity through the movement of our limbs. If an athlete stands erect and then moves a leg forward to take a step, the athlete's center of gravity shifts in the same direction. If the athlete moves the leg plus an arm, the center of gravity shifts forward even farther.

The distance that an athlete's center of gravity shifts from one position to another depends on how much of the athlete's body mass is moved and how far it's moved. Legs usually have a lot of mass, so they cause a greater shift in the center of gravity than moving one arm by itself. Flexing at the waist shifts the center of gravity, as does tilting the head. The shift of an athlete's center of gravity always relates to the amount of mass and the distance that it is moved (see figure 2.2).

What happens if an athlete hoists a heavy barbell above her head? In this situation, consider the combined center of gravity of the athlete's mass plus that of the barbell. Hoisting a heavy barbell to arm's length above the head shifts the combined center of gravity of athlete and barbell a considerable distance in a vertical direction. In addition, the athlete has raised the mass of her arms above her head. The longer and more massive the athlete's arms and the more massive the barbell, the farther the combined center of gravity will move (see figure 2.3).

If the athlete lets go of the barbell, then her center of gravity immediately reestablishes itself relative to her body position. The barbell, of course, will have a center of gravity of its own.

Is it possible to shift the center of gravity outside the body? Yes, and the more flexible the athlete the easier it is. A diver performing a toe touch in a piked position reaches forward with the arms and flexes at the waist. This causes the athlete's center of gravity to move forward to a position where it is no longer within the body (see figure 2.4).

A gymnast performing a high back arch or a back walkover also shifts the center of gravity into a position where it is temporarily outside the body. The center of gravity moves in relation to the shift in mass of the legs, upper body, and arms. The more extreme the arch, the greater the shift of the center of gravity (see figure 2.5).

A gymnast's back arch position is much like the draped layout position a flop high jumper uses when clearing the bar. A very flexible athlete

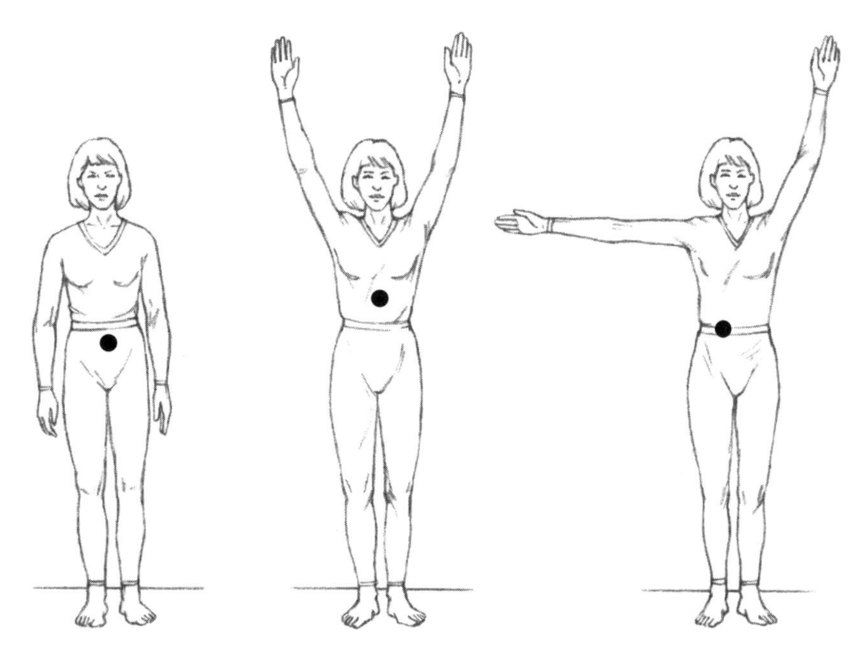

FIGURE 2.2 An athlete's center of gravity shifts as the body's position changes.

FIGURE 2.3 Combined center of gravity of an athlete with a heavy barbell overhead.

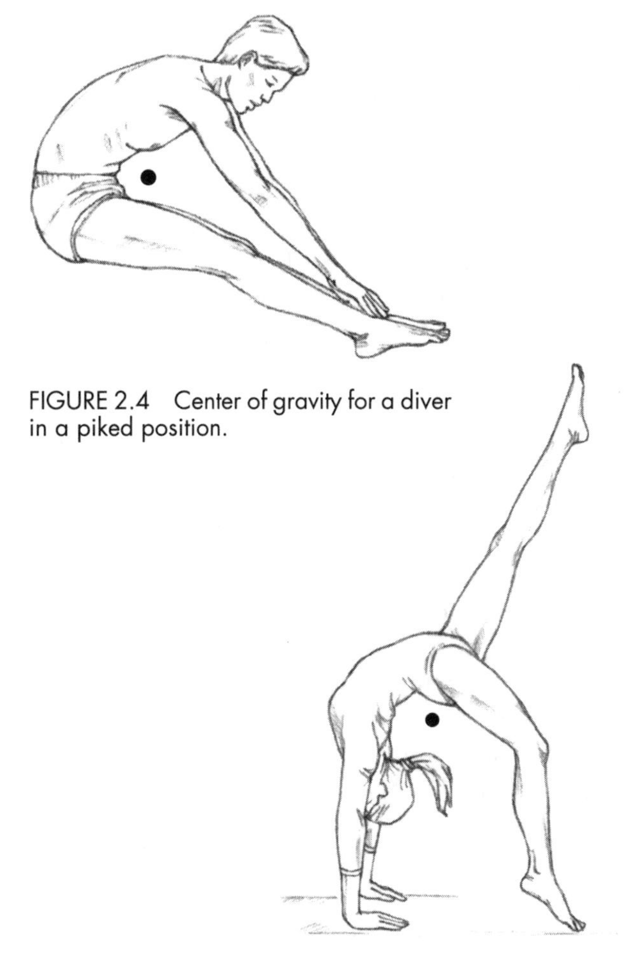

FIGURE 2.4 Center of gravity for a diver in a piked position.

FIGURE 2.5 Center of gravity during a back walkover.

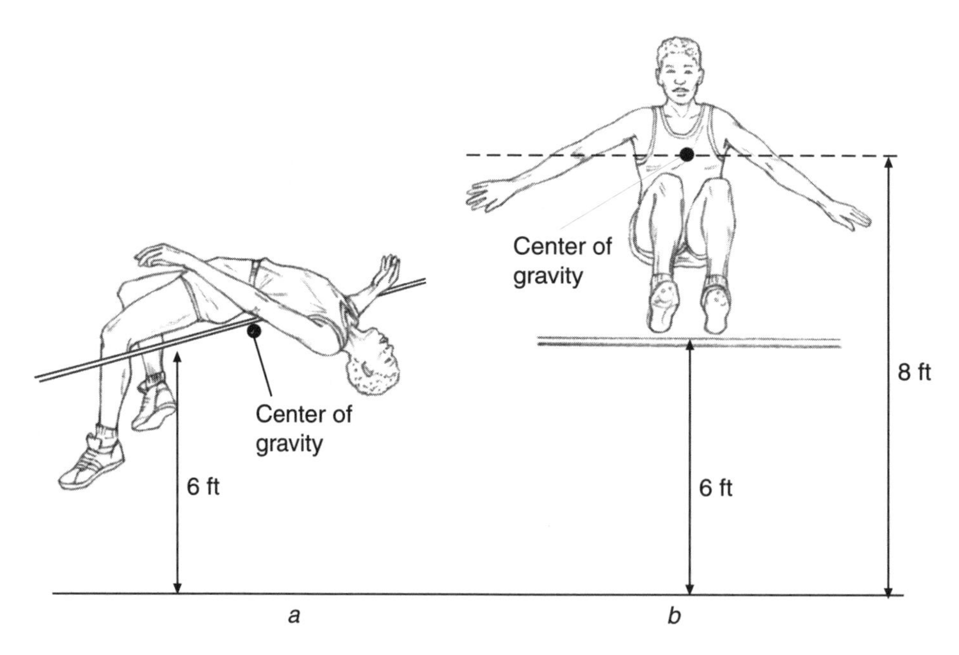

FIGURE 2.6 Height that an athlete's center of gravity must be raised to clear a 6-ft bar using *(a)* the flop technique or *(b)* a squat jump.
Adapted, by permission, from G. Dyson, 1986, *Mechanics of athletics*, 8th ed. (London: Hodder & Stoughton Educational), 168.

can make his center of gravity pass under the bar while his body snakes its way over the top. The benefit of this draped bar clearance technique can be easily understood if the same athlete clears the bar using a squat jump rather than the flop technique. If the athlete clears 6 ft (1.83 m) using the flop technique, he would need to raise his center of gravity close to 8 ft (2.44 m) to clear the same height using a squat jump (see figure 2.6).

Just where an athlete should position his center of gravity depends on the demands of the sport skill the athlete is performing. Athletes in wrestling and judo constantly reposition their center of gravity to increase their stability relative to the actions of their opponents. In the set position in sprint races, athletes shift their center of gravity in the direction they want to sprint so that there's no time wasted getting up and out of the blocks when the gun goes off. Gymnasts balancing on the beam make sure that their center of gravity stays centered above the beam; in track and field throwing events, athletes position their center of gravity so that they can apply the greatest amount of force over the greatest time frame to the implement. In all sport skills, quality performances require precise positioning of the athlete's center of gravity. When you get to chapter 5, you will explore this topic in more detail.

How Gravity Affects Flight

In events in which the athlete is in flight for a short time (e.g., high jump, long jump, gymnastics, figure skating, trampoline, and diving), the flight path of the athlete's center of gravity is set at takeoff. Some force will be applied in a vertical direction and some in a horizontal direction. The combined effect, or **resultant,** of these two forces sets the athlete's takeoff trajectory. In flight, the earth's gravity pulls at the athlete's center of gravity just as it does on the surface of the earth, and this downward tug gives the athlete's flight path its familiar parabolic (i.e., curved) shape. Once in the air, it's impossible for the athlete to alter the flight path that was set at takeoff. So if a diver makes the mistake of thrusting straight up from the board (i.e., placing all of her thrust in a vertical direction), there's nothing she can do to avoid coming back down on the board. Likewise, a flight path that is close to the horizontal means that the diver shoots out across the pool without enough height (and therefore not enough time) to perform all the twists and somersaults required in the dive. Gravity fights against the small amount of vertical thrust that the athlete put into the dive. Crazily waving the arms and legs around like kids at the local swimming pool will not change this flight path.

In contrast to springboard and tower divers, trampolinists want to rise vertically above the middle of the trampoline and then drop down to the trampoline in the same way. They prefer not to "travel" along the bed of the trampoline. Springboard divers train on the trampoline, but they must learn to move away from the diving board sufficiently to avoid contact. Even extraordinary divers like multiple gold medalist Greg Louganis have been punished for a flight path that did not allow sufficient clearance from the end of the board.

How the Reaction Force of the Earth Acts on Athletes

Because of the attraction of their mass to other objects that have mass, athletes standing on the surface of the earth pull upward on the earth. The more massive they are, the greater their pull. At the same time they are pulled down toward the earth's core by the earth's gravitational force. Athletes press down against the surface of the earth with a force equal to how much they weigh. So athletes pull up on the earth with the same force that the earth pulls down on them. In addition, where they come in contact with each other, the earth reacts to the athletes' weight by pushing against them with an equal and opposite force. On the surface of the earth, the pull down and the push up cancel each other out.

The force pushing up against the athletes is commonly called a **ground reaction force.** And like all the other equal and opposite forces that occur in the scenario we've described, the ground reaction force is an example of Isaac Newton's third law, which tells us that every action has an equal and opposite reaction. If you push, press, or hit something, it's going to do the same back to you. An easily visible action and reaction is the recoil of a rifle when it is fired. Less visible is the huge mass of air that must be deflected toward the earth so that a jetliner can take off and rise above the earth's surface. And more difficult to visualize is the outcome when an unhappy player punches a locker after a poor game. The player gets punched back by the locker! Don't worry if this latter example is difficult to picture. In subsequent chapters we'll enlarge on Newton's three laws of motion and the phenomenal contribution he has made to our understanding of the world around us.

The magnitude (i.e., size) of the earth's reaction force pushing against an athlete depends on how much the athlete pushes against the earth's surface. So the earth's reaction force depends not only on how much the athlete weighs but also on whatever movements the athlete makes. For example, in figure 2.7 an athlete is landing at the end of a long jump. In this situation the athlete exerts considerable force on the earth. The earth responds with an equal force against the athlete in the opposing direction. Soft sand

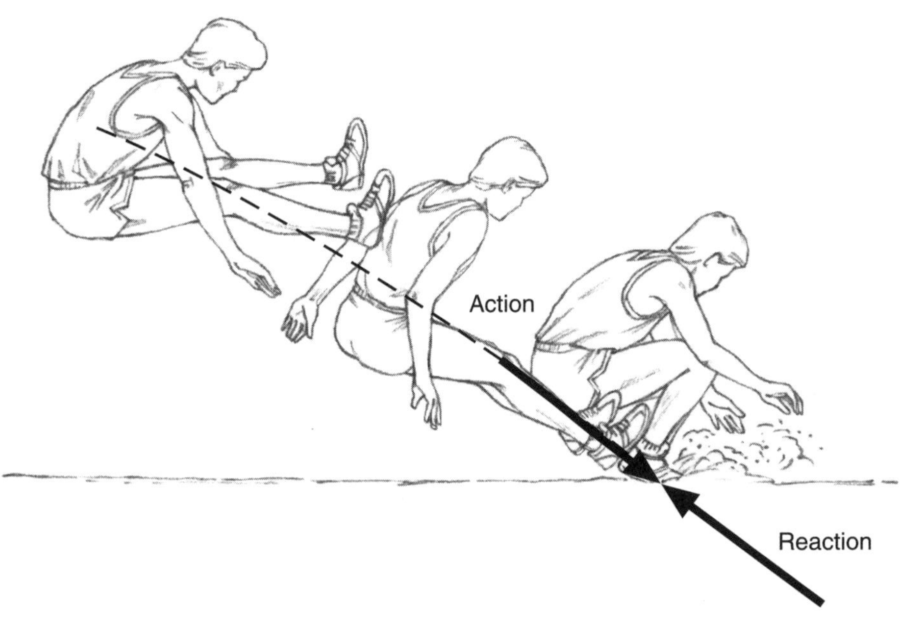

FIGURE 2.7 Action and reaction in the long jump. The action is the athlete landing; the reaction is the earth pushing back equally and in opposition to the athlete.

makes this force-counterforce comfortable and injury free for the athlete.

The forces of an athlete pushing down and the earth pushing back are crucial in determining how much friction occurs between the athlete and the earth's surface. Friction is necessary for traction, and traction is essential for movement.

In all sports, the amount of friction and traction required depends on what the athlete wants to do. Sometimes an athlete wants friction and traction to be maximal and at other times minimal. Excellent examples of this variation in frictional requirement occur in snowboarding. Snowboarders talk of varying the pressure that they exert on the board. This increases and decreases friction with the snow and assists in carving turns and adjusting to variations in terrain. Pressure can be varied by flexing and extending the legs. Extending the legs temporarily increases pressure on the board. Once extension ceases, the snowboarder's body (because of its inertia) continues to rise for an instant. At that instant the snowboarder becomes un-weighted. A reduction in friction with the snow (and a reduction in the ground reaction force) allows quick directional movements to be made. Interestingly enough, the same process can occur by flexing the legs. When a snowboarder flexes her legs, her body accelerates downward, temporarily reducing (not increasing) the pressure on the board. When leg flexion ceases, pressure on the board increases. How much can a snowboarder increase and then reduce the pressure on the board? Both depend on how vigorously and how quickly extension or flexion is carried out. These same mechanical principles apply to skiers and there's no better place to see them in action than in a moguls competition!

In flight, an athlete will notice much more readily how every action produces equal and opposite reaction. Move part of the body one way and another part visibly moves in the opposing direction. Divers, gymnasts, trampolinists, and trapeze artists make full use of this law of equal and opposite reactions when they somersault and twist. If you extend your arms in front of you and swing them hard to the left while standing erect on the surface of the earth, you don't see any reaction—friction by way of the soles of your shoes joins you to the surface of the earth, and the earth itself reacts by rotating to the right.

Of course the mass of the earth is so enormous that its movement to the right is infinitesimal, and we simply say that the earth "absorbed" the reaction. Is it possible to see the force-counterforce of action and reaction on the ground? Yes. Sit upright in a swivel chair with your feet off the ground. Extend both arms horizontally in front of your body. Then swing both arms vigorously to the left. You'll find that your legs and lower body rotate to the right. Your arm movement is the action, and the movement of your legs and lower body is the reaction. Figure 2.8 shows this action and reaction performed on a special turntable designed to have very little friction so it moves easily in relation to the movements that you perform. The turntable simulates a situation that would occur if you were in flight. An explanation of how athletes in flight make use of rotary action-reaction will occur in chapter 4.

Force

Whenever an athlete performs a sport skill, the athlete primarily produces force within the body by contracting the muscles. The muscles contract and pull on tendons and the tendons on the bones. The force produced by the athlete then competes against the external forces produced by gravity, ground reaction force, friction, air re-

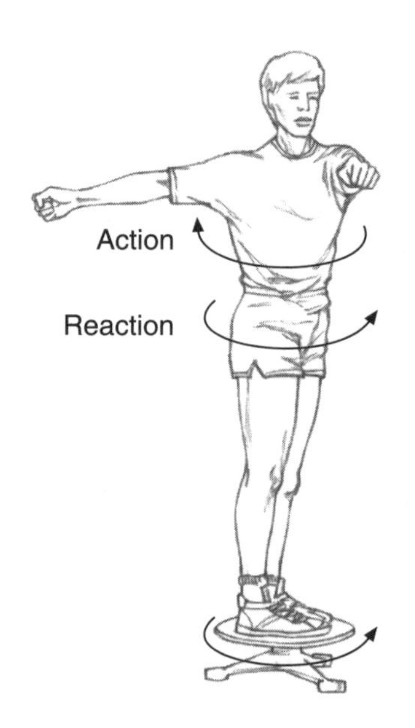

FIGURE 2.8 Action and reaction demonstrated on a turntable.

sistance, and in many sports, the contact forces provided by opposing players.

What exactly do we mean by force? You cannot actually see a force, but you can see and experience its effects. A **force** is a push or a pull that changes or tends to change the shape or the state of motion of an athlete or an object. Here's an example to explain what we mean by "tends to."

Imagine a weightlifter attempting to lift a barbell from the floor in a vertical direction. The athlete reaches down and pulls on the bar. If he pulls hard enough and applies sufficient force, the barbell is hoisted upward. But what happens if he doesn't apply enough force to move the barbell? In this situation you could say there is a tendency for the athlete to set the barbell in motion—it is closer to moving with the athlete pulling on it than if he didn't pull on it at all.

If another athlete adds his force in the same direction as that of the original lifter, maybe their combined force would be sufficient to move the barbell off the ground. The tendency toward movement caused by the first athlete is turned into action by help from the second. The barbell moves. In this scenario you must assume that both athletes pull in the same direction. A totally different effect occurs if the second athlete pulls the barbell sideways rather than upward.

Force Vectors

In the weightlifting scenario, we imagined that two lifters combined their muscular force to lift a barbell. The combination of their forces totaled a certain amount and was aimed in a particular direction. When the direction and amount of the applied force are known, the combination of these two items is called a **force vector.** The term *vector* simply means a quantity that has direction. In the case of the weightlifter, a certain amount of force was vectored in a vertical direction.

In mechanics, force vectors are often represented diagrammatically by arrows. The head of the arrow indicates in what direction the force is acting, and the length of the arrow represents the amount of force being applied.

In our weightlifting example, if one athlete lifts vertically and the other pulls the bar horizontally, the result is that the two athletes pull the barbell partially upward and partially to the side. Depending on the amount of force applied by each athlete, the barbell moves in the direction of the **resultant force vector.** The resultant force vector in this case is the equivalent of two forces that simultaneously pull the barbell in different directions. A representation of what occurs when two forces are applied in different directions against an object is shown in figure 2.9, which shows two athletes pulling on a large resistance. Athlete A applies 10 units of force toward the north. Athlete B applies 10 units of force toward the east. Arrow A represents 10 units of force and arrow B represents 10 units of force. Each unit of force is given the same dimension (one centimeter per unit, or, if preferred, one inch per unit). A parallelogram is drawn with opposite sides and angles equal. (In this particular

Newton's Universal Law of Gravitation

Newton's universal law of gravitation tells us that if we were to move away from the surface of the earth, our weight would diminish inversely by the square of the distance that exists between us and the earth. For example, if astronauts were twice the radius of the earth (i.e., 2 × 3,958 miles) in distance from the earth's core, they would weigh 1/4 of what they weighed on the surface of the earth. At three times the radius of the earth in distance from the earth's core (i.e., 3 × 3,958), they would weigh 1/9 of their weight on earth. Four times the distance and they would weigh 1/16 of their weight on earth. This progressive "reduction" in gravitational attraction tells us that the earth's gravitational attraction never fully disappears. It just gets less and less the farther away from the earth that you travel.

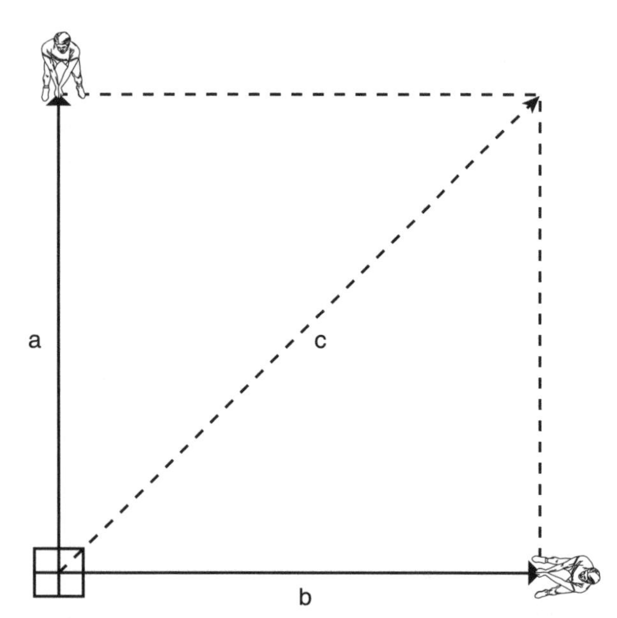

FIGURE 2.9 A parallelogram of forces. Arrow *a* shows the force applied vertically by the athlete; arrow *b* shows the force applied horizontally by the athlete; arrow *c* shows the resultant force vector.

case, each side will be 10 cm [or in.] in length and each angle will be 90 degrees.) The diagonal C gives us the resultant force vector, which will be the direction of force resulting from the combined efforts of athletes A and B. Careful measurement indicates that C is 14.14 units in length and represents 14.14 units of force. This also means that a single force of 14.14 units (e.g., C) acting at 45 degrees from the horizontal can be split into two forces—A, which has 10 units of force, acting vertically, and B, which has 10 units of force, acting horizontally. To find the outcome of several forces acting on a single object, you would need to know (1) the number of forces involved, (2) the magnitude of each individual force, and (3) the direction that each force is applied.

When an athlete performs a sport skill, several forces usually act at the same time. Let's look at these forces at work in the shot put event. Think of elite athletes putting a shot at a release angle of about 42 degrees to the horizontal. To give the shot some height, athletes must apply force in that direction. So the athletes apply some (but not all) of their force in a vertical direction. To get the shot moving horizontally, they apply force in that direction as well. The combination of horizontal and vertical forces gives the shot its 42-degree trajectory.

Obviously the shot-putters cannot apply all of their force in just a vertical or horizontal direction. If an athlete is foolish enough to put all his force in a vertical direction, the shot will go straight up and come straight down. This is hardly the desired result in a competition that is won by achieving the greatest horizontal distance. On the other hand, if all of the thrower's force is directed horizontally, the shot will hit the ground long before it has time to cover the optimal distance. The ideal trajectory angle is partway between horizontal and vertical.

During flight the earth's gravitational force pulls the shot directly downward (perpendicularly). Gravity only fights against the vertical force vector that the athlete applied to the shot. Gravity is not interested in the horizontal force vector. In addition to the force of gravity, air resistance provides a force that battles the forward motion of the shot. The result of this war of forces determines the distance the shot travels (see figure 2.10).

There are many examples in sport of athletes combining forces to produce a desired result. Top-class soccer players know from experience how long it takes a soccer ball to travel a particular distance. They assess the speed of the forwards on their team as they sprint into open field positions. When the player with the ball makes a downfield pass, several factors must be taken into consideration, including the direction and strength of the wind and the velocity of the forward sprinting to receive the pass. If the ball is kicked with the correct amount of force and given the right trajectory, it drops at the feet of a forward who is running flat out. The same principles apply to a quarterback who wants to hit a receiver cutting across the field, or a basketball player attempting to hit a teammate who has broken away from the opposition. In all cases, the passers perform a mental vector analysis to make sure that the ball arrives at a particular spot at the same time as their teammate.

An Athlete's Movement

An athlete can move in three different ways. Movement can be linear (i.e., in a straight line),

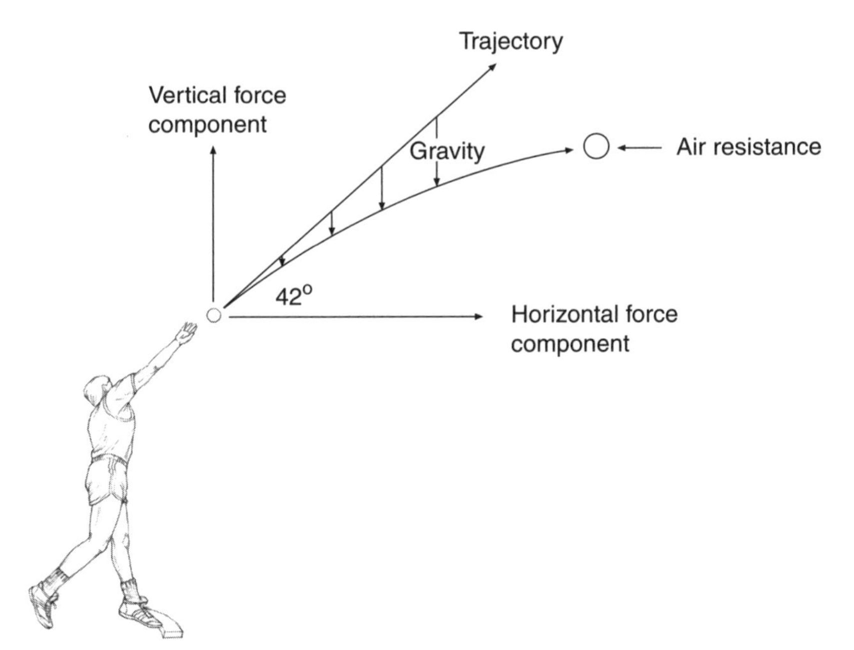

FIGURE 2.10 Forces influencing the trajectory of a shot in the shot put event.

angular (i.e., in a circular, or rotary, fashion), or a mix of linear and angular motion, which we simply call general motion.

In sport, a mix of linear and angular movement most commonly occurs. Angular movement plays the dominant role because most of an athlete's movements result from the swinging, turning action of the athlete's limbs as they rotate around the joints.

Linear Motion

Linear motion describes a situation in which movement occurs in a straight line. Linear motion can also be called **translation** but only if all parts of the object or the athlete move the same distance, in the same direction, and in the same time frame. As you can imagine, translation rarely occurs in an athlete's movement because some parts of an athlete's body can be moving faster than other parts and not always exactly in the same direction. For example, an athlete in the 100-m sprint wants to travel the shortest distance from the start to the finish. The shortest distance is a straight line. Yet sprinting is produced by a rotary motion of the limbs as they pivot at the athlete's joints, and the athlete's center of gravity rises and falls during each stride.

Angular Motion

Angular motion is described by many names. Coaches talk of athletes rotating, spinning, swinging, circling, turning, rolling, pirouetting, somersaulting, and twisting. All of these terms indicate that an object or an athlete is turning through an angle, or number of degrees. In sports such as gymnastics, skateboarding, basketball, diving, figure skating, and ballet, the movements used by athletes include quarter turns (90 degrees), half turns (180 degrees), and full turns, or "revs" (i.e., revolutions), which are multiples of 360 degrees. Slam dunk competitions are a great example of basketball players showing off their "360s."

To produce angular motion, movement has to occur around an axis. You can think of an axis as the axle of a wheel or the hinge on a door. An athlete's body has many joints and they all act as axes. The most visible rotary motion occurs in the arms and legs. The upper arm rotates at the shoulder joint, the lower arm at the elbow joint, and the hand at the wrist. The hip joint acts as an axis for the leg, the knee for the lower leg, and the ankle for the foot. Movements like walking and running depend on the rotary motion of each segment (e.g., foot, lower leg, and thigh) of an athlete's limbs as they rotate around the joints.

General Motion

All human motion is best described as general motion that is a combination of linear and angular motion. Even those sport skills that require an athlete to hold a set position involve various amounts of linear and angular motion. A gymnast balancing on a beam and the aerodynamic crouch position during the acceleration prior to take-off in ski jumping are good examples. In maintaining balance on the beam, the gymnast still moves, however slightly. This movement may contain some linear motion but it will be made up primarily of angular motion occurring around the axes of the athlete's joints and where the gymnast's feet contact the beam. The ski jumper holding a crouched position attempts to reduce air resistance to a minimum to accelerate as much as possible prior to takeoff. Sliding down the inrun holding a crouched position is a good example of linear motion. But the athlete never fully maintains the same body position throughout, and the inrun is not straight throughout, so any motion that the ski jumper makes will be angular in character. Perhaps the most visible combination of angular and linear motion occurs in a wheelchair race. The swinging, repetitive angular motion of the athlete's arms rotates the wheels. The motion of the wheels carries both the athlete and the chair along the track. Down the straightaway, the athlete and chair can be moving in a linear fashion. At the same time the wheels and the athlete's arms exhibit angular motion (see figure 2.11). This combination of angular and linear motion is an example of general motion.

Projectiles

There are many events in which athletes and objects are projected (or propelled) through the air to be caught or to land on the ground or in water. Golf balls, basketballs, baseballs, and javelins fly through the air, and athletes in ski-jumping, trampoline, trapeze, and diving become **projectiles** themselves. These sports all require an athlete to manipulate, control, or assess the flight path that occurs. For example, an archer angles the bow and pulls the strings back just the right amount so that the arrow flies to the bull's-eye. A high jumper aims for height, distance, and rotation so that the bar is successfully cleared. A diver looks for a flight path that gives adequate time in the air to perform all the required twists and somersaults yet still line up for a splashless entry. A goalkeeper assesses the velocity and flight of the ball or puck to make a successful save, and a tennis player tries to land the ball in the correct place in the opposing court yet where the opponent is unable to return it successfully.

In events that incorporate flight, several factors are "vectored" in to influence the character of the flight path. An athlete takes off at an angle in jumping events. A baseball is given a trajectory angle during a pitch or when the ball is thrown or hit. Jumpers vary the speed they use during

FIGURE 2.11 A wheelchair athlete exhibits a combination of angular and linear motion.

FIGURE 2.12 A baseball pitcher delivers the ball at a particular height (a) and gives the ball a particular velocity (b) and angle of release (c).

Reprinted, by permission, from, S.J. Hall, 1995, *Basic biomechanics*, 2nd ed. (St. Louis, Mosby-Year Book, Inc.), 313.

takeoff, and as for the baseball, the speed at which the ball comes off the bat or is released from the pitcher's or fielder's hand varies as well. Finally, you need to consider the height at which the athlete takes off or the height at which a baseball is hit or thrown (see figure 2.12).

Trajectory

To understand how all these factors are interrelated, let's consider the flight of the baseball. We'll start by eliminating gravity and air resistance, and then we'll get a pitcher to throw a baseball so that it is released at an angle of 35 degrees above the horizontal. To produce the 35-degree **trajectory,** the pitcher must apply slightly more force to the ball in a horizontal direction than in a vertical direction. Without the presence of gravity and air resistance, a baseball released at an angle of 35 degrees will fly indefinitely on and upward at the speed the pitcher applied to the ball at the instant of release. (This is Newton's first law of inertia at work!)

In reality, we all know that gravity pulls a baseball toward the earth, and without considering the effects that spin causes, we also know that air resistance will counteract the forward motion of the ball as it rises and falls in the air. Gravity and air resistance change the flight path of the baseball from its trajectory, set by the pitcher at 35 degrees, to the familiar curved flight path in which the ball rises to a certain point and then arcs back toward the earth's surface. Gravity battles the ball's rise upward from the surface of the earth. It finally stops the ball's upward motion and changes it so that the ball falls back toward the earth. Remember that gravity pulls perpendicularly toward the earth's surface and so does not counteract the force that the pitcher applied to the ball in a horizontal direction. Gravity only battles the force applied in a vertical direction.

To see how angle of release, speed of release, and height of release combine to influence the distance (or range) that a baseball travels during flight, let's look at each of these factors in turn.

Angle of Release

When a pitcher hurls a baseball, the shape of the ball's flight path depends on the angle at which the athlete releases the ball. The speed with which the athlete throws the ball then determines the size of the flight path. So the ball's flight path can be of a different shape, and each shape can vary in size. If you discount air resistance, the shape of the ball's flight path will be one of three types:

1. If the pitcher hurls the baseball straight up, the ball goes directly upward and is pulled straight down again by gravity. The flight path is a straight line. Gravity decelerates the ball on the way up and accelerates the ball on the way down.
2. If the pitcher hurls the ball at an angle between vertical and horizontal, an angle of release that is above 45 degrees (i.e., closer to the vertical) gives the ball a trajectory in which height dominates over distance.
3. If the pitcher hurls the ball at an angle below 45 degrees (i.e., closer to the horizontal), the flight path is long and low. Distance dominates over height.

Speed of Release

What happens if an object's speed of release is varied? If the pitcher hurls a ball straight up, an increase in the speed of release is obviously going to make the ball go higher. The **apex** of the flight path (the highest point) is raised as the speed of release is increased. It's the same in a vertical jump. The faster the athlete's takeoff speed, the higher the athlete rises.

When the pitcher hurls a ball at an angle between vertical and horizontal, any increase in the ball's speed of release increases not only the height of the ball but also how far it travels.

Height of Release

The third important factor that influences the flight path of a ball is the height of release relative to the height of the surface on which it lands. Golfers most often hit the ball from ground level and frequently have to land it on a fairway or green that is at another level. Baseball pitchers and batters release and hit the ball above ground level. In track and field, giant shot-putters release the shot well above shoulder level, with the shot then landing at ground level. Some shot-putters are taller than others, so body type can increase the height of release.

If it were possible for a shot-putter to release a shot at ground level and the ground was perfectly horizontal, then 45 degrees would be the best release angle for the greatest distance. In this situation the athlete puts equal amounts of force in a vertical and horizontal direction. The force applied in a vertical direction is used to battle the downward pull of gravity. However, when an athlete releases a shot above ground level, as all shot-putters do, to get the greatest horizontal distance the athlete must lower the trajectory angle and release the shot at slightly less than 45 degrees. Elite athletes release the shot at an angle that ranges from 35 to 42 degrees (see figure 2.10). This is a large trajectory angle if you compare it with the angle used by ski jumpers when they take off. (Think of the takeoff of the ski jumper in the same manner as the release of the shot. The ski jumper, like the shot, is a projectile. Both fly through the air and both athlete and shot are projected for maximum distance.) Ski jumpers take off from a ramp and land on a surface that slopes downward, similar to their flight path. So they thrust themselves off at an angle that is close to horizontal.

Long jumpers take off from ground level and want to travel as far as possible, so you might guess that their takeoff angle would be 45 degrees. But this is not the case. These athletes actually take off at an angle between 20 and 22 degrees (see figure 2.13), and the angle of takeoff is even less for triple jumpers. Both types of jumpers would be forced to cut their speed down the runway to take off at an angle of 45 degrees. No long jumper wants to do this because a reduction in approach speed drastically reduces the distance traveled in flight and the distance jumped. So long jumpers compromise between velocity at takeoff and takeoff angle. Velocity is the more important factor, and as a result, the takeoff angle is reduced from 45 degrees to 20 to 22 degrees.

Velocity, height, and angle of takeoff (or release) are all interrelated. Altering one component inevitably causes changes in the others. Aerodynamic and environmental factors must

Ball Speed

Humans may not be able to run as fast as animals but when athletes hit with a bat, racket, or club, the speeds that result are phenomenal. Baseball and softball pitches have been measured at over 100 mph and volleyball spike-serves can leave an elite athlete's hand at around 75 to 80 mph. A serve in tennis can be in excess of 130 mph, a slap shot in hockey can reach up to 125 mph, a golf drive can be around 150 mph, and a badminton shuttle can be smashed at 168 mph. Winners in this list are jai alai balls, which have been measured at speeds over 180 mph.

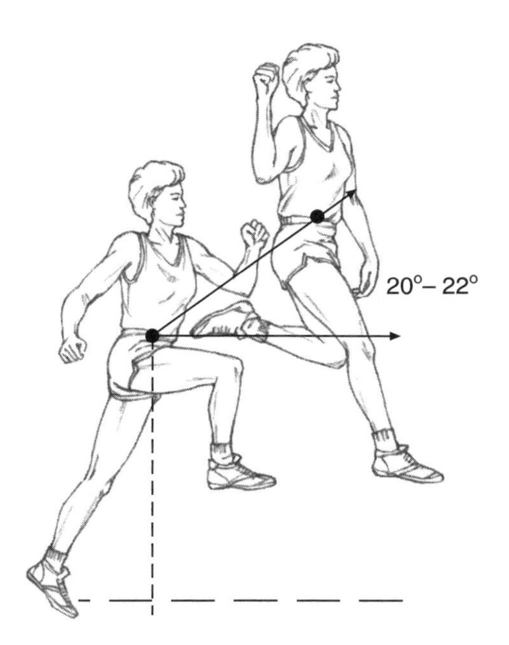

FIGURE 2.13 Takeoff angle in the long jump.
From E. Kreighbaum and K.M. Barthels, *Biomechanics: A qualitative approach for studying human movement.* Copyright © 1990. All rights reserved. Adapted by permission of Allyn & Bacon.

also be taken into consideration. The javelin and discus, for example, are very much affected by the way the wind is blowing. Headwinds approaching from a favorable angle and blowing at an optimal velocity (about 15 to 20 mph for the discus) can dramatically increase the distance the athlete throws. With a headwind, the thrower not only reduces the angle of release but also lowers the leading edge of the discus relative to air flowing past it. The angle of the discus relative to the airflow is called the **angle of attack.** If the angle of attack is too great, the discus stalls in flight and the distance is dramatically reduced. In chapter 6 you'll read more about aerodynamic factors and how they affect the flight of discuses, javelins, baseballs, and athletes such as ski jumpers.

SUMMARY

- An athlete's body weight results from both the earth's gravitational pull on the athlete's body and the pull of the athlete's body on the earth. The earth's gravitational attraction varies according to location.
- All objects that have substance have mass. An athlete's body has mass, commonly termed body mass.
- Mass is directly related to inertia: the more mass, the more inertia.
- Inertia is characterized by resistance and persistence. All objects resist movement because of their inertia. Once movement is initiated, an object's mass (and inertia) is expressed by a tendency to continue moving at a uniform speed in a straight line. This would occur if forces such as gravity, air resistance, and friction did not intervene. Massive athletes require the application of great force to get moving and the application of great force to stop or change their direction. The characteristics of inertia are described in Newton's first law.
- Speed is a scalar measure indicating how fast an object is traveling at a particular instant in time. An athlete who runs 100 m in 10 sec has an average speed of 10 m/sec. Average speed does not tell us the athlete's top speed.
- Acceleration and deceleration refer to the rate that speed changes. An athlete who increases speed at 5 ft/sec for every second in the sprint is accelerating at 5 ft/sec/sec, or 5 ft/sec^2. Acceleration and deceleration are indicated with one distance unit (i.e., 5 ft) and two time units (i.e., sec/sec, or more commonly, sec^2).
- Velocity indicates both speed and direction.
- The earth's gravitational acceleration is approximately 32 ft/sec^2 (9.8 m/sec^2 in the metric system). Slight variations in gravitational acceleration occur relative to location on the earth's surface.

- The center of gravity is where the earth's gravitational pull is concentrated in an object or in an athlete's body. Changes in body position and the density of body parts shift an athlete's center of gravity.
- Density refers to the amount of substance (or mass) contained in a particular space. Bones and muscles are more dense than body fat.
- An athlete standing on the surface of the earth is pulled by gravity toward the earth's core. The earth reacts against the downward force exerted by the athlete by pushing upward with an equal and opposing force. This is called the earth's reaction force (or ground reaction force).
- Force is a push or pull that changes, or tends to change, the state of motion of an object. Athletes primarily use muscle contractions to apply force. Among the external forces affecting an athlete are gravity, air resistance, friction, the earth's reaction force, and forces applied by objects and opponents.
- Athletic skills are made up of a mix of angular (rotational) and linear (translational or straight-line) movement. This mix of movement is called general motion. Because bones rotate at the joints, movements performed by athletes in sport skills are predominantly rotational.
- In many sports, objects or athletes are projected or propelled into the air. Their trajectories depend on their velocity, height, and angle of release. The forces exerted by gravity and air resistance help determine the resulting flight path.
- With no air resistance, a trajectory angle of 45 degrees produces the greatest distance for objects projected from ground level on a horizontal surface. When the object is projected from above ground level, an angle less than 45 degrees produces the greatest distance.

KEY TERMS

acceleration
angle of attack
angular motion
apex
center of gravity
deceleration
density
force
force vector
gravitational acceleration
gravitational attraction
ground reaction force
inertia
linear motion
mass
matter
negative acceleration
positive acceleration
projectile
resultant
resultant force vector
scalar
speed
trajectory
translation
uniform acceleration
uniform deceleration
velocity
weight

REVIEW QUESTIONS

1. Is it possible for an athlete with no change in body mass to weigh a certain amount at one place on the earth's surface and a different amount at another place?
2. A shot-putter carrying a shot lets it slip out of her hand and fall toward the earth. If action equals reaction, then the earth pulls on the shot with the same force that the shot pulls on the earth. If this is the case, why does the shot accelerate toward the earth?

3. The acceleration of gravity is 32 ft/sec^2 (i.e., 32 ft/sec/sec), or 9.8 m/sec^2 in the metric system. Why do we use one distance unit (i.e., 32 ft) and two time units (i.e., sec/sec) when we refer to gravity's acceleration?
4. What happens in a mechanical sense when a slalom skier weights and unweights her skis? Why does a skier carry out this maneuver?
5. What reasons relating to inertia did the text cite for elite squash and badminton players being lean and lightweight?
6. Explain how a quarterback can make a mental "vector analysis" to assure that the football arrives at a particular spot at the same time as the receiver cutting across the field.
7. Explain in everyday language Isaac Newton's first and third laws.

PRACTICAL ACTIVITIES

1. **Center of gravity.** With a pair of scissors, cut a 1-ft-square piece of cardboard (or use 1/8 in. plywood, if a saw and drill are available) into an irregular shape. Punch four or five holes around the perimeter that are large enough for a pencil to slide through. Tie a string two to three ft in length to the middle of the pencil, and at the other end tie any small weight (such as an eraser). Put the pencil through any hole in the cardboard. Allow the cardboard and the eraser at the end of the string to swing freely on the pencil. (The eraser at the end of the string will act as a plumb line.) With a felt pen, draw a line where the string drops across the cardboard. Repeat this action for all the holes punched around the perimeter of the cardboard. Draw a large circular mark where the lines intersect. Explain why this circular mark gives you the center of gravity of the cardboard. Using the cardboard alone and holding it by any edge, spin it (thin side uppermost) vertically up in the air in front of you. Why does the cardboard always spin around its center of gravity?
2. **Newton's first law of inertia.** Obtain a four-wheeled cart that has a large enough platform to carry a similar but smaller four-wheeled cart. Align the wheels so they all face the same direction. With your hand on the bottom cart, accelerate it quickly to the left. The smaller cart will roll to the right. Next, simultaneously accelerate both carts to the left and then suddenly stop the bottom cart. The smaller cart will continue to the left. Write an explanation of what occurred, demonstrating your knowledge of the characteristics of Newton's first law of inertia.
3. **Newton's third law of action and reaction.** Stand on a skateboard and throw a medicine ball to a partner who is standing on the floor. As the medicine ball is thrown in one direction, you will move in the other. Explain how the outcome is related to Newton's third law.
4. **Ground reaction forces.** Stand on a scale and take note of your weight. Then perform the following actions: (1) Starting in a position with extended arms beside your body, raise your arms quickly and vigorously outward and upward above your head; (2) Starting in a position with your arms extended above your head, swing your arms vigorously outward and downward toward the sides of your body. Check what happens to your weight as indicated by the scale when you perform these actions. Does your weight increase or decrease? Referring to ground reaction forces and Newton's third law, explain what occurs.

Getting a Move On

When you finish reading this chapter, you should be able to explain

- the factors that influence an object's or an athlete's acceleration;
- what is meant by impulse and how impulse relates to acceleration and deceleration;
- how an athlete's actions cause reactions that are equal and opposite;
- what is meant by momentum and how objects or athletes gain or lose momentum;
- the mechanical definition of work;
- the difference between power and strength;
- the relationship and differences between kinetic energy, gravitational potential energy, and strain energy;
- how momentum is conserved and kinetic energy dissipated;
- the factors that influence the rebound of balls; and
- the factors that control friction as objects slide or roll across contacting surfaces.

This chapter gets us involved with athletes and objects in motion, and that's why it's called "Getting a Move On." The mechanical principles we discuss in this chapter build on those we've already talked about in chapter 2. As you develop your knowledge about mechanics, you'll notice how all these principles tie in to one another. Equally important, you'll see that every action your athletes make in a sport skill involves several mechanical principles that occur simultaneously. Good technique in sport is based on making the best use of these mechanical principles. So let's get a move on by seeing what happens when athletes apply force with their muscles and put themselves and objects such as bats and balls in motion.

Forces at Work in a Sprint Start

Imagine an elite sprinter in a set position. The gun goes off and the sprinter drives out of the blocks. In this situation the sprinter applies muscular force and extends his legs against the blocks. The blocks, of course, are attached to the earth. The force (i.e., the push) that the sprinter applies against the blocks is the action. The reaction (the push back) comes from the earth pushing equally and in the opposite direction by way of the blocks against the sprinter (see figure 3.1). This parallels the action-reaction situations that we discussed in chapter 2, where skiers weighted and unweighted their skis and where a long jumper landing in the pit applied force against the earth and the earth pushed back equally and in the opposite direction against the jumper.

The force produced by the sprinter's muscles overcomes the inertia of the athlete's mass and the sprinter begins to accelerate. If there were no oppositional forces acting against the sprinter, then the athlete's legs extending against the blocks would cause the sprinter to continue moving indefinitely in the direction that his muscular force propelled him. Gravity, friction, and air resistance apply the brakes to this endless motion.

The acceleration of any sprinter's body mass is proportional to how much muscular force the athlete applies and the time frame in which the force is applied. It is also inversely proportional to the athlete's mass. This means that if two sprinters apply the same muscular force to their bodies for the same amount of time, the less massive of the two athletes accelerates more. Likewise, if two sprinters have the same mass and apply force for the same amount of time, the athlete who applies more force within that time frame will accelerate more. This scenario is a great example of the second of Isaac Newton's three laws of motion. This second law is called Newton's law of acceleration and as a simple formula is written $F = ma$ (force = mass $\times$ acceleration).

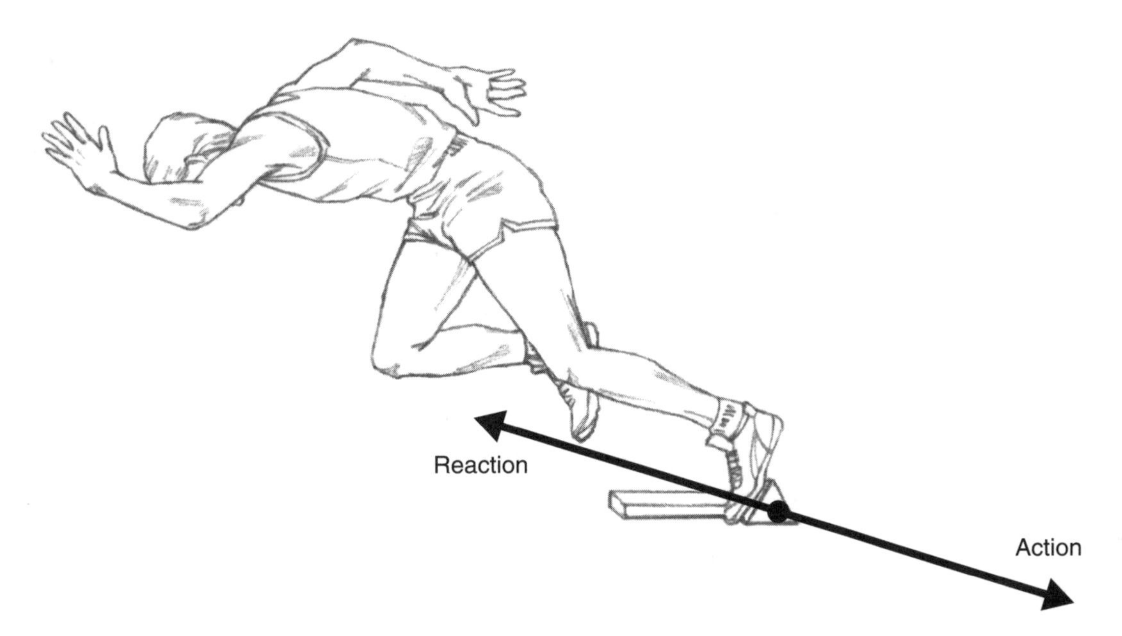

FIGURE 3.1 Action and reaction in a sprint start. The athlete applies force against the block. The earth (via the block) applies an equal and opposite force against the athlete.

By extending his legs powerfully at the start, a sprinter contracts his muscles to push against his body mass and simultaneously against the mass of the earth via the blocks. The athlete moves in one direction, and the earth (plus the blocks) moves a negligible amount in the opposing direction. The earth and the athlete move in opposite directions relative to their mass and their inertia.

You can picture the relationship between the sprinter and the earth by compressing a spring between a 16-lb shot (resting on the ground or another hard surface) and a table tennis ball. The shot is the earth, the spring is the athlete's muscles, and the table tennis ball is the athlete. If you let go of the shot and the table tennis ball at the same time, the spring expands. The table tennis ball accelerates in one direction, and the shot rolls a short distance in the opposing direction or maybe barely rolls at all. Now visualize increasing the size of the table tennis ball so that it's equal to the sprinter and increasing the size of the shot so that it's equal to the earth. This is why the sprinter moves in one direction relative to his mass (and his inertia) and the earth moves an immeasurable amount in the opposing direction relative to the earth's mass and its inertia. It's easy to see why the sprinter does the moving!

Momentum

An athlete who is moving is an example of mass on the move. Because the athlete's body mass is moving, we say that the athlete has a certain amount of momentum. **Momentum** describes the quantity of motion that occurs. How much momentum an athlete possesses depends on how massive the athlete is and how fast the athlete is traveling at the time. Increase the athlete's mass, velocity, or both, and in each situation you increase the athlete's momentum. Momentum's simple formula is $M = mv$ (momentum = mass × velocity).

Sport commentators frequently talk of a team gaining momentum, and by this they mean that one team is starting to dominate over the other. Politicians also talk of their campaigns gathering momentum. In mechanical terms, momentum has a far different meaning. It always takes into consideration both the velocity and the mass of an athlete or a moving object. A massive athlete sprinting at the same velocity as an athlete with less body mass has more momentum. Likewise, an athlete with minimal body mass sprinting at phenomenal velocity can have more momentum than an athlete who is more massive but moving slowly. But, if there's no movement, there's no momentum, no matter how massive an athlete might be.

To make up for a large difference in mass, athletes with little body mass must sprint at a much higher velocity to match the momentum of a more massive athlete. As an example, if a 300-lb (136-kg) lineman ambled through 100 m in 20 sec (which is really slow!), a 150-lb (68-kg) running back would have to flash through the same distance in 10 sec to produce the same momentum.

Momentum occurs any time an athlete or an object moves, and it plays a particularly important role in sports and situations where collisions occur. An easy way to think of momentum is to see it as a weapon that an athlete can use to cause an effect on another object or an opponent. A puck hit with immense velocity by a hockey player can have enough momentum to drive a goalie backward. Even though a puck is light and has little mass, NHL players can hit it over 100 mph. When the puck hits the goalie, both puck and goalie (plus pads, gloves, mask, skates, and stick) for an instant become a combined mass. The puck slows and loses some of its momentum. Because the goalie is driven backward, he gains momentum in that direction. A similar example occurs when a tennis ball is served at such velocity that it knocks the opponent's racket backward.

Hockey players skate at tremendous speed and can generate great momentum. They demonstrate their momentum in the bone-rattling body checks that characterize their sport. Offensive and defensive linemen in the NFL, with their immense size and great acceleration over 40 yd (36.6 m), also generate tremendous momentum. Like the hockey player, the lineman who has the most momentum at impact is likely to dominate in a collision with an opponent.

We have seen that if you increase either the velocity or the mass of an object or an athlete, momentum increases as well. If there's no velocity, then there's no momentum. The best way for an athlete to increase mass is to "pack on" quality muscle mass rather than fat. The extra muscle mass then provides the power to help the athlete move faster and maneuver more efficiently.

Remember that not all sport situations require maximum momentum. Many skills require momentum to be carefully controlled. For example, a punter is often required to make the ball go out of bounds as close as possible to the end zone. The momentum given to the ball has to be exact so that it travels a precise distance. This means that the momentum of the punter's kicking leg must be the right amount as well. Similarly, an outside shooter who fires at the basket from three-point range aims to put the basketball through the hoop and not have it ricochet haphazardly out to midcourt. The trajectory, spin, and momentum given to the ball must be exact the moment it is released.

Impulse

Muscular force has to be produced when athletes want to get moving or they want to accelerate an object such as a soccer ball and give it momentum. The force that athletes apply always takes time. When athletes apply force to an object over a certain time, we say that the athletes have applied an **impulse** to the object. Of course, athletes can also apply an impulse to their own bodies or to another athlete.

How force and time are combined depends on the physical capabilities of the athlete. An athlete who is strong and flexible can apply more force over a greater range (i.e., time frame) than an athlete who is weaker and less flexible. Equally important, the combination of force and time depends on the needs of the skill. Some skills, such as some of the punches used in boxing, require tremendous force to be applied over a short distance and a short time frame. Other skills require a lesser force to be applied over a large time frame. The variations in the use of force and time are limitless. Let's look at some examples to see how the requirements of sport skills vary the demand for force and the time of its application.

Impulse in a Javelin Throw

A javelin throw is an example of incredible force being applied to an implement over a long time frame. Power and flexibility are required in this event. After an approach run, an expert thrower accelerates the javelin by pulling it from way behind his body and releasing it far out in front. Long arms are beneficial, but more important is a backward body lean when entering the throwing position. In this way, the athlete applies force to the javelin over a long period. To a spectator it may not seem that the athlete is accelerating the javelin for very long because the whiplike action of a good javelin throw seems to happen so quickly. But when an elite thrower is compared with a novice, the distinction between the two athletes is easy to see, even to a person who is not familiar with the event. To start with, the elite thrower is much more powerful and applies more force at high speed to the javelin. Second, greater flexibility and careful technical training allow the elite athlete to accelerate the javelin over a longer time frame. Because of this, the impulse applied to the javelin by an expert thrower is far greater than that applied by a novice, and as a result the javelin moves at a tremendous velocity when it is released (see figure 3.2).

Impulse at Takeoff in the High Jump

High jump is similar to javelin in that both events require the athlete to generate considerable velocity. The javelin is a projectile hurled by the thrower, and a high jumper becomes a projectile propelled upward into the air by muscular force. Because the high jumper wants to go as high as possible, you might think it would be beneficial to apply as much force as possible at takeoff over the longest available time frame. So why don't we see athletes lowering down onto a fully flexed leg and then thrusting upward until the jumping leg is fully extended? Surely this would maximize the total force applied by the

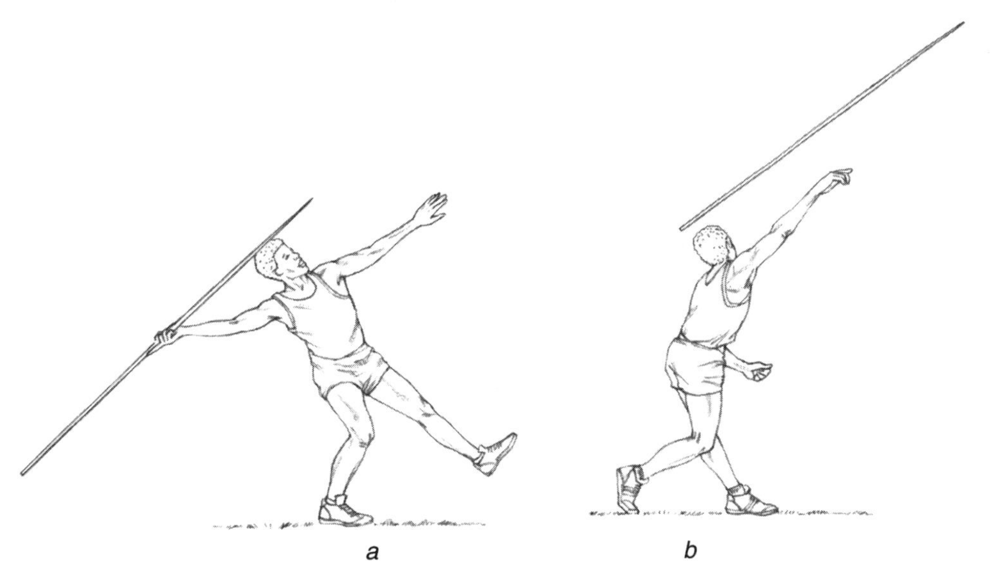

FIGURE 3.2 Impulse in a javelin throw. Elite athletes are able to apply force over a long time frame by *(a)* leaning back and pulling the javelin from behind the body and *(b)* releasing it in front of the body.

Rules Limit High Jumpers

The rules of high jumping look not for an athlete who can jump the highest in any style, but reward the athlete who can jump the highest off one foot (and then cross a bar). If a two-foot takeoff were allowed, the world record of just over 8 ft would surely be beaten by a gymnast using a high-speed run-up, a round-off, several back handsprings to gain speed, and a triple back somersault to cross the bar. Today's elite gymnasts reach heights of 9 to 10 ft on the second of the three somersaults. Forty years ago, using a single back somersault, gymnasts cleared a bar set at 7 ft 6 in.

leg muscles. Unfortunately, this is not the case. Starting from a fully flexed position, the athlete cannot develop maximum force because the leg muscles in the jumping leg are in an anatomically poor position to drive the athlete upward. What you'll find instead is that all great high jumpers start their upward thrust from a jumping leg that is flexed no more than the equivalent of a quarter squat.

If high jumpers cannot use a jumping leg that is fully flexed at the knee, is there any other way they can extend the time frame over which they apply force? Yes, there is. Like an elite javelin thrower, all great jumpers lean backward as they plant the jumping foot before takeoff. Straightening up from a backward lean allows the athlete to spend more time applying force to the ground, which in reaction thrusts the athlete upward (see figure 3.3). The same technique is used by volleyball players when they jump to spike and block, by soccer players when they jump to head the ball, and by basketball players when they leap to block or perform a layup. The use of a backward lean before takeoff helps all athletes get higher in the air. Watch slow-motion videos of great ballet stars such as Mikhail Baryshnikov or Rudolf Nureyev when they jump upward. The correct mechanics of getting up in the air in ballet are no different from those required of a high jumper, basketball player, or volleyball player.

Impulse and Cadence in Sprinting, Speedskating, and Rowing

Sprinting, speedskating, and rowing are similar in that athletes repeat the same actions in a cyclic, repetitive fashion throughout the race. These skills are not like the high jump (in which the athlete gets a rest after a jump) or the volleyball spike (in which other skills like blocking, digging, or setting can occur immediately after a spike). Elite athletes in track repeat their running action for the duration of their race. Rowers repetitively pull on the oars, and speedskaters thrust and glide at high speed around the rink. Throughout their races these athletes vary the amount of force and the time that they apply force with their muscles. Because they are accelerating and overcoming their own inertia, rowing eights use a higher cadence with more strokes per minute at the start than they do farther down the course. Each pull on the oar is quick and powerful, but over a short range or distance. Likewise, sprinters and speedskaters use short, quick strides as they accelerate from the start. Once moving at high velocity, they reduce their stride rate but each stride is longer. Why? Because great force applied quickly and repeat-

a

b

FIGURE 3.3 Impulse in a flop high jump takeoff. Elite high jumpers lean back before takeoff *(a)*, which allows them to spend more time applying force to the earth *(b)*. The earth, in reaction, thrusts the athlete upward.

edly over a short distance, or a short range of motion, is the most efficient way of overcoming inertia. It's the best way for a rowing eight or a sprinter to accelerate and get up to top velocity as quickly as possible. Unfortunately, a high stroke or stride rate burns up a lot of energy and, although efficient when accelerating, is inefficient once moving at high velocity. Once up to speed, sprinters and speedskaters reduce their stride rate and extend more fully with each leg thrust. Rowers pull over a larger distance with each stroke using a greater range of motion. So even though stride rate and stroke rate may be reduced, great force is applied over an increased range of motion at a lower cadence. This "change-down" in cadence helps the athletes maintain their velocity without running out of energy.

Using Impulse to Slow Down and Stop

Let's now look at impulse in a different light. In this section you'll see how impulse is used to slow down and stop an object. Here's an example. A field hockey player hits a ball and it rolls across the turf. Friction with the turf coupled with a small amount of air resistance slowly brings the ball to a halt. The small forces of friction and air resistance are applied to the ball over a long time (and distance), and the result is that they progressively reduce the momentum of the ball to zero. At any instant in time the force applied to the ball is small, but it is applied progressively over a long time.

Compare the forces working on the field hockey ball with those working on a basketball player who leaps for a slam dunk and afterward lands stiff-legged on the floor. The player's mass, dropping from a height, slams into the floor and comes to a halt in an instant. The reaction force of the floor (i.e., the earth) hits back at the athlete with the same force that the athlete hits the floor. Unfortunately, the time during which the athlete's body must absorb this force is extremely short, so the shock and stress on the athlete's body are phenomenal. What do most athletes do naturally to counteract this? They flex at the ankles, knees, and hips. When you coach, you tell your athlete, "Bend your legs as you land!" In a mechanical sense, your coaching advice tells the athlete to extend the time during which her body receives and absorbs the force applied by the ground. At any instant over this longer time frame, the force applied to the athlete's body will be less.

In many sport situations, it's difficult to extend the time to soften the impact. A first baseman reaching to catch a ball will be fully extended and cannot always draw his hand back to prolong the time of contact with the ball. But the glove helps. Its soft padding extends the time of contact, which keeps the ball from applying all its force in an instant to the player's hand. Fielders in the game of cricket don't wear gloves like those used in baseball. Yet a cricket ball is similar in size but harder than a baseball. To stop the sting of catching a hard-driven ball, fielders in cricket reach out to catch the ball and, at the instant of contact, quickly draw their hands backward. This increases the time frame over which the force of the ball is applied to their hands. Timing the catch and withdrawing the fielder's hands takes considerable practice.

In addition to enlarging the time frame that force is applied to their bodies, athletes are taught to enlarge the area of **impact** (i.e., the place where forces are applied) as much as possible. A runner's slide into home base is a good example. The slide not only gets the runner's legs below the reach of the opponent's tag but also extends the time during which friction with the ground brings the athlete to a halt. In addition, the sliding action puts a large area of the runner's body in contact with the ground. Visualize the pain and discomfort of a runner who dives at the bag headfirst and stops in an instant on the point of his nose! Although this comical scenario seldom, if ever, occurs, it illustrates a situation in which all the forces produced by the athlete are reduced to zero in an instant and in a very small and sensitive area.

Athletes in many sports are taught specific techniques to extend the area and time that forces act on their bodies. By doing this they avoid injury and reduce to comfortable levels the pressure exerted on them. Ski jumpers perform a "telemark landing" after flying through the air. Flexing their legs with one leg forward and one leg back lengthens the time that the impact force of landing is applied to their bodies (see figure 3.4). Hockey players try to ride out the force of a check from an opponent, and athletes in judo use breakfall techniques that enlarge the area and the time frame that the force of impact with the mat (i.e., the earth) is applied to their bodies. Perhaps one of the most famous examples of this mechanical principle was Muhammad Ali's legendary technique of rolling with the opponent's punch. As his adversary threw a punch, Muhammad rolled his head and body backward. In this way the force of the opponent's punch was extended over a long time frame and so had less effect. Imagine the difference if Muhammad had stepped forward into a punch thrown by his opponent. The time frame of contact would have been reduced to an instant, and the effect of the punch would have been much greater.

Athletes do not always have to rely on special techniques to avoid injury when they are involved in impact situations. They get help from equipment that is designed to extend the time and enlarge the area over which external forces are applied to their bodies. Helmets, padding, gloves, crash pads, foam-rubber-filled landing pits, and even parachutes do this job. The total force applied to the athlete's body does not change, but by extending the time and the area of application, the force applied at any one instant and in any one place on the athlete's body is significantly reduced. Air bags in cars use the same principle.

Conservation of Linear Momentum

When we talk of the "conservation of linear momentum," we mean that in any interaction (e.g., collision) between two objects (such as a baseball being struck by a bat or two hockey players slamming into each other), the total amount of linear momentum of both objects after the collision will be the same as the total amount that existed beforehand. So if two football players bring a total of 100 units of linear momentum into a tackle when they collide with each other, then after the tackle, 100 units of linear momentum will still exist. Linear momentum is not gained or, surprisingly, lost; we say it is "conserved." The law of **conservation of linear momentum** is directly related to Newton's third law, which says that every action has an equal and opposite reaction. Using two football players, Jack and Pete, here's how the law of conservation of linear momentum works:

FIGURE 3.4 Reducing the impact of landing in ski jumping.

- If Jack tackles Pete in a game of football, Jack and Pete exert equal and opposite forces on each other during the tackle for the same period of time. If Jack applies a force against Pete for 1 sec during the tackle, then Pete does exactly the same in return to Jack. It may not look this way when you watch a receiver being hit the instant he receives the ball, but mechanically, this is what occurs.
- From the above explanation, we can see that Jack and Pete both experience equal and opposite impulses. Equal and opposite impulses means that the product of Jack's force multiplied by the time that he applies his force in the tackle is applied back to him in the opposing direction by Pete [$F \times t$ (Jack) = $F \times t$ (Pete)].
- Since the impulses that Jack and Pete apply to each other are equal in amount and in opposing directions, then the change in linear momentum of both Jack and Pete must also be equal and opposite. As an example, if Pete's linear momentum is increased by a certain amount during the tackle, Jack's linear momentum must be decreased by the same amount.
- The combined linear momentum of Jack and Pete has not changed in any way as a result of the tackle. Linear momentum has been transferred from one athlete to the other, but there has been no gain or loss in the total linear momentum of the two players. Consequently, we say that total linear momentum of the two players has been conserved.

Absorbing the Impact

Air bags in cars act like a pole-vaulter's landing pad: Both are designed to absorb impact. A pole-vaulter's landing pads are permanently filled with absorbent material that cushions the athlete's landing. Air bags in cars have to fill at high speed to absorb the impact of a driver or passenger who is thrown forward in a collision. Using the same principle as the air bags in cars, air bubbles cushion springboard and tower divers in training. Pressurized air is released from the bottom of the diving tank. The air expands as it rises and provides an elevated area of frothy water at the surface. There's less punishment if athletes fail a dive because there's a bed of watery air bubbles to land on. Divers can fail in a dive from the 10 m tower knowing that at over 30 mph, most of the sting of hitting the water has been eliminated.

Collisions occur throughout sport, and it makes no difference if one object is moving (e.g., a golf club) and the other (a golf ball sitting on a tee) is not moving. The ball gains linear momentum and the club loses some of the linear momentum it had before impact with the ball. Collisions cannot create or dissipate linear momentum. All that happens is a transfer of linear momentum from one object to another.

Mechanical Work

In everyday use, work usually involves some kind of activity not as pleasurable as "play." Working out in the weight room, even though it can be fun, suggests labor and hard work. In mechanics, when **work** is done it specifically means that an object or athlete has applied force over a particular distance. Because distance is involved, the object or the athlete moves from one position to another. You can see that force applied over a particular distance must be related to impulse, where force is applied over a certain length of time. In a mechanical sense, work and impulse are related when a resistance is moved over a certain distance. An athlete applies force to the javelin for a certain amount of time and over a particular distance. So in applying an impulse to the javelin, the athlete is also performing work on the javelin. In our field hockey example, where the grass slowly brought the field hockey ball to a stop, the grass was progressively applying force to the ball. This is an expression of the impulse of stopping, and it is also an example of mechanical work being done. In both cases, force is applied to the javelin and the field hockey ball over a large time frame as well as over a large distance.

Isometric exercises are good examples of how muscular force is applied for a certain amount of time against a static, immoveable object. For example, a bar can be positioned at head height in a rack and an athlete can push against it for 10 seconds. If the bar doesn't flex or move, then no mechanical work is done. It doesn't matter how vigorously the athlete's muscles contract or how much physiological work is done. If a resistance is not moved over a certain distance, no mechanical work is done.

Power

Power refers to the amount of mechanical work done in a particular time period. In everyday life we use **horsepower** as a measure of the power of machines and engines such as those used in the Indy 500. One horsepower is the ability of a machine (or a human) to move 550 lb a distance of 1 ft in 1 sec. In the **metric system,** power is measured in watts (745.7 watts = 1 horsepower).

How is power used in athletic contests? Using a weightlifting example, imagine two athletes lifting barbells of the same weight. One takes 2 sec to lift the barbell overhead, and the other takes 1 sec. They both lift the barbell the same distance. In this comparison, the latter athlete is more powerful. Why? Both athletes moved the same weight over the same distance and performed similar amounts of mechanical work. But the second athlete took less time and so is considered more powerful.

Here's another example illustrating power. Let's imagine that two athletes (we'll call them Scott and Rick) race against each other over 100 m. They cross the line in a dead heat in 10.0 sec. And let's say that on the day of the race, Scott is more massive than Rick. This means that Scott is more powerful because he moved more mass over the same distance (100 m) in the same time frame (10.0 sec) than Rick. This description also indicates that power differs from **strength** (the ability of a muscle to exert force) because strength does not necessarily imply the application of force with speed. Perhaps on this basis, the sport of power lifting, which tends to measure strength rather than power in squats, the bench press, and the deadlift, should be called "strength lifting."

In many sports, power is tremendously important because a slow application of force will not get the job done. This is particularly the case in throwing and jumping events, the snatch and the clean and jerk in Olympic weightlifting, and a sport like gymnastics where skills such as back and front somersaults cannot be performed slowly. Successful performance in these events demands that great force be applied over a particular distance very quickly.

Energy

In everyday life, an energetic person is one who has the capacity for action. In mechanics, the term **energy** specifically means the ability of an athlete or an object to do mechanical work (i.e., to apply force over a distance against a resistance). Mechanical energy comes in three forms: kinetic energy, gravitational potential energy, and strain energy.

Kinetic Energy

The word *kinetic* means that motion is involved, and **kinetic energy** is the capacity of an object and an athlete to do mechanical work by virtue of being on the move. The more mass they have, and in particular the faster they move, the greater their capacity to do mechanical work. Any object or athlete that is moving will always have both momentum and kinetic energy. We'll see how momentum and kinetic energy are related but different later in this chapter.

Gravitational Potential Energy

Gravitational potential energy is a form of stored energy—energy that is potentially available and ready to be put to use. Objects and athletes have gravitational potential energy when they are raised above a planet's surface and are primarily influenced by the gravity of that planet. Since the sports we are considering take place on or close to the earth's surface, gravitational potential energy means energy that objects and athletes have by being raised even a short distance above the earth's surface. The greater the height and the greater their mass, the more potential energy they have. (Height above the surface of the earth = distance, mass = resistance, and the force that accelerates them = gravity.) Unlike objects that have kinetic energy, objects or athletes can have gravitational potential energy just by being positioned at a distance above the earth's surface. They don't necessarily need to be moving.

Mechanical work has to be performed against the pull of gravity to get an athlete or an object up above the surface of the earth so that they have gravitational potential energy. Great examples in this respect are athletes who climb up to the 10-m tower to somersault and twist on their way down to the pool. The athletes perform mechanical work climbing the ladder and raising their body mass upward. With every rung they climb, they increase their gravitational potential energy. Roller coasters are no different. Roller coaster cars and their occupants are pulled by a machine to the top of the first hill on the track, and at that point the car and its riders possess maximum potential gravitational energy for the height they are at. Once the roller coaster car (or a diver) accelerates downward toward the surface of the earth, gravitational potential energy is converted primarily to the energy of motion (kinetic energy). We say *primarily* because not all of the kinetic energy is expressed as motion. Some kinetic energy is converted into noise and heat.

Strain Energy

Strain energy, like gravitational potential energy, is a form of stored energy. Objects have the capacity to store strain energy if they have the ability to restore themselves back to their original shape after being squashed, pulled, twisted, or pushed out of their original (resting) shape. Mechanical work must be done to put an object in this condition, and once distorted, the ability of an object to regain its original shape as fast as possible is a measure of its strain energy. An archery bow springing back to its original shape after being flexed is an example of strain energy. A golf ball also springs back to its original shape after being squashed by a driver, and likewise the leaf and coil springs used on cars and motorcycles spring back to their original shape after being flexed. All are examples of strain energy.

Kinetic Energy, Gravitational Potential Energy, and Strain Energy in Pole Vaulting

Kinetic, gravitational potential, and strain energy are all well demonstrated in the pole vault. In this event an athlete sprints flat out carrying the pole. The kinetic energy developed during the approach (because the athlete's mass and the mass of the pole are in motion) is used to flex the pole and "load it" with strain energy.

When a vaulting pole flexes at takeoff, the athlete has performed mechanical work on the pole by applying force to it over a particular distance. If the athlete runs slowly during the approach, the pole is flexed less and less strain energy is stored in it. Like an archer's bow that is pulled back only slightly, a vaulter's pole that is not flexed the optimal amount cannot do very much "work" when it straightens out and drives the athlete skyward. Top-class pole-vaulters always combine the skills of a gymnast and a sprinter. The faster they run and the taller and more powerful they are, the higher they can hold the pole. A phenomenal athlete like Sergei Bubka of Ukraine, who has vaulted over 20 ft (just over 6 m) both outdoors and indoors, is so powerful and such a good sprinter that he can hold extremely "high" on a long, stiff pole that is specially rated for his body weight (176 lb, or 80 kg). If you can flex this kind of pole, you store a tremendous amount of strain energy in the pole (see figure 3.5), which is returned to you by rocketing you up toward the crossbar.

FIGURE 3.5 In the pole vault, a flexed pole stores strain energy, which is later used by the vaulter.

During the vault, the flexed pole straightens out and drives the athlete up toward the bar. The strain energy stored in the pole by the athlete now does mechanical work by propelling the athlete upward, and because he is rising in the air, he gains gravitational potential energy. At the high point of the vault the athlete rises no more. His kinetic energy is momentarily zero because, for an instant, he is not moving. But the athlete is a long way above the surface of the earth and so has maximal gravitational potential energy for the height he has risen. As the athlete accelerates toward the earth, gravitational potential energy is progressively replaced by kinetic energy. By the time the athlete hits the landing pad, he is moving at top speed for the distance fallen and so has plenty of kinetic energy. The athlete depresses the landing pad, making the contacting surfaces slightly warmer. The depth that he squashes into the landing pads, the noise of impact, and the heat that is developed are expressions of the work done by the athlete's kinetic energy. Once the athlete is back on the surface of the earth and not moving, his gravitational potential energy is zero, and he has no kinetic energy either.

Many other examples of kinetic, gravitational potential, and strain energy occur in sport. An athlete bending a springboard in diving is an example of work being performed to the board to load it up with strain energy. In trampoline, which became an official Olympic sport in the 2000 Sydney Olympics, athletes use a pumping action to progressively increase the stretch in the springs of the trampoline. In this way the stretched springs are given more and more strain energy. Strain energy then performs mechanical work by propelling these athletes high into the air. Trampolinists will have zero kinetic energy when, for a brief instant, they are motionless at the top of their flight path. As with pole-vaulters, their gravitational potential energy will be maximal for the height they have risen. After they drop downward, their kinetic energy (which progressively increases during the fall) will be primarily spent in stretching the springs of the trampoline bed. Some will be lost as noise and heat, but the trampolinists make up for this loss by repeating their muscular pumping action when they are in contact with the bed of the trampoline. Muscular thrust coupled with the kinetic energy of the drop to the trampoline bed stretches the trampoline springs once again for the next flight up in the air. When you compare trampolinists to divers, you can understand why trampolinists don't want to travel along the bed of the trampoline during their routines. They want to stay in the middle. Divers, on the other hand, must travel away from the board to avoid hitting it on the way down to the water.

How Momentum and Kinetic Energy Are Related

All objects on the move have both momentum and kinetic energy. The more mass they have and the faster they move, the greater their capacity is to apply force over time (which we call impulse). Isaac Newton's third law of action and reaction always gets involved in these situations and tells us that the impulse applied against another object or another athlete is "mirrored" back against the person or thing applying the impulse. Let's look at momentum and kinetic energy in impact situations (i.e., collisions and tackles) to see how the concepts of momentum and kinetic energy are related, and also how they differ.

The easiest approach is to regard kinetic energy as the ability of a moving object to do work on whatever it collides with and in return to do work on itself. The moving object can be anything at all. It can be an athlete running into an opponent, or an archer's arrow piercing a target. A moving object has the ability to do mechanical work, and it can apply force over a particular distance to whatever it hits. The athlete drives into the opponent, the pole-vaulter compresses the landing pads, and the arrow buries itself in a target.

The formula for kinetic energy is $KE = 1/2mv^2$, where m is the mass of the object and v is its velocity. Kinetic energy is directly proportional to any increase in mass, but much more important, it increases according to the square of the velocity. These characteristics indicate that if you leave the mass of a moving object unchanged but double its velocity, the object's kinetic energy increases fourfold. Triple its velocity and its kinetic energy increases to nine times the original. Increasing the kinetic energy of an object ninefold will give it the ability to do nine times the mechanical work on whatever it hits. Take the archer's arrow as an example: An archer fires an arrow at a target. The arrow sinks in a certain distance. The archer then picks an arrow that is twice as massive and has it hit the target at the same velocity as the original arrow. The result is that it sinks in approximately double the depth of the original arrow. When the archer goes back to the original arrow but fires it so it's traveling twice as fast when it hits the target, the arrow sinks into the target approximately four times deeper. Three times as fast would drive it in approximately nine times deeper. In each example, we need to use the word *approximately* because not all of the arrow's kinetic energy is used in burying its head in the target. Some of the arrow's kinetic energy is dissipated as vibrations, noise, and some as heat, which slightly warms the arrowhead and the pierced target area.

Whenever an athlete is on the move, he has both momentum and kinetic energy. Let's consider tackling in football to see how momentum and kinetic energy are always related but at the same time quite different. Imagine a 300-lb lineman entering a tackle at a velocity of 4 ft/sec. At this velocity the lineman has 300 × 4 = 1,200 units of momentum. A 150-lb safety sprinting into a tackle at 8 ft/sec has the same amount of momentum (i.e., 150 × 8 = 1,200) (see figure 3.6). If each of their opponents weighs 200 lb and runs at 6 ft/sec, then the opponents also have 1,200 units of momentum. In the tackle, both the 300-lb lineman and the 150-lb safety are equally effective in stopping their 200-lb opponents. In each case, 1,200 units of momentum runs into 1,200 units of momentum. If there is no rebound and they stick together like clay, then in both tackles the athletes come to a stop on the spot where they collide.

Let's now look at the kinetic energy brought into the tackle by the 300-lb lineman and the 150-lb safety. Because the safety sprints at 8 ft/sec, he goes into the tackle twice as fast as the lineman, who enters his tackle at 4 ft/sec. If the lineman and the safety had the same mass, the safety would have four times the kinetic energy of the lineman because he runs twice as fast. (Remember, if you double the velocity, it increases kinetic energy fourfold [i.e., 2 × 2 = 4].) But at 150 lb he has half the mass of the 300-lb lineman. So his kinetic energy is 4 divided by 2, or twice as much as the lineman.

How is double the amount of kinetic energy expressed in the tackle? Think of the archer's arrow piercing the target, or an athlete flexing the pole at the start of a vault. In his tackle, the 150-lb safety is going to do twice as much work on his opponent as the 300-lb lineman. Think of him driving himself twice as "deep" (just like the archer's arrow) into his opponent's body. The result of this extra ability to do work causes a lot of pain and maybe broken bones for his 300-lb opponent—and for the 150-lb safety as well, because Newton says, "For every action, there is an equal and opposite reaction."

Kinetic energy figures in all situations where something is on the move. Perhaps the most important lesson to be learned about kinetic energy is its effect when your car skids. A skid is an example of mechanical work being done by an object (you and your car). So imagine that you're driving your car at 10 mph and you suddenly have to slam on your brakes. And let's say that you skid 5 ft. What would be the approximate distance of your skid if you were traveling at 20 mph? In this situation you've doubled your velocity, which squares its effect on the car's kinetic energy and its ability to do mechanical work (in the skid). Consequently, your car's stopping distance will be 20 ft (i.e., four times longer than the original of 5 ft). An increase in the car's velocity to 30 mph gives you 3 × 3 × 5 = 45 ft and 40 mph produces 4 × 4 × 5 = 80 ft. All of these distances are approximate because kinetic energy is lost in generating heat and noise, and because we "guesstimated" the length of the original skid and we assume that all other conditions stay the same throughout. However, an increase in velocity dramatically increases the kinetic energy of an object and its ability to do mechanical work. So the faster you drive your car, the more important it is to leave larger and larger distances between you and the car ahead of you and, if possible, drive a car with antilock brakes, which minimize skidding.

How Kinetic Energy Is Dispersed

Now let's return to our football players, Jack and Pete. Imagine Pete running with the football and Jack approaching to tackle him. They are both

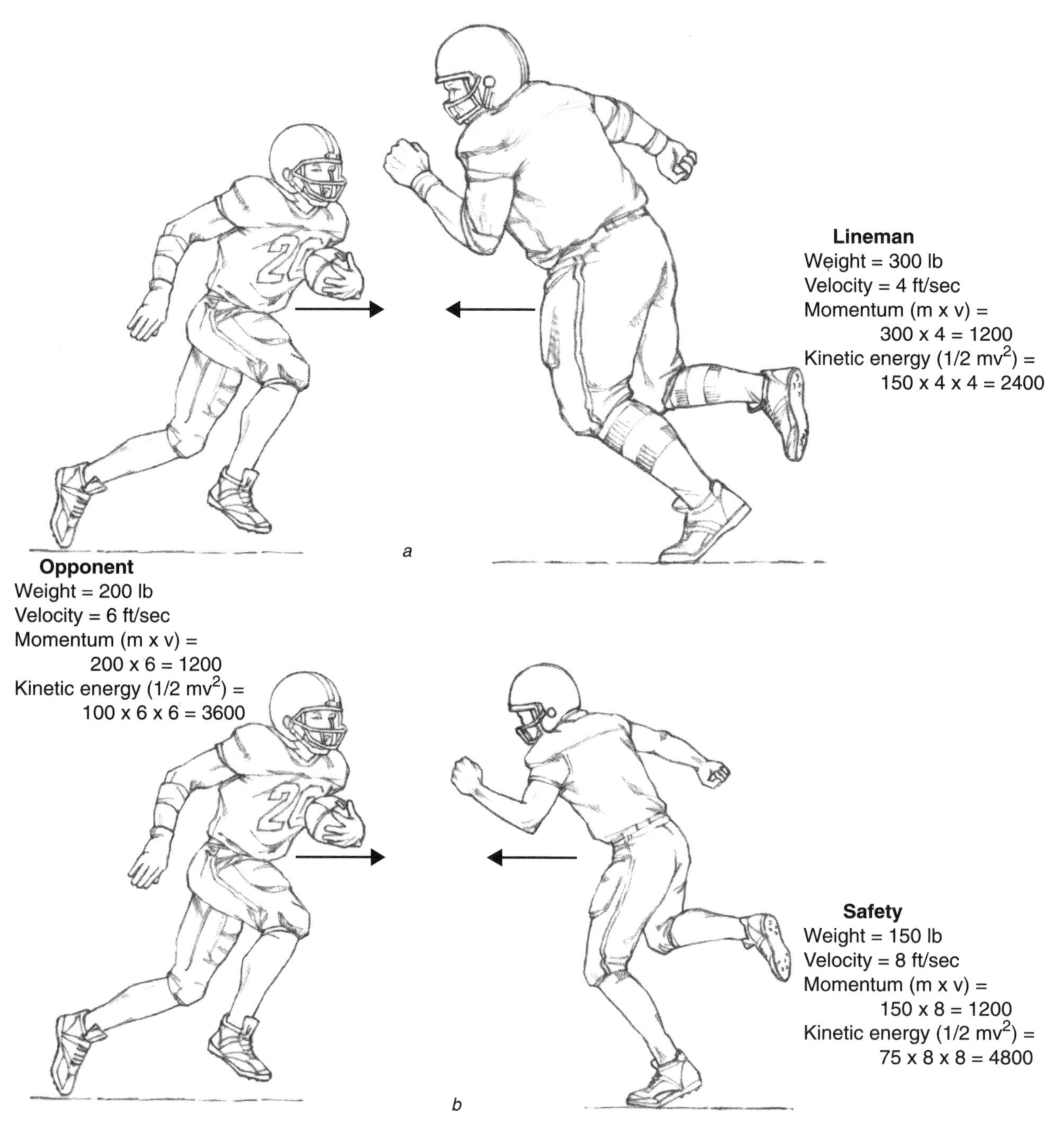

FIGURE 3.6 Momentum and kinetic energy in a tackle. The lineman *(a)* and the safety *(b)* each have the same momentum, but the safety has twice as much kinetic energy.

moving, so they both possess momentum and kinetic energy. Is kinetic energy conserved by the two players in the tackle in the same way as momentum? No, because the kinetic energy that the two players bring into the tackle doesn't remain totally with the players but is spread around, or dispersed, in many different ways. Some kinetic energy is certainly used by Jack in doing work on Pete and likewise by Pete in doing work on Jack. They squash into each other and maybe they rebound as well. Their kinetic energy does the "squashing" and the elasticity of their bodies (which is strain energy) can cause them to bounce apart. But kinetic energy is also used in generating noise and heat during the tackle. Noise and heat occur in all sport collisions. You hear the crack of bat on ball in baseball, ball against ball in the game of pool, and hand against ball in a volleyball serve. The two colliding objects also get a bit warmer. Noise and heat always take some of the kinetic energy of two colliding objects. Consequently, in a collision momentum is conserved by the colliding objects, whereas their kinetic energy is not. In car accidents, metal is bent and crushed, and there's tremendous noise and heat generated by the kinetic energy of the colliding vehicles. Most modern cars now have built-in "crushability." They are designed to dissipate the kinetic energy involved in the impact with minimal effect on the occupants. The same design is used for high-speed race cars.

Law of Conservation of Energy

Although kinetic energy gets used up in the deformation of objects (and athletes) and into heat and sound, all that's really happening is that one form of energy is being changed into another. For example, imagine a trampolinist who makes a mistake in her routine and misses the trampoline and falls hard onto a wooden floor. We know that as the trampolinist falls faster and faster, her gravitational potential energy gets converted into kinetic energy, and that her kinetic energy is maximum just prior to impact. At impact, there is no more gravitational potential energy and, since there appears to be no more movement, it seems that kinetic energy has disappeared too. But has it really? After impact, the kinetic energy of the trampolinist has been converted into the movement of millions of molecules, some expressed as noise, some as heat, some as the movement of the floor, and some as the deformation of the trampolinist's body. What has happened is that one form of energy—kinetic—has been exchanged for other forms of energy with no real loss. In reality, energy is conserved.

Rebound

When bats hit balls or balls bounce off floors, a collision, or impact, occurs. When the objects separate, and one (or both) moves away from the other, we call it a **rebound.** What actually happens during and after the collision depends on many interacting factors. Let's look at what happens to balls when they are hit and as they bounce.

Often both the ball and the object it collides with are moving (e.g., a racket hitting a ball). Sometimes one object can be moving and the other can be momentarily stationary, such as a bowling ball hitting a pin. After the collision the bowling ball slows down slightly and loses momentum. The pin accelerates and gains momentum. In another common situation, a moving object (e.g., a squash ball or basketball) collides and rebounds from an immense stationary object such as a wall or the floor.

Coefficient of Restitution

The coefficient of restitution (COR) is a measure of the ability of an object like a ball to spring back to its original shape after being hit by a club, bat, or racket, or after bouncing off a floor or a wall. It's essentially a measure of "bounciness" or "resilience." All ball sports have specific rules controlling just how much bounce a ball is allowed to have. Rules also take into consideration the composition and manufacture of the club, bat, and racket (and strings). Too much or too little rebound and the whole character of the sport is changed. The highest COR rating is 1 and the lowest is 0. Children's Super Balls® and golf balls have tremendous bounciness and have a COR between 0.8 and 0.9. Tennis balls have a COR of 0.72. Baseballs are required to have a COR of between 0.51 and 0.57, and basketballs must have a COR of between 0.76 and 0.80. Squash balls have a COR of between 0.34 and 0.57 and are available in a wide range of rebound and hang-time ability. This is intended to accommodate the various abilities of different players. In addition, all balls vary in their bounciness according to temperature.

In the previous examples, the angle at which the collision occurs can vary. Both objects can hit head-on (such as a basketball bounced straight down onto the floor), or they can glance off one another at an angle (such as a carom shot in pool or a bounce pass in basketball).

Rebound and Elastic Recoil

An important factor that determines what happens after a collision is the degree of elastic recoil that objects have. (By elastic recoil, we mean the force by which objects push back to their original shape.) Some objects have very little recoil, or **elasticity,** and stick together like clay after colliding. A Hacky Sack® footbag offers virtually no bounce at all. Likewise, an athlete who falls out of control onto the ground hardly bounces back up into the air. The human body in this condition has very little elasticity.

Golf balls, on the other hand, are well known for their elastic recoil. They become deformed (squashed out of shape) when hit by a club. Like a spring that is compressed, the ball stores strain energy, which is transformed into kinetic energy as it regains shape. Just like the pole-vaulter's pole, strain energy produces kinetic energy, which then makes a contribution toward the velocity of the golf ball as it comes off the club face. Kinetic energy also contributes to the noise of impact and increases the warmth of the ball slightly after it has been hit. The same thing happens to a baseball when it is hit. A baseball is flattened to almost half its size when it is hit for a home run. As the ball leaves the bat, it springs back to its original shape. The energy produced by this action adds velocity to the flight of the ball. This extra energy helps great baseball hitters put the ball out of the park or in the upper row of the bleachers. The elastic recoil of any object also depends on the nature of the object with which it collides. A football can rebound with considerable velocity off artificial surfaces. This differs dramatically from the rebound that occurs when the ball drops on wet, muddy turf.

Rebound and Temperature

Temperature makes a difference in the way balls rebound after a collision. Heat causes the air inside a ball to expand, and this increases the ball's ability to rebound. In squash, players spend time before a match rallying back and forth so that the ball is hot and will bounce correctly. A cold squash ball, in contrast, is virtually a dead ball—with little or no bounce—making it much more difficult for players to return a shot. Elite squash players are required to play with a ball designed to bounce very little, even when it's warmed up and the air inside it is hot. This challenges the players according to their ability. Novices, on the other hand, use a ball designed to bounce much higher after it's warmed up. The ball's additional time in flight gives novices more time to get into position and make their strokes.

Angle, Velocity, and Frictional Forces in a Rebound

Many factors determine what happens to a ball after it collides with another object. In addition to those items already mentioned, rebound depends on the ball's angle and velocity, whether the ball is spinning, and finally how much friction occurs between the ball and whatever it is hitting. These factors play a big part in table tennis. The rough, spongy surface of the bat imparts tremendous spin on the ball. Even though the ball and the table have smooth surfaces, the effect of the spinning ball hitting the table at high velocity is dramatic. The direction of rotation of a topspin makes the ball accelerate off the table at a low angle. A backspin causes the reverse to occur. Its spin is in opposition to the ball's movement, and the ball can slow down or even reverse direction.

The direction of spin plays a vital role in many sports. In billiards, the spin (or "English") given to the cue ball affects what happens to the ball that the cue ball strikes and what happens to the cue ball afterward. Similarly, backspin applied to a golf ball by the angle, or loft, of the club face helps give the ball lift and enables it to stay in the air longer. (You'll read more about lift in chapter 6.) Backspin can also prevent a ball from rolling off the far side of the green. In basketball, good shooting requires that the athlete give the ball a backspin at the instant of release. This is done by making the ball roll off the fingertips as it leaves the shooter's hand. A basketball shot with backspin that hits the rim or the backboard not only loses velocity when it hits but also is more likely to drop in the basket because of the backward rotation. Rick Barry, one of the greatest foul shooters ever to play basketball (.900 average), was well known for his underhand shooting technique. He could throw the ball with considerable backspin in this manner, and the backspin was instrumental in rotating the ball down into the basket.

Most modern basketball players shoot fouls the same way they shoot during the game. Even though they shoot overhand, the finger flick applied at the end of their release produces backspin. The backspin then helps sink the ball in the basket. A topspin, on the other hand, is more likely to cause the ball to rebound back out onto the court.

Rebound in Tennis

The game of tennis is in many ways more complex than table tennis because of the number of factors that affect the rebound of the ball. For example, a tennis court can be clay, grass, or a hard surface. All these surfaces have a different effect on the way the ball rebounds. Courts can be indoors or outdoors, and environmental conditions such as temperature, humidity, and wind velocities alter play. Tennis rackets vary in size, shape, weight, and flexibility; and strings differ in type and in tension. Even the tennis ball itself changes the way it reacts. Tennis balls bounce higher after they are warm and after some nap (i.e., fuzz) is worn off. When they have been out of their pressurized cans for a long time, the balls age and lose their bounce irrespective of how much use they've had.

Friction

Friction occurs when an object moves or tends to move while in contact with another object. This force is present when a jogger's running shoes make contact with the road and when a bowling ball rolls down the surface of the lane. In both these examples friction occurs between two solid (i.e., nonliquid) surfaces. However, friction also results when a ski jumper or javelin flies through air or when swimmers move through water. In these situations both the air and the water act as fluids, and they produce fluid friction; we'll examine fluid friction in chapter 6 and concentrate in this chapter on friction that occurs between solid surfaces.

The demands for friction vary dramatically from one sport to the next and from one set of environmental conditions to another. An athlete may want plenty of friction at one instant and only minimal friction at another. In cross-country skiing the correct choice of wax relative to the snow conditions allows sufficient friction for traction but not so much friction that an athlete cannot glide on the skis when necessary. In football, making sudden changes of direction depends on friction for good traction. Unfortunately, on artificial turf friction between an athlete's shoe and the turf can sometimes be too good. During tackles the athlete's foot can get trapped in one position. The result can be torsion (i.e., twisting) injuries to the knee and ankle.

Tennis professionals compete on three different surfaces: hard courts, grass, and clay. Hard courts vary in texture, and the more grit used

Technology Changes the Tennis Racket

The sweet spot is that part of the tennis racket that returns the ball with the greatest velocity and with the least shock and vibration to the player. Modern tennis rackets have increased in size from 75 sq. in. for the old wooden rackets to 95 to 100 sq. in. for modern composites. On wooden rackets, the sweet spot was close to the base of the racket face. Modern composite rackets have a larger sweet spot, which is higher on the racket face. In addition, composite rackets are stiffer so that they transfer more energy to the ball. With a higher sweet spot, the ball is hit more reliably, the player feels less shock, and the ball comes off the racket faster.

in the top dressing, the more friction will occur between the ball and the court surface. Increased friction slows down the ball as it rebounds off the surface. Grass provides very little friction. The short grass at Wimbledon is recognized as a fast surface. It benefits power players who serve the ball over 100 mph. Clay is the complete opposite of grass. Rough clay courts, such as those at Roland Garros Stadium in Paris, are extremely slow surfaces. These surfaces reduce the effectiveness of the serve and volley game of power players who dominate on faster grass and hard courts. On clay, defenders have more time to get to the ball and to counterattack. The strategy of clay court specialists is to hit passing shots over and past power players. Thomas Muster of Austria was a king of the endurance battles that occurred on clay court surfaces. He built up a 46-3 win/loss record on clay, bringing his 1995-1996 total to an amazing 111-5. Unfortunately he was not able to replicate this level of success when he played on other surfaces.

Types of Friction

Three types of friction occur between solid surfaces. The first is **static friction,** which exists between the contacting surfaces of two resting objects and provides the resistive force opposing the initiation of motion. The second is **sliding friction,** a resistive force that develops when two objects slide and rub against each other. The third is **rolling friction,** which produces a resistive force when objects—such as balls and wheels—roll over a supporting or contacting surface. In sport, it's possible for all three of these frictional forces to oppose the motion of a single object. For example, a field hockey ball at rest resists movement initially because of the static friction existing between the ball and the turf. Once a player hits the ball, it may simultaneously slide and roll, in which case both sliding and rolling friction resist its motion. However, only rolling friction remains once the ball is fully rolling and no longer sliding.

Let's examine static and sliding friction by looking at a football player shoving a blocking sled, a common piece of equipment designed to slide on turf and provide both static and sliding resistance against a player. We'll look at rolling friction later in the chapter.

Static and Sliding Friction

The frictional force generated by a blocking sled at rest comes from static friction. If an athlete pushes against the sled with minimal effort, the sled remains motionless. In this situation the static friction is greater than the force produced by the athlete. If the athlete increases his thrust against the sled, the frictional force opposing the athlete reaches a critical level, called the sled's maximum static friction. Should the athlete further increase the force of his push, he will overcome the sled's maximum static friction and the sled will start to slide. Static friction is now replaced by a frictional force between the base of the sled and the turf called sliding, or kinetic, friction. Sliding friction is always less than maximum static friction. In other words, it's easier to keep an object moving than to start it moving.

Now that we know the difference between static and sliding friction, let's look at the factors that influence the amount of static and sliding friction that can occur.

Force Pressing the Two Surfaces Together

Using the blocking sled as an example, the force pressing the base of the sled to the turf is equal to the weight of the sled pushing down and the reaction of the earth pushing up. These two forces press the contacting surfaces of sled and turf together. If a coach gets on the sled, both the weight of the sled pressing downward and the reaction force of the earth pressing upward increase. The frictional force opposing an athlete attempting to push the sled will be greater.

The importance of weight (or mass) pressing two surfaces together cannot be overemphasized. More mass means more pressure thrusting the contacting surfaces of turf and sled together. For maximal friction, the weight of the sled plus whatever weight is added must act perpendicularly to the supporting surface. If a coach hangs on the sled with his body angled (so that he is not standing upright), only part of his weight contributes to pushing the sled down onto the turf.

Actual Contact Area Between the Two Surfaces

Actual is the important word here because friction can occur only where there is contact

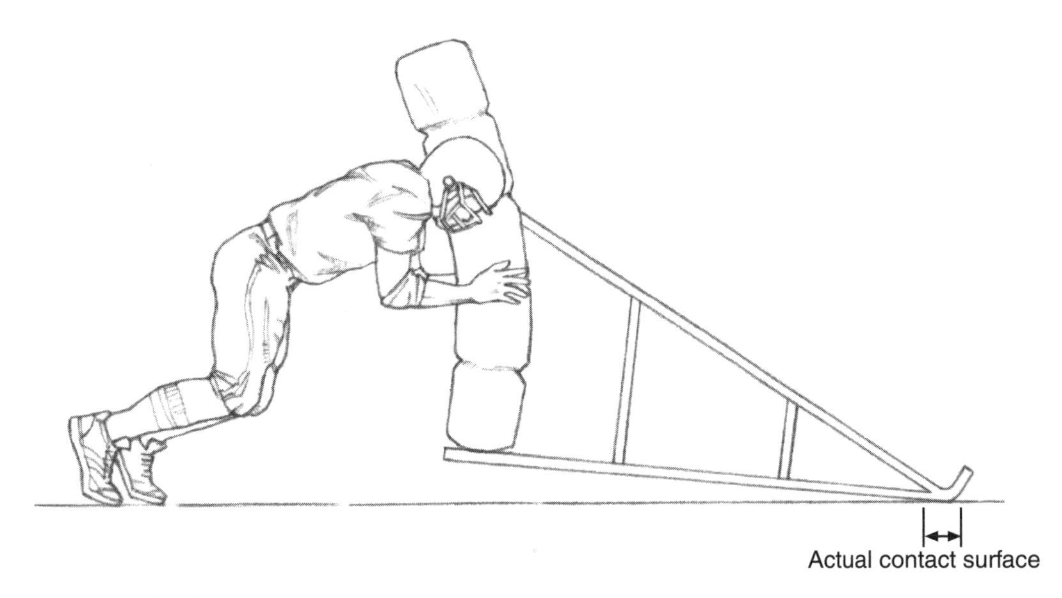

FIGURE 3.7 Lifting a football blocking sled reduces the actual contact surface.

between two surfaces. Where part of the base of the blocking sled does not contact the ground, friction cannot occur. If an athlete lifts the sled by pushing forward and upward, then no friction occurs where the sled loses contact with the ground (see figure 3.7). Likewise, if a soccer player's studs are the only part of the boot contacting the ground, then only the surface of the studs contributes to the contacting area; the rest of the sole makes no contact and generates no friction.

The following example illustrates the importance of union between the actual contact area (i.e., the parts of the surfaces in contact) and the forces pressing the two surfaces together. Imagine two sleds that weigh the same but have bases of different sizes. An athlete pushes against one sled and then in exactly the same manner (and with the same force) against the other. The sled with the larger base will be no more difficult to push than the sled with the smaller base, because the weight of the bigger sled is spread over a bigger area and presses onto the turf proportionally less for every square inch of its base. Consequently, the same friction is produced by both sleds.

Nature and Type of Materials That Are in Contact

The nature and type of materials in contact refer not only to the material on the base of the sled but also to the type of surface that the sled contacts. Imagine that the base of the sled is rough and pitted and it is forced to slide over a muddy, sticky surface. The frictional force opposing the player would be greater with this sled than the force produced by a sled with a smooth base sliding over a hardened, level surface. The many different materials used in sport (and everyday life) all produce varying levels of friction as they contact one another. At one end of the scale are the long steel blades of a speedskater gliding on ice: A thin film of water created by the blade pressing on the ice produces a lubricant that reduces friction to an extremely low level. At the other end of the scale, court shoes with gum rubber soles pressing against a rubberized surface create an extremely high level of friction.

Relative Motion Between the Two Surfaces

The relative motion between the base of the blocking sled and the ground refers to whether the athlete is applying force to initiate movement or whether the sled is already moving. Remember that static friction produces a higher level of resistance than sliding friction. We all know from experience that it's easier to keep an object sliding than to start it sliding.

Rolling Friction

Rolling friction occurs when a round object, such as a ball or wheel, rolls across a contacting, or supporting, surface. Rolling friction is common in bowling, billiards, golf, field hockey, cycling, and soccer. The resistive force generated by rolling friction is significantly less than that of sliding friction. Indeed, the use of the wheel is commonplace because of the minimal levels of friction produced as a wheel rolls over a supporting surface. This extremely low-level friction results from the ease with which a rounded surface detaches itself from the surface it is rolling over. Rolling friction varies according to the nature of the surfaces in contact, the pressure pushing the surfaces together, and the diameter of the rolling object.

Sprint cyclists use narrow tubular tires, inflating them to phenomenal pressures that average 120 psi (pounds per square inch). The result is that even with the weight of the cyclist, hardly any tire surface contacts the track, and rolling friction is reduced to a minimum. Even so, the rubber of the tire in contact with the smooth velodrome surface provides excellent traction. It is because of rolling friction that fat-tire mountain bikes feel so sluggish on the road, particularly if their tires are underinflated (see figure 3.8). On the other hand, wide, knobby mountain bike tires are designed to produce great stability and traction on rough terrains. High-pressure, narrow tubular tires are virtually useless in these conditions.

Every sport involves friction in some form or another, from the static and sliding friction in gymnastics, ice skating, and wrestling to the rolling friction in ball games and wheeled sports such as cycling, auto racing, and in-line skating.

How a Rolling Wheel Makes Use of Static Friction

Many cars are equipped with antilock brakes that allow the wheels to continue to rotate while progressively applying the brakes. This is much better than wheels that suddenly lock and start to skid. A skidding wheel has less traction (grip on the road surface) than a wheel that is nonskidding. Antilock brakes stop you from skidding while slowing your car down. Sensors in your car are designed to respond to a sudden reduction in rotation of the wheels. This allows them to continue rolling rather than locking up and causing a skid. With antilock brakes, you stop faster and you can continue to steer. This is something you cannot do very well when you are skidding. A skidding tire is slowed by sliding friction between the contact points of each tire and the road surface. A rotating tire can be considered to have hundreds of contact points that are each stationary for one instant after the other. So, a rotating tire is slowed down by static friction, which is more powerful than sliding friction.

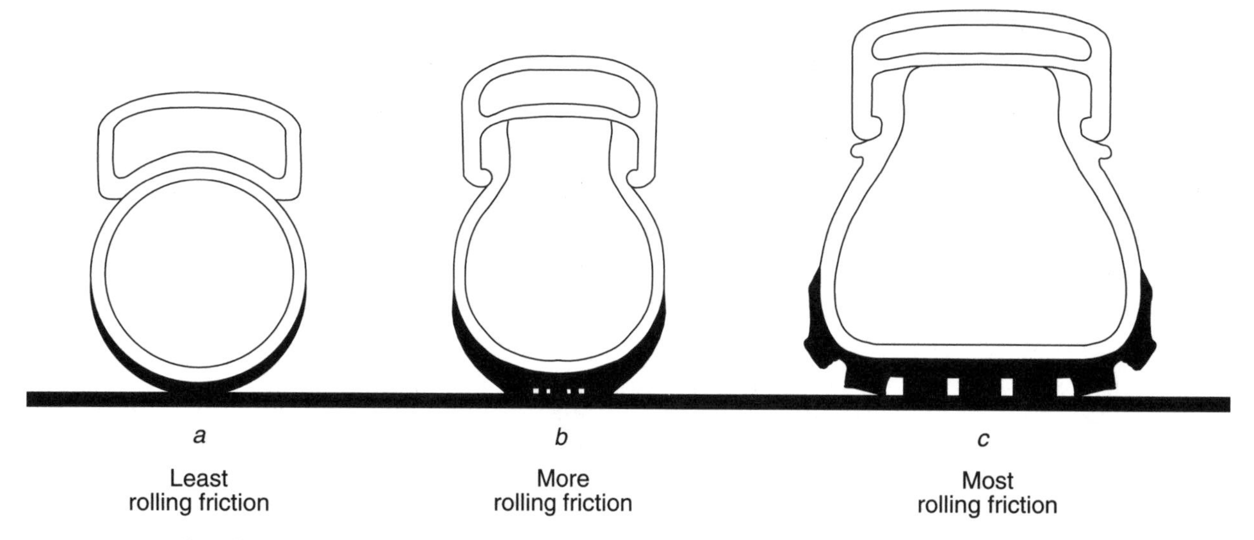

FIGURE 3.8 Rolling friction increases with the amount of bicycle tire in contact with the road. The least rolling friction occurs with the fully inflated high-pressure tubular tire *(a)*. Moderate rolling friction occurs with the fully inflated clincher tire *(b)*, and the most rolling friction occurs with the underinflated mountain bike tire *(c)*.

SUMMARY

- The acceleration of athletes or objects is proportional to the force applied against them and the time that this force acts. Increase force or time, or both, and acceleration will increase. Force multiplied by the time that it acts is called impulse.
- The acceleration of athletes or objects is inversely proportional to their mass. For a given impulse, any increase in mass reduces acceleration. These first two bulleted points summarize Isaac Newton's second law of acceleration.
- All actions cause reactions. Action and reaction are colinear forces, being both equal and acting in opposite directions. The law of action and reaction is Isaac Newton's third law.
- Momentum describes quantity of motion. An increase in mass or velocity, or both, increases momentum.
- In a collision between two objects (or athletes) that is isolated from other influences, linear momentum is conserved. The total amount of linear momentum that the objects possess after the collision is the same as the total linear momentum before the collision. The law of conservation of linear momentum is derived from Newton's third law.
- Collisions vary from those that are considered elastic, in which the colliding objects rebound from each other, to those that are considered inelastic, in which the colliding objects do not rebound from each other.
- In a collision between two objects (or two athletes) kinetic energy is not totally retained by the two colliding objects. Kinetic energy is used in deforming the colliding objects, and is also dissipated as heat and noise.
- Force that is absorbed over a large time frame and over a large area helps prevent injury when athletes come to a stop or when they stop moving objects.
- Mechanical work is force multiplied by the distance through which the force is applied.
- Power is the rate at which work is done. One horsepower equals 550 lb moved 1 ft in 1 sec (1 horsepower = 746 watts).
- There are three types of mechanical energy: kinetic energy, gravitational potential energy, and strain energy. Kinetic energy is the energy an object possesses by virtue of its motion. Gravitational potential energy is the energy an object possesses when at a distance above the earth's surface. Strain energy is an object's capacity to do mechanical work when it recoils after being pulled, pushed, or twisted out of its normal shape. Strain energy is also considered a form of potential energy.
- The rebound of a ball depends on the elastic recoil of both the ball and the object with which it collides. The velocity of the ball, the angle of impact, and factors such as temperature and friction affect the manner in which a ball rebounds.
- When two objects slide against each other, static friction resists the initiation of motion and sliding friction resists the sliding motion that occurs. Sliding friction is always less than static friction.
- Four factors affect static and sliding friction: the forces pressing the contacting surfaces together, the actual contact area between the two surfaces, the nature and type of the materials in contact, and the relative motion between the two surfaces.
- Rolling friction occurs when a round object rolls against a contacting surface. It is significantly less than sliding friction. Rolling friction is influenced by the forces pressing contacting surfaces together, the nature and type of the materials in contact, and the diameter of the rolling object.

KEY TERMS

conservation of linear momentum
elasticity
energy
friction
gravitational potential energy
horsepower
impact
impulse
kinetic energy
metric system
momentum
power
rebound
rolling friction
sliding friction
static friction
strain energy
strength
work

REVIEW QUESTIONS

1. A soccer ball is kicked at tremendous velocity toward the goal. The goalie standing on the line catches the ball and is driven backward into the goal; the ball loses momentum because it slows down. What gains momentum?
2. Imagine you've been invited to play baseball, but there's no glove for you to use. A batter drives the ball toward you at incredible velocity. You reach out toward the ball but don't draw your arm back as you catch it. Why will this be such a painful experience?
3. Two football players collide with each other. Explain why their linear momentum is conserved and describe what happens to the kinetic energy that the two players bring into the tackle.
4. At the apex of a vault, when for an instant a pole-vaulter is going neither up nor down, what happens to (a) the athlete's kinetic energy, (b) the athlete's potential energy, (c) the strain energy of the pole, and (d) the force of gravity?
5. A four-man bobsled reaches the bottom of the course after a poor run. The brakeman applies the brake, and the sled slides a certain distance. With conditions the same on the second run, the sled is sliding twice as fast at the time that the brakeman applies the brake. About how much farther will the sled slide?
6. A gymnast shifts from a two-hand handstand to a one-hand handstand. The pressure that the athlete experiences between her hand and the floor increases as she shifts from the two hands to one. Why?

PRACTICAL ACTIVITIES

1. Newton's second law of acceleration. Obtain a small block of wood (4 in. × 3 in. × 2 in.) and screw a hook in one end. Rest the block of wood on the surface of a table. Tie a string to the hook and let it drape over the edge of the table; attach a weight to the string so that it pulls toward the floor. Initially, use a weight that is just sufficient to move the block across the surface of the table. Then repeat with progressively heavier weights. As you increase the weight (which represents the force acting on the block) the block will increase its acceleration. Explain how this simple experiment shows the relationship between the force applied to a specific mass (i.e., the block of wood on the table) and its acceleration. In your explanation, demonstrate that you understand the relationship between force, mass, and acceleration.

2. Work, potential energy, and kinetic energy. Bend down and pick up a block of wood from the floor. Lift it vertically and put it on the edge of a table. Then push the block off the table so that it falls back down to the ground. With this scenario in mind, answer the following: (a) When were you performing mechanical work on the block? (b) When was the gravitational potential energy of the block maximal? (c) When was the kinetic energy of the block zero? (d) When was the kinetic energy of the block maximal? (e) How was the kinetic energy of the block dissipated once it hit the floor?
3. Impulse applied to initiating motion. Watch a video of a speedskating competition. (a) What kind of stride length and stride cadence do the speedskaters use at the start? (b) What kind of stride length and stride cadence do the skaters use during the remainder of the race? Give mechanical reasons for the difference in stride pattern. Comment on the amount of force and the time that force is applied during each stride in these two phases of the race.
4. Impulse applied to arresting motion. Have two of your friends hold a bedsheet, with one gripping two corners at the left and the other gripping two corners at the right. The sheet should look like the curved sail on a galleon or like a fireman's safety net. Throw an uncooked egg into the sheet. The egg will be stopped without breaking. Why? Your explanation should refer to the impulse of stopping and the mechanical principles involved in stopping safely without injury.
5. Starting and sliding friction. Place a hook in the end of a fairly heavy block of wood (4 in. × 3 in. × 2 in.) and tie some string to the hook. Attach a small spring scale to the other end of the string. Place the block on the wooden surface of a table. Holding the free support of the spring scale, pull the block across the table. Note the measurement on the spring scale immediately before the block starts to slide across the table. Second, note the measurement on the scale once the block is sliding across the table. Perform the same measurements with the block resting on other surfaces, such as various types of carpeting or rubber. Note the measurements for the starting and sliding friction and write down your conclusions.

4

Rocking and Rolling

When you finish reading this chapter, you should be able to explain

- the factors necessary to initiate and vary angular motion;
- how athletes apply torque and how torque varies angular motion;
- how athletes make use of first-, second-, and third-class levers;
- why rotating objects initially resist rotation and then want to continue rotating once they've been set in motion;
- how the resistance of an object to rotation can vary;
- what angular momentum is and how it is determined;
- how athletes make use of the principle of conservation of angular momentum in such sports as diving, gymnastics, and track and field; and
- how specific movements, like the body tilt and cat twist, allow athletes to mix somersaults and twists while in flight.

Rocking and Rolling is about rotation, and it's the largest chapter you'll find in this book. The size of this chapter is an indication not only of the importance of rotation but also of its presence in all sport skills.

The first half of this chapter explains the basic principles that relate to rotation. The second half will be of great interest to those of you wanting to coach gymnasts, divers, and figure skaters. Here we explain the basic mechanics of twisting and somersaulting. Even if you don't intend to coach these sports, you will benefit from reading this material because the mechanical principles involved in gymnastics, diving, and figure skating occur in other sport skills as well.

Rocking and *rolling* are two of the many terms that coaches use when they refer to angular motion. We also talk of athletes rotating, circling, revolving, spinning, somersaulting, twisting, pirouetting, turning, and swinging. All these terms refer to angular motion.

If you watch closely, you'll see that rotation occurs in all sports, including those that require an athlete to stay as still as possible. In Olympic pistol and rifle shooting, for example, athletes try to eliminate all unnecessary movements. They train to slow down their heart rate and, amazingly, to pull the trigger between heartbeats so that the thump of the heart doesn't jog the barrel of the gun. Even in this slow and patient sport, when the athlete squeezes the trigger the bones of the index finger move in a rotary fashion. It's the same when an archer flexes her arm to pull a bow. As the bowstring is drawn back, the forearm and upper arm flex and rotate toward each other. Both shooter and archer hardly move at all, but rotation occurs nevertheless.

Far different from archery and shooting are the sports of gymnastics, diving, ski aerials, and figure skating. Dramatic rotary skills performed in flight characterize all these sports. Many of these skills combine somersaults and twists. Somersaults most commonly occur around the athlete's **transverse axis** (from side to side, or from hip to hip) and twists around the **longitudinal axis**, or long axis (head to feet). Gymnasts also cartwheel and side-somersault around their **frontal axis** (from front to back) (figure 4.1 shows you these axes).

Because they compete in so many different events, gymnasts perform the greatest variety of rotational skills. In contact with stable apparatus (e.g., the floor, beam, vault, bars, and pommel horse) and highly unstable apparatus (i.e., the rings), gymnasts perform somersaulting and twisting skills around axes formed by their feet, hands, hips, shoulders, and even their knees.

The prize for the greatest number of somersaults and twists performed at any one time belongs not to gymnasts but to ski aerialists. These daredevil athletes use a "kicker" (i.e., a

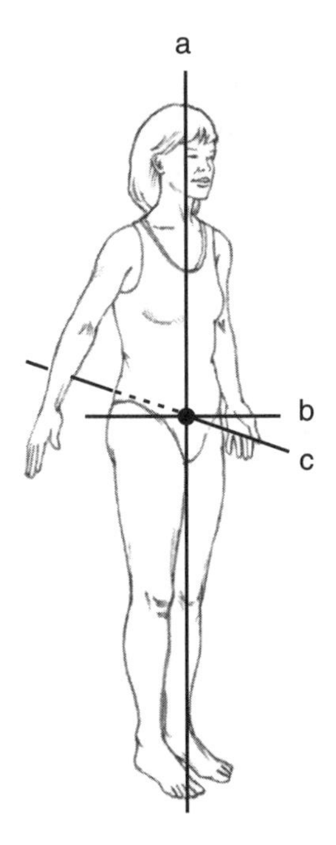

FIGURE 4.1 Major axes in the human body include *(a)* long axis, *(b)* transverse axis, and *(c)* frontal axis.

ramp) that throws them high in the air, and with their flight time extended by a downward sloping "outrun" they combine as many as three somersaults with five twists.

Angular Motion

Angular motion means that an athlete or an object rotates, spins, swings, or twists through an angle of a certain number of degrees. If a gymnast makes one full revolution around a bar, she has rotated through an angle of 360 degrees. In golf, from backswing to follow-through, a golfer can swing a club from 10 to 15 degrees in a putt to considerably more than 360 degrees in a drive. Basketball players who perform "360s" in slam dunk competitions spin one full revolution around their long axis before slamming the ball through the hoop. Now, "720s" are being performed in the "half-pipe" by snowboarders.

Whether rotation occurs through several revolutions or through an arc of a few degrees, the same mechanical principles apply. It's good to understand these principles because they will help you teach technique based on sound sport mechanics.

Components of Lever Systems

Levers move in a rotary fashion and are used in all sports. To understand how levers work in the human body and to see what part they play in sport, let's first look at the components of a lever system (shown in figure 4.2).

A **lever** is a simple machine that transmits and changes mechanical energy from one place to another. It usually incorporates objects such as the bones in an athlete's body or the pole used in pole vaulting, which is made to rock, or rotate, around an axis and in so doing produces angular motion. In the case of athletes, their bones rotate at the joints. In pole vault, the pole rotates around the point where it contacts the ground, and the pole-vaulters rotate where they grip the pole at the upper end. An athlete's muscles, bones, and joints work together as lever systems. The muscles pull on the bones and the bones rotate at the joints. The joints act like a hinge on a door, a **fulcrum** (i.e., an axis about which a lever rotates) on a scale, or an axle of a wheel.

In a lever system, force is applied at one location on the lever and a resistance applies its own force at another. The action of the applied force attempts to make the lever rotate in one direction. The force produced by the resistance tries to make the lever rotate in the opposing direction. Force and resistance battle each other. In an athlete's body, force is primarily produced by muscular contraction. The weight of the athlete's limbs plus the weight of whatever the athlete is trying to move produces resistance. Figure 4.3 shows the position of force and resistance in a dumbbell curl, which is an exercise most athletes use.

The perpendicular distance from the axis to where the force is applied is called the **force arm.** Likewise, the perpendicular distance from the

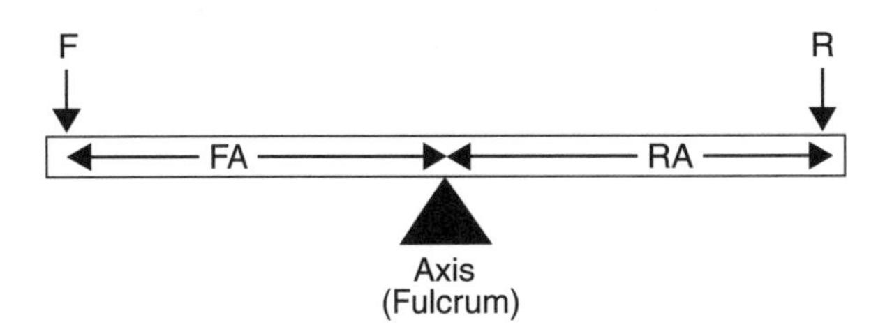

FIGURE 4.2 Components of a lever.

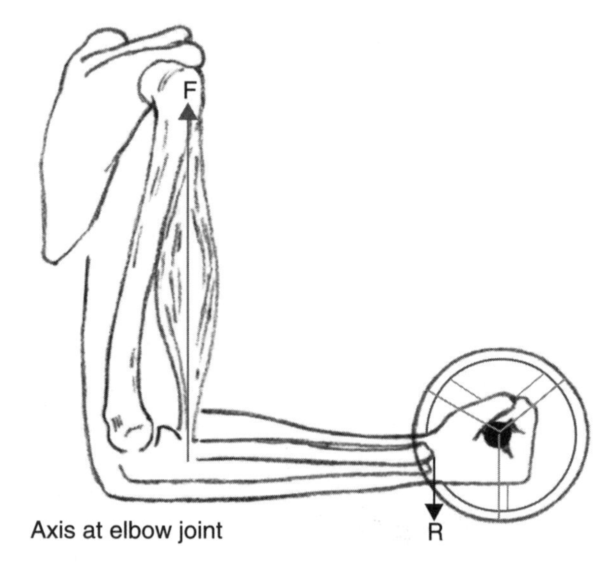

FIGURE 4.3 A lever system in the human body. Resistance is the combined weight of the dumbbell and the athlete's forearm.

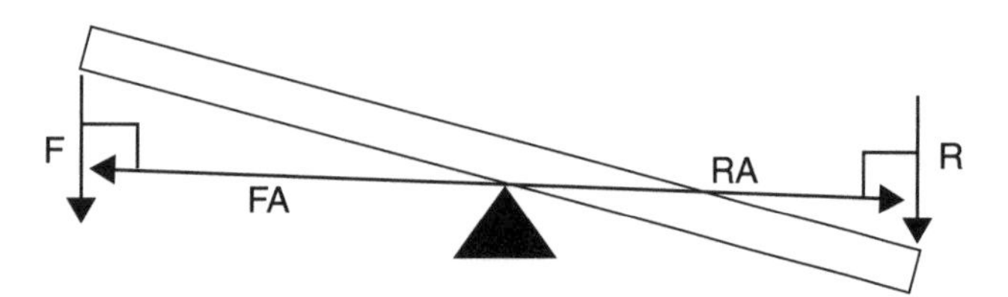

FIGURE 4.4 The force arm and the resistance arm on a lever that has rotated a few degrees. The force arm is the perpendicular distance from the force to the axis. The resistance arm is the perpendicular distance from the resistance to the axis.

axis to where the resistance applies its own force is called the **resistance arm.** Figure 4.4 shows the force arm and the resistance arm on a lever that has rotated a few degrees. Notice that the force arm forms a right angle (90 degrees), or is perpendicular, where it contacts the arrow indicating the direction that the force is being applied. The same situation occurs with the resistance arm. Later in this chapter you'll see that the length of the force arm relative to that of the resistance arm is important in determining what advantage an athlete gains when using a lever.

Torque

Because all levers rotate around an axis, they always produce a turning effect, which we call **torque.** In auto mechanics, a torque wrench is designed to apply a precise turning effect to a bolt. In weight training, a dumbbell curl requires the biceps muscle to pull on the forearm and produce a turning effect in an upward direction. How much torque occurs depends on the amount of force produced by the biceps multiplied by the length of the force arm. The force arm is the perpendicular distance from the biceps' attachment on the forearm to the axis (i.e., the elbow joint) (see figure 4.5). The dumbbell plus the weight of the athlete's forearm generate their own torque as gravity pulls them down.

To understand how we can increase the turning effect of torque, let's have an athlete become a mechanic and use a wrench to loosen a bolt. When the athlete pulls on the wrench, he applies torque to the bolt. Whether the athlete is successful in overcoming the rotary resistance of the bolt depends on how much force he exerts, how far from the bolt he applies force (i.e., the length of the force arm), and at what angle he pulls on the wrench. A 90-degree angle of pull is most efficient.

In figure 4.6a, the athlete applies 10 units of force at 5 units of distance from the bolt (which acts as the axis). The torque produced is 50 units. In figure 4.6b, the athlete applies the same amount of force 10 units of distance from the bolt. In this case the turning effect, or torque,

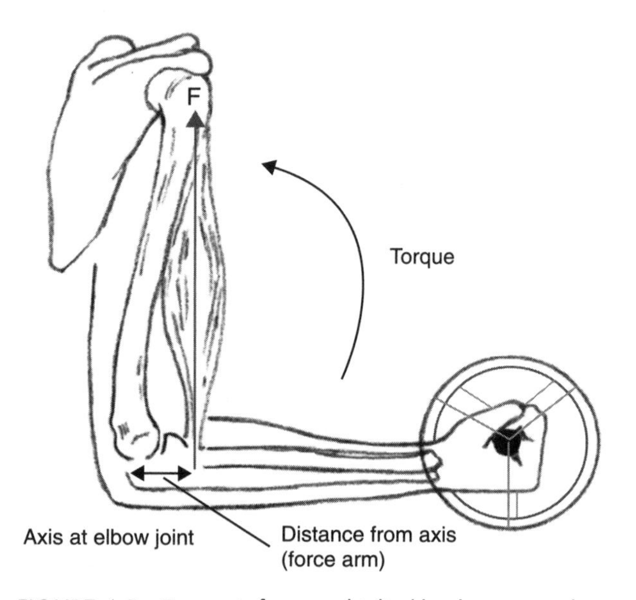

FIGURE 4.5 Torque is force multiplied by the perpendicular distance from the axis to where force is applied.

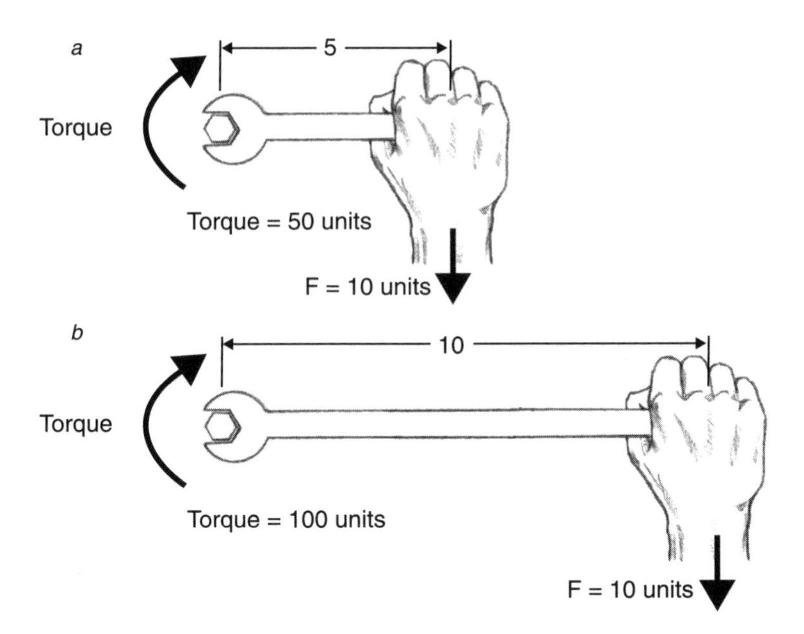

FIGURE 4.6 Torque is doubled when force is applied twice the distance from the axis. In *a* force is applied 5 units from the axis, producing a torque of 50 units. In *b* force is applied 10 units from the axis, producing a torque of 100 units.

applied to the wrench has been doubled to 100 units because the force arm has been made twice as large. If the athlete wants to produce even more torque, the options available are (a) to apply more force at 90 degrees to the wrench, (b) to increase the size of the force arm, or (c) to combine (a) and (b).

From this example you'll notice that a large force arm is an important requirement if an athlete wants to produce a large amount of torque. Now let's look again at the biceps curl in figure 4.5. Notice that in producing the turning effect of torque, the biceps has to work with a very small force arm. Later we'll explain how this characteristic (which is common throughout the human body) can be both a disadvantage and an advantage.

The principles followed by the athlete in applying torque to a bolt apply to all lever situations that occur in sport. In essence, what happens is a battle between the turning effect of two opposing torques—torque produced by an athlete's muscular forces and torque produced by any type of resistance, such as the weight of the athlete's limbs plus whatever the athlete is holding (e.g., a discus or a barbell). In a sport such as wrestling or judo, resistance and torque can also be generated by your opponent.

Sport skills involve many ways to apply the turning effect of torque. To understand these differences, let's look at how lever systems can vary.

Types of Levers

Levers are divided into three groups called first, second, and third class. This classification is based on how the force, resistance, and axis are positioned on the lever relative to each other. In an athlete's body, third-class levers are most common. However, in the performance of sport skills you'll find that athletes frequently use all three classes of levers.

There's an easy way to remember the relationship of force, resistance, and axis for each of the three lever systems. This is by using the acronym *ARF*, with the *A* associated with a first-class lever, *R* associated with a second-class lever, and *F* associated with a third-class lever. On a first-class lever, *A* (the axis) is positioned between the resistance and the force. On a second-class lever, *R* (the resistance) is positioned between the axis and the force. Finally, on a third class-lever, *F* (the force) is positioned between the axis and the resistance.

First-Class Levers

In a **first-class lever** the axis is positioned somewhere in between the force and the resistance (see figure 4.2). The force arm is the perpendicular distance from the axis to where force is applied to the lever, and the resistance arm is the perpendicular distance from the axis to where the resistance applies its force to the lever. Force and resistance arms can be equal in length, as they are in figure 4.2, or they can be unequal. If the force arm is longer than the resistance arm, then the lever favors force output, meaning that this lever arrangement multiplies the force exerted by the athlete. As a result, on the opposing side of the fulcrum, a greater force is applied to the resistance. If the force arm is shorter than the resistance arm, then the lever favors speed and range of movement at the expense of force. What is lost in force produces a gain in speed and distance, and what is gained in force output only occurs with a loss in speed and distance. So there's always a compromise. The following examples show how this compromise occurs in sport.

Leg Press As a First Class Lever

First-class levers are used frequently in the design of weight training machines. Figure 4.7 shows a simplified illustration of a leg press machine. The athlete applies force with her legs on one side of the axis and the weight stack as the resistance applies its own force on the other. On some leg press machines, the athlete will find two sets of foot pedals to push against. One set is purposely positioned lower on the machine than the other. An athlete who struggles to lift the weight stack using the upper pedals will find it much easier using the lower pedals. Why? Because on the lower set of pedals, the force arm (which is the perpendicular distance from the axis of the machine to the line of force applied by the athlete's legs) is much longer than on the upper set. As a result, the same amount of force applied by the athlete's legs to the lower set of pedals produces greater torque. In figure 4.7, the turning effect of torque produced by the athlete in a clockwise direction must overcome the torque generated by the weight stack acting in a counterclockwise direction. But if something is gained by the athlete in using the lower set of pedals, then something has to be lost or given

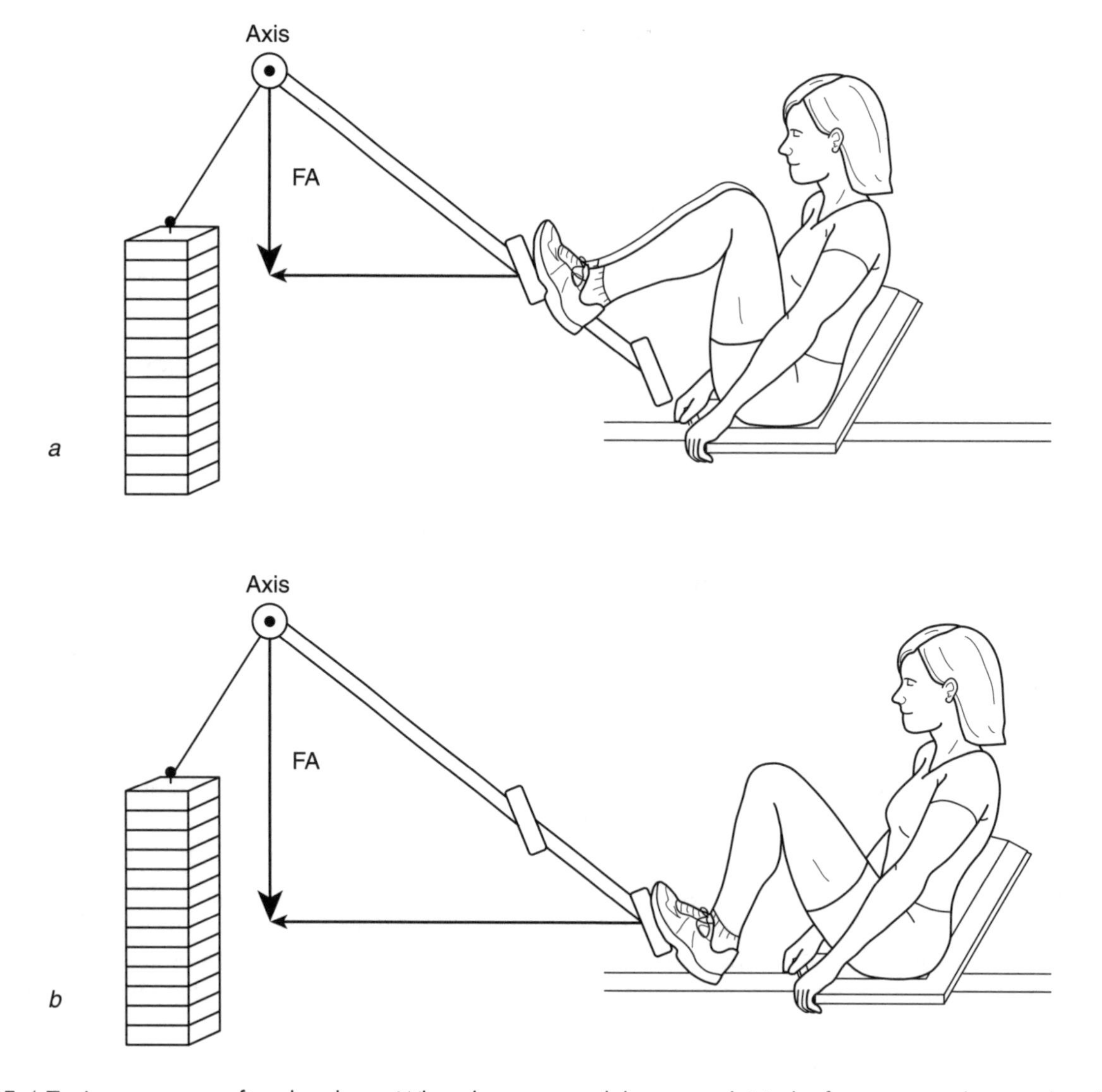

FIGURE 4.7 Leg press as a first-class lever. When the upper pedals are used *(a)*, the force arm is shortened and more force is required to raise the weight stack.

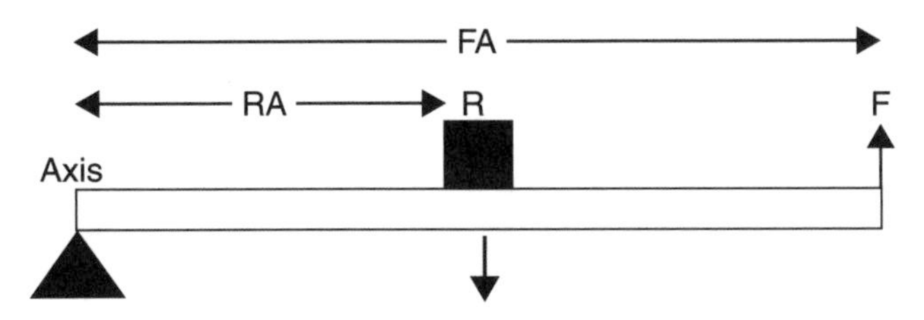

FIGURE 4.8 Second-class lever.

up. The compromise in this situation is that the athlete must push the lower set of foot pedals through a bigger arc of movement than when using the upper set.

Second-Class Levers

A **second-class lever** is characterized by having both force and resistance on the same side of the axis, with the force arm always longer than the resistance arm (see figure 4.8). The applied force (which could be the force applied by you) acts in one direction (e.g., counterclockwise), and the resistance tries to move in the other direction (i.e., clockwise). If you apply sufficient force, both force and resistance move in the same direction. Second-class levers favor the output of force at the expense of speed and range of movement. The larger the force arm in relation to the resistance arm, the greater the force output. An athlete who uses a second-class lever applies less force over a large distance at a greater speed in order to shift a heavier resistance a small distance at a slower speed. The following example demonstrates these characteristics.

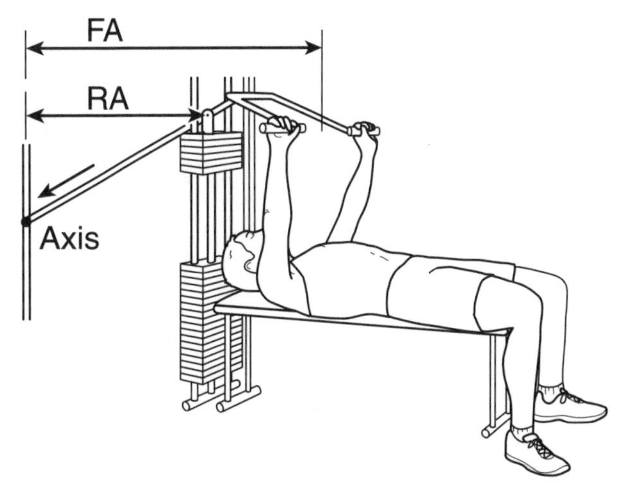

FIGURE 4.9 Bench press on a multipurpose weight training machine as a second-class lever.

Bench Press As a Second-Class Lever

A second-class lever arrangement is shown in figure 4.9, where an athlete performs a bench press on a multipurpose weight machine. Notice that the athlete pushes upward on handles attached to a single bar; the bar rotates at an axis to the left of the machine. The weight stack, supported by rollers that run on the upper surface of the same bar, is positioned closer to the axis than the handles on which the athlete is pushing. The force exerted by the athlete moves upward in an arc, and the rollers allow the weight stack to be lifted vertically.

When the athlete's arms are at full extension, the perpendicular distance of the force arm is reduced in length relative to the perpendicular

Use of Cams on Weight Training Machines

Many types of weight training machines make use of oddly shaped cams. Cams are basically pulley wheels with off-center axes and have a profile (i.e., shape) that depends on the specific exercise for which they were designed. Joined to the circumference of the cam is a cable or chain that is often attached to the weight stack. As the athlete uses the machine, the cam is forced to rotate. The rotation of the cam varies the length of the radius from the cam's axis to where the cable contacts its circumference. This variation in radius changes the resistive torque produced by the weight machine so that the athlete has to work harder at some joint angles than at others. This type of "accommodating resistance" takes into account the fact that the human body is stronger and more efficient at certain joint angles than at others.

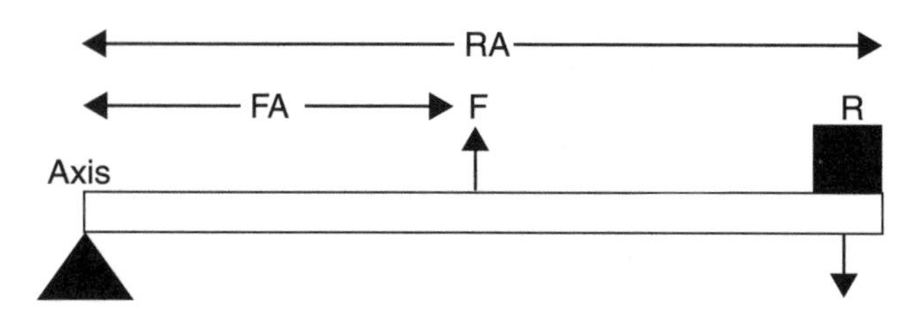

FIGURE 4.10 Third-class lever.

distance of the resistance arm. In other words, as the athlete straightens his arms and becomes anatomically more efficient in applying force with the muscles of his arms, chest, and shoulders, the weight machine counters this anatomical change by reducing the length of the force arm relative to the resistance arm.

Third-Class Levers

In a **third-class lever** the axis is at one end of the lever and the applied force is always closer to the axis than the resistance (see figure 4.10). As in the second-class lever, if the applied force is great enough to move the resistance, then both applied force and resistance move in the same direction. This distinguishes second and third-class levers from first-class levers, where applied force and resistance move in opposing directions. Third-class levers always move the resistance through a larger range and at a greater speed than that moved by the force. On the other hand, the force that is applied is greater than that applied by the resistance.

Biceps Curl As a Third-Class Lever

A biceps curl is a great example of a third-class lever because it shows the typical relationship that exists between most of the muscles, bones, and joints in an athlete's body. It also demonstrates how your muscles must exert considerable force to move even a light resistance.

In figure 4.11 an athlete is holding a dumbbell so that it neither rises nor falls. The counterclockwise turning effect (torque) produced by the athlete's biceps equals the torque produced by gravity acting in a clockwise direction on the mass of the athlete's forearm and the dumbbell. In this example we have set the ratio of resistance arm to force arm at 10:1. If the resistance is 10 units and the distance of the resistance to the elbow joint is 10 units, then the clockwise

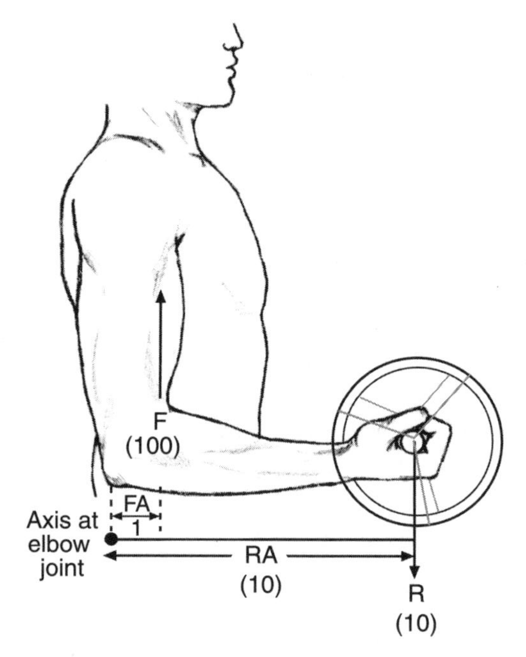

FIGURE 4.11 Biceps curl as a third-class lever.

turning effect of the resistance is 100 units ($10 \times 10 = 100$). If the force arm is only 1 unit of distance, the force necessary to hold the forearm and dumbbell horizontal is 100 units, or 10 times that of the resistance.

We must mention one additional characteristic here. With the arm held horizontal, the biceps normally pulls on the forearm at an angle slightly less than 90 degrees. This arrangement helps to rotate the forearm to the upper arm and also pulls the forearm toward the elbow joint, which helps to stabilize this joint. (Physiologically, this helps keep the bones at the elbow joint held together.) In figure 4.12 we have purposely exaggerated the angle of pull of the biceps so that the division of the force produced by the biceps is more apparent. In this figure, the contraction of the biceps pulling the forearm toward the upper arm is approximately 25 degrees from the vertical. Any angle of pull by the biceps that is not perpendicular to the forearm means that some of the force of muscle contraction is directed elsewhere and so plays no part in producing torque. In figure 4.12, *a* represents the line of action (contraction) produced by the biceps, *b* indicates that part of the force produces torque and lifts the forearm, and *c* shows that some force is directed toward the elbow joint.

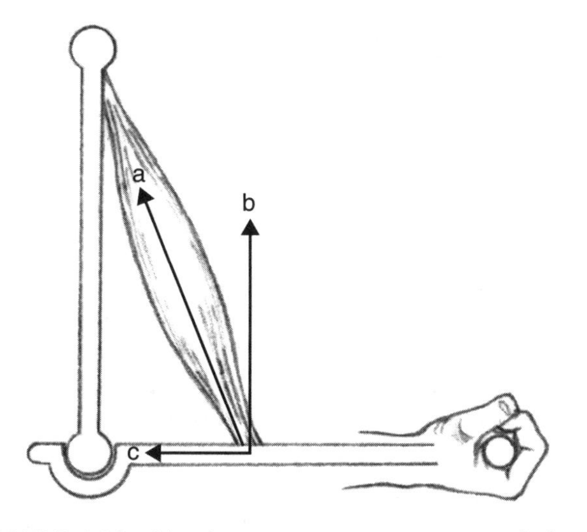

FIGURE 4.12 Muscle a is contracting at an angle less than 90 degrees. Its force can be split into force b, which acts perpendicularly to the forearm, and force c, which acts toward the elbow joint. (The angle of the muscle in this figure is exaggerated for the sake of clarity.)

Third-class levers in the human body are commonly characterized by muscles that attach near the joint and whose contraction produces tremendous torque. But, this torque is transmitted along very long levers. The longer the lever, the smaller the resistance that can be moved. A force of 100 units produced by the biceps muscle is immediately reduced because the muscle pulls on the forearm at an angle other than 90 degrees. The turning effect occurring where the muscle is attached to the forearm is further reduced by the length of the forearm. By the time it reaches the athlete's hand it has been reduced considerably. However, an athlete does gain from this arrangement, as you'll see in the following section.

Advantages and Disadvantages of Limb Length

If an athlete contracts his biceps in the curl illustrated in figure 4.11 so that its attachment on the forearm moves in an arc 1 unit in length, the dumbbell travels around an arc 10 times larger. This increase in range of movement, shown schematically in figure 4.13, occurs because the resistance arm is 10 times longer than the force arm. Furthermore, if it takes 1 sec for the athlete's biceps to complete its arc of 1 unit, the dumbbell moves through an arc 10 times larger in 1 sec also. Thus, the longer the athlete's arm, the faster and farther the dumbbell moves.

These characteristics illustrate the compromise we all experience as a result of the way our bodies are designed. We are all at a disadvantage

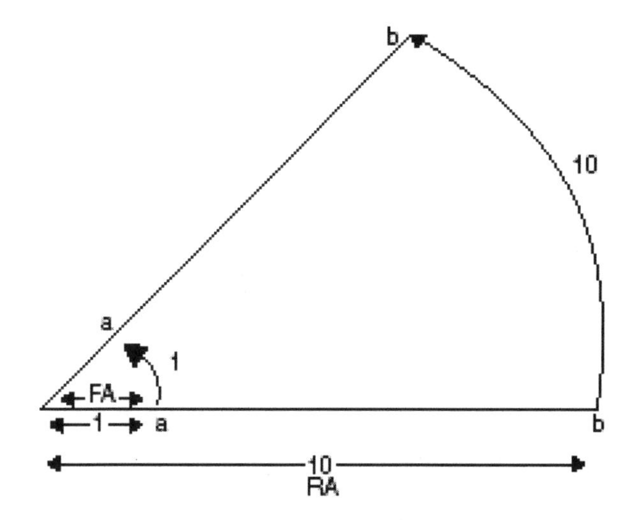

FIGURE 4.13 Point b moves 10 times farther and 10 times faster than point a on the force arm.

Rowers Use Hatchet Blades to Apply More Force

Elite rowers use oars with huge blades that look like giant meat cleavers. Called "hatchet blades," these oars are shorter from the oarlock to the blade than standard oars. The mechanical principle behind the new design is that for the same effort from the rower, the new blade travels more slowly through the water but applies more force. Moving slower, the blade slips less in the water but propels the shell faster. Do these blades present any problems? According to many coaches, hatchet blades can cause stress injuries because the rowers have to pull against a stiffer and less mobile resistance.

because our muscles must exert tremendous force to move a much lighter resistance. Obviously, if we don't possess the necessary strength we cannot move even a light resistance. And to make matters worse, the longer our limbs, the greater the force required from our muscles. So what do we gain from this arrangement? Think of an athlete with long arms. Long arms mean that an athlete's hand travels at great speed over a huge distance when the arm is swung in an arc. Providing an athlete can generate sufficient force in the muscle, a long limb moves a light resistance (such as a ball, club, or discus) over an immense range at tremendous speed. This explains why discus throwers in the Olympic Games are such huge long-armed athletes. The more powerful an athlete's muscles and the longer the arms, the better. Short limbs, on the other hand, give up less force along their length than do long limbs. You'll see what we mean by this statement in the following fantasy competition between a 5-ft 6-in. male gymnast weighing 130 lb and a giant 7-ft professional basketball player who weighs 300 lb. Let's imagine for the sake of this scenario that these two athletes have similar strength in their pectoral and latissimus muscles. We now ask both athletes to attempt an iron cross on the rings—an extremely difficult gymnastic skill.

An iron cross requires an athlete to contract the latissimus and pectoral muscles. These muscles do most of the work in pulling the arms and hands downward on the rings and simultaneously, as part of their contraction, pulling the body up. This action stops the athlete's body from dropping down out of the cross position.

A basketball player with the physique that we've described is immediately at a disadvantage in this skill. Long arms are long levers and in a mechanical sense form huge resistance arms. The resistance is the basketball player's body weight pulling downward. The axis of rotation is at his shoulder joint, and his pectoral and latissimus muscles, which provide force, have to work with a very short force arm. (Of course, by being human both basketball player and gymnast suffer from this problem!) For our basketball player, the force produced by muscle contraction in the pectoral and latissimus muscles, multiplied by a small force arm, faces off against a resistance of 300 lb of body weight multiplied by a huge resistance arm—the enormous length of the basketball player's arms. For our basketball player to hold an iron cross and not drop down toward the floor, the turning effect (torque) produced by force multiplied by force arm must equal the torque produced by the resistance multiplied by the resistance arm (see figure 4.14). A gymnast with shorter arms has a much better chance of success in this skill. Short arms in a mechanical sense mean short levers and small resistance arms. The torque produced by the gymnast's lighter body weight multiplied by these short resistance arms is considerably less than that of a heavyweight basketball player with exceptionally long arms.

Let's consider the issue of body weight for a moment. Our basketball player has 300 lb of body weight acting as the resistance. Even though his pectoral and latissimus muscles collectively battle 300 lb of body weight pulling downward, an iron cross would demand strength in these muscles that you'd find only in a "bionic man." The contractile force that the basketball player's

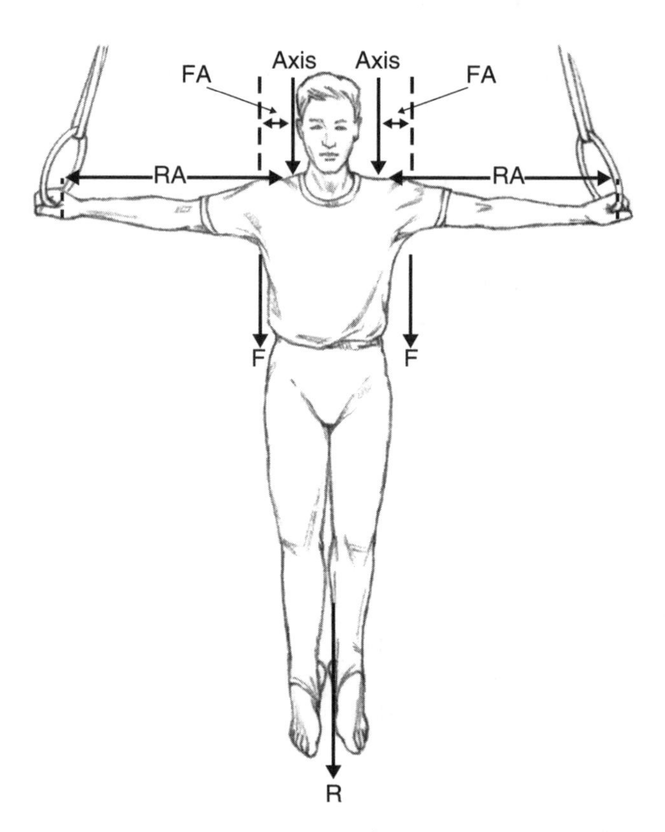

FIGURE 4.14 Force arms and resistance arms in the iron cross.

muscles would have to produce for an iron cross could even tear the muscles away from where they attach to the bones!

The basketball player's muscles get little help from the short force arm they must work with, and they fight immense resistance (his body weight) multiplied by a huge resistance arm (the length of his arms). These characteristics are reason enough why athletes with the physique of basketball players seldom if ever venture into the sport of gymnastics.

Most male gymnasts are around 5 ft 6 in. tall and weigh between 120 and 130 lb. Females are even smaller, averaging 5 ft and weighing no more than 110 lb. So gymnasts not only have the advantage of short arms, but they also benefit from having minimal body weight. Their body weight is often less than half that of basketball players in the NBA. Success in an iron cross (and other difficult gymnastics skills) is much more likely when you are built like a gymnast. This explains why elite gymnasts are super strong, have short limbs, and diet to control their body weight. A couple of extra pounds (i.e., a kilogram) can mean the difference between success and failure in many gymnastics skills.

Now let's take our basketball player and gymnast onto the track to compete against each other in a discus competition. When they throw a discus we find that a taller, heavier, long-limbed athlete has a distinct advantage. A spin across the ring by our long-armed basketball player gives the discus incredible velocity at the instant of release because it is pulled around a circular pathway made enormous by the length of his arms. Proof of the importance of arm length for a discus thrower was demonstrated in the 2000 Sydney Olympics, where the men's discus gold and silver medalists were both enormous men with arm spans reaching 7 ft 4 in. (2.2 m) from fingertip to fingertip.

Our gymnast loses in the discus event. If both gymnast and basketball player were to spin across the ring at the same rate, a discus in the hand of a long-armed athlete such as a basketball player would travel faster. What could our gymnast do to compensate? He could spin at such a phenomenal rate that he's a blur as he goes across the ring. But this is very difficult to accomplish, and today few small athletes compete seriously in the discus event.

Long arms are a great advantage in an event like the discus and a disadvantage in most gymnastics skills. It depends on what is required by the sport. In some sport skills it is better to be heavy and to have long arms and legs; in other sports these characteristics become a disadvantage.

Initiating Rotation

Athletes produce the turning effect of torque any time they apply force at a distance from an axis. If they increase the distance or the force that they apply, the turning effect becomes greater. The application of torque can make objects and athletes rotate.

In ball games like tennis or volleyball, the server wants to spin the ball on some occasions and avoid spinning it on others. If volleyball players wish to serve a floater, they make sure that the force they apply to the ball passes directly through the ball's center of gravity, which in flight is the ball's axis of rotation. When this happens, the ball floats across the net without spinning (see figure 4.15).

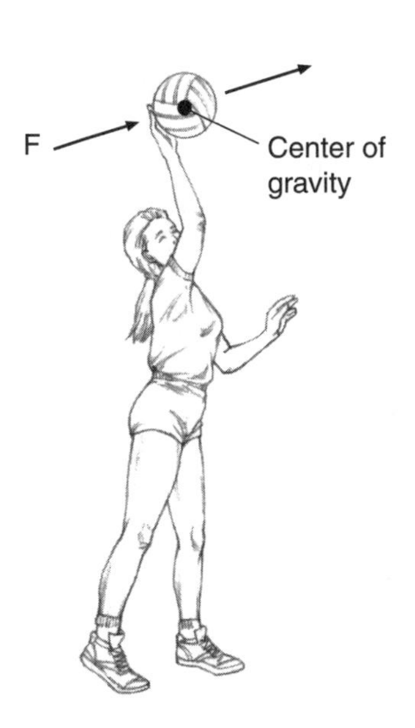

FIGURE 4.15 No spin is imparted to a volleyball when force is directed through its center of gravity.

Reprinted, by permission, from K. Luttgens and K. Wells, 1992, *Kinesiology: Science basic of human motion*, 8th ed. (Times Mirror Higher Education Group), 314. Reproduced with permission of The McGraw-Hill Companies.

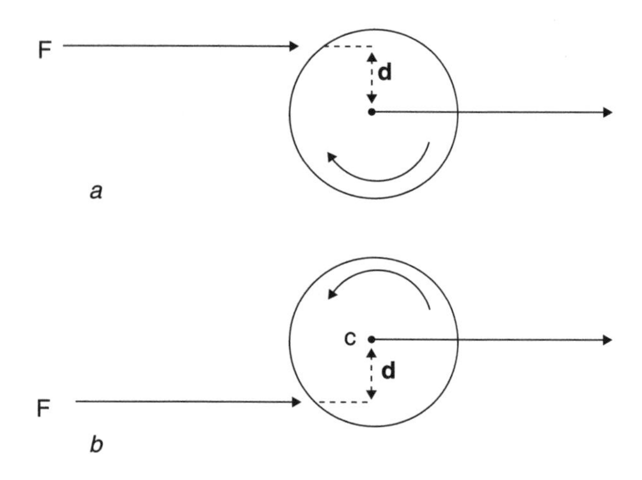

FIGURE 4.16 Application of topspin *(a)* and backspin *(b)* to a ball.

What must happen if servers want to spin the ball? Then they apply the force from their hand at some distance from the ball's center of gravity. Increasing this distance magnifies the spin. Figure 4.16a shows force applied well above the ball's center of gravity. When this happens the ball receives a topspin, and as a result it arcs downward in flight. In golf, a chip shot requires golfers to apply force well below the ball's center of gravity. The angle of the club face and the stroke technique produce a backspin, which causes the ball to lift in flight. The amount of backspin influences how much the ball will rise. Depending on the surface on which it lands, the ball can stop dead or, if shot to the far side of the green, can roll back toward the pin.

The amount of spin given to any ball depends on how much force is applied and how far it is applied from the ball's center of gravity. The greater the force and the larger the distance from the center of gravity, the greater the torque and the greater the spin. If you want to apply maximum spin to a football, you grip it closer to the middle where the ball is fattest. Here the force applied by your fingers to the circumference of the ball is farthest away from the ball's center of gravity and its long (i.e., longitudinal) axis. In this way you apply maximum torque, which gives the ball the greatest amount of spin. The drawback with this position is that you put less of your force into throwing the ball for distance.

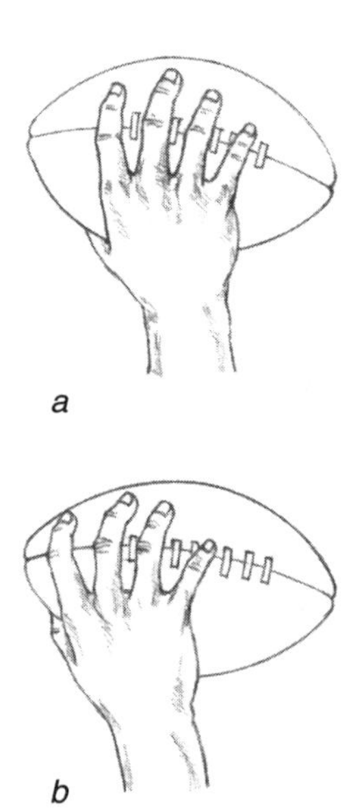

FIGURE 4.17 Varying spin and directional force of a football. Gripping the middle of the ball *(a)* provides maximal spin and less force for distance. Gripping the end of the ball *(b)* provides less spin and maximal force for distance.

On the other hand, if you want to apply less spin and concentrate on applying maximum force to the ball, you grip it closer to the end (see figure 4.17, a-b). Most quarterbacks compromise by gripping the ball halfway between these two positions.

How Athletes Make Themselves Rotate

Athletes who want to rotate in the air employ the same mechanical principles to themselves as those used to apply spin to a volleyball. For example, a trampolinist who bounces on a trampoline with his center of gravity directly above the upward thrust of the trampoline bed rises vertically without rotating. The thrust of the trampoline pushes directly upward through his center of gravity. Like the volleyball, the trampolinist moves vertically (and without rotating) in the same direction as the thrust of the trampoline (see figure 4.18a).

If the trampolinist positions his center of gravity so that it is no longer directly above the upward thrust of the trampoline bed, then there is a tendency to rotate. The more the athlete shifts his center of gravity out of line with the thrust of the trampoline (e.g., by leaning

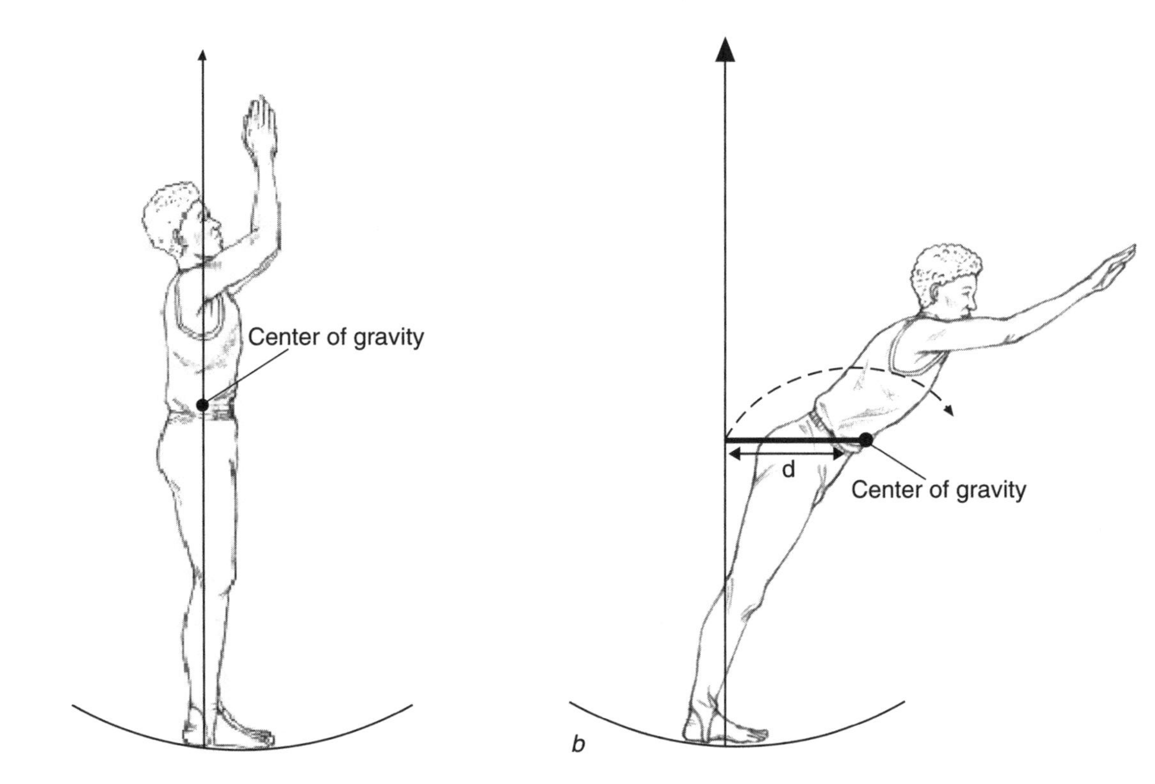

FIGURE 4.18 A gymnast rises vertically when the vertical thrust of the trampoline is through the center of gravity *(a)*. When the thrust from the trampoline does not pass through the gymnast's center of gravity *(b)*, the torque produced causes the gymnast to rotate.

Adapted, by permission, from H. Braecklin, 1974, *Trampolinturen II* (Germany: Limpert Verlag), 29.

forward or backward), the greater the turning effect (i.e., torque) applied to his body and the greater the tendency for rotation to occur (see figure 4.18b).

How Gravity Can Assist With Rotation

Earth's gravitational force (which pulls perpendicularly to the earth's surface) can be used to produce torque and cause rotation. In figure 4.19 a gymnast is performing a front giant, which is one 360-degree revolution around the high bar. In this skill the gymnast is accelerated toward the earth by gravity's pull on the downward portion of the circle. The gymnast is then decelerated by gravity on the upward portion.

As the gymnast holds on to the bar and rotates toward the earth, gravity applies torque to his body. The earth's pull is concentrated at his center of gravity and acts perpendicularly toward the earth's surface at every instant in the giant swing. The distance (d) of the gymnast's center of gravity from the bar (the axis around which the gymnast rotates) is greatest when his body is fully extended in a horizontal position. The greatest turning effect (i.e., torque) applied by gravity occurs as the gymnast passes through this position. Conversely, no torque is produced by gravity when the gymnast is directly above or directly below the bar (see figure 4.19).

In football, a tackle around the ankles can suddenly turn a running back's ankles (or feet) into an axis. The running back rotates around that point. If the running back is sprinting flat out when the tackle occurs, the linear motion of the athlete's body is dramatically changed into rotation. As the player rotates toward the ground, gravity assists in accelerating the player downward.

The taller the running back, the greater the distance of the athlete's center of gravity from his feet. Therefore the turning effect is also greater. This is one reason big men fall "heavier" than those not as tall.

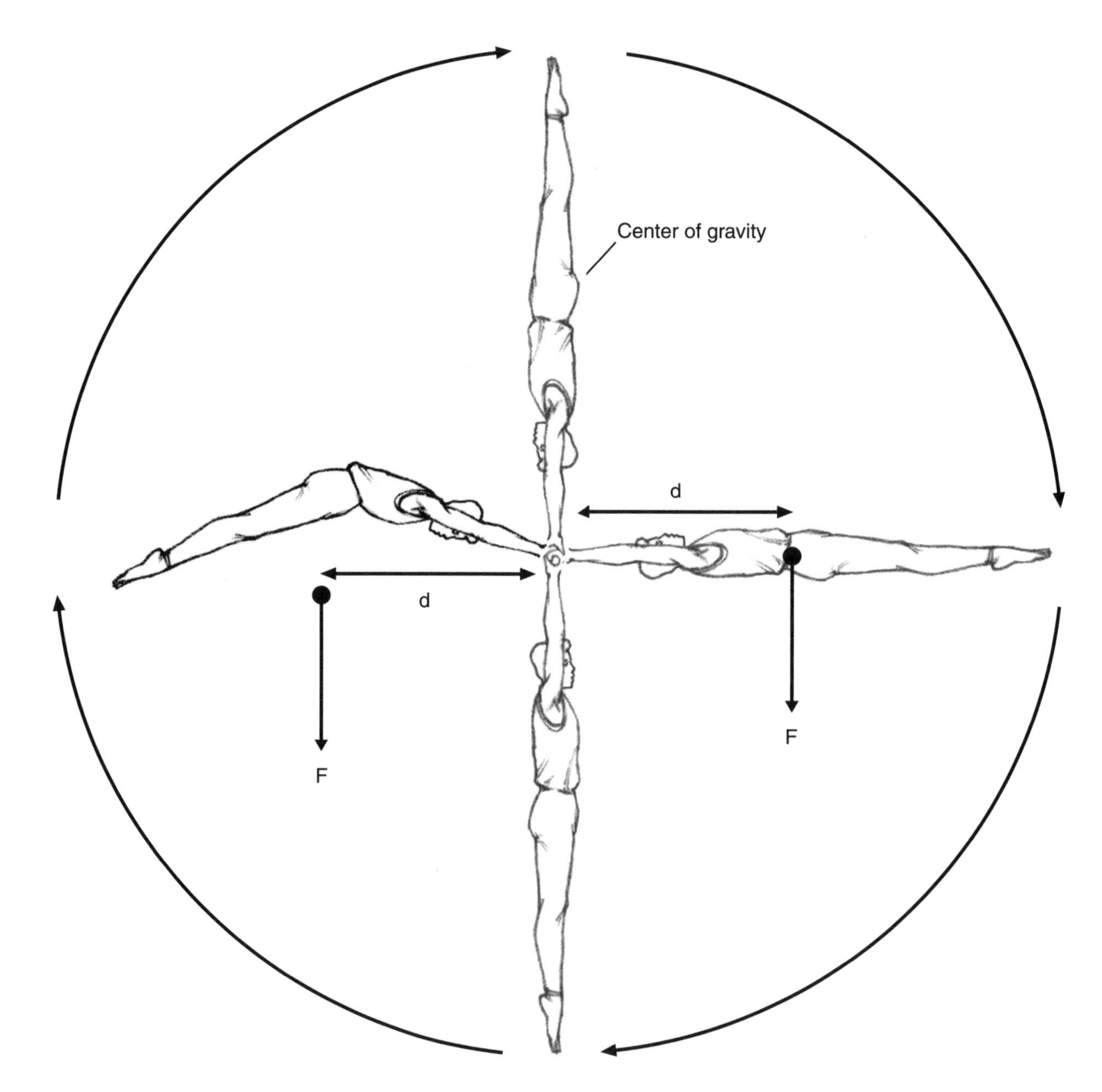

FIGURE 4.19 Torque produced by gravity in a front giant around the high bar.

Angular Velocity

Angular velocity is a term used to describe the rate of spin of an athlete or an object. This phrase also describes the rate of swing of a bat or a club. However, there's a difference between the angular velocity and the speed of an object or an athlete as they rotate. To understand this difference, let's revisit our gymnast performing giants (i.e., 360-degree rotations) around the high bar and ask him to rotate around the bar with a perfectly rigid body. (In reality, gymnasts must flex at the hips and at the shoulders at specific phases in a giant around the high bar, but for this example, we'll assume that the gymnast remains rigid throughout this skill.)

If you time the gymnast as he rotates around the bar, your stopwatch might show that he makes one complete revolution in a clockwise direction every second. You now know the angle, or the number of degrees (i.e., 360 degrees for each revolution), that the gymnast performs in a particular time and in a particular direction.

If you know the number of revolutions, the time frame, and the direction that the gymnast is rotating, then you know his angular velocity, or rate of spin. (Angular refers to angle, degrees, or revolutions, and velocity means speed with direction.) All parts of the gymnast's body make one complete circuit around the bar in a clockwise direction in 1 sec.

Let's now turn our attention to the gymnast's body as it rotates. If you watch his hips as they follow their circular pathway, and then watch his feet, you'll come to the following conclusions:

- The gymnast's hips travel around a much smaller circle than his feet. His feet, which are farther from the bar, travel around a bigger circle.
- Both the gymnast's hips and feet complete their different size circles in the same time frame. So the gymnast's hips and feet have the same angular velocity.
- If the gymnast's feet are twice the distance from the bar as his hips, they travel around a circle that is twice as big and must be moving twice as fast. Their speed is twice that of the hips.

This information tells us that although all parts of the gymnast's body have the same rate of spin (i.e., angular velocity), the farther away from the bar, the faster the gymnast's body parts will move. Figure 4.20 shows that in a front giant the gymnast's feet *(a)* move around faster than his hips *(b)*, which in turn move faster than his shoulders. The gymnast's fingers gripping the bar *(c)* move slowest of all, yet his fingers have the same angular velocity as all other parts of his body. Can you guess by looking carefully at figure 4.20 which factors influence how fast the gymnast's feet will travel? Their speed depends on how many revolutions per second the gymnast

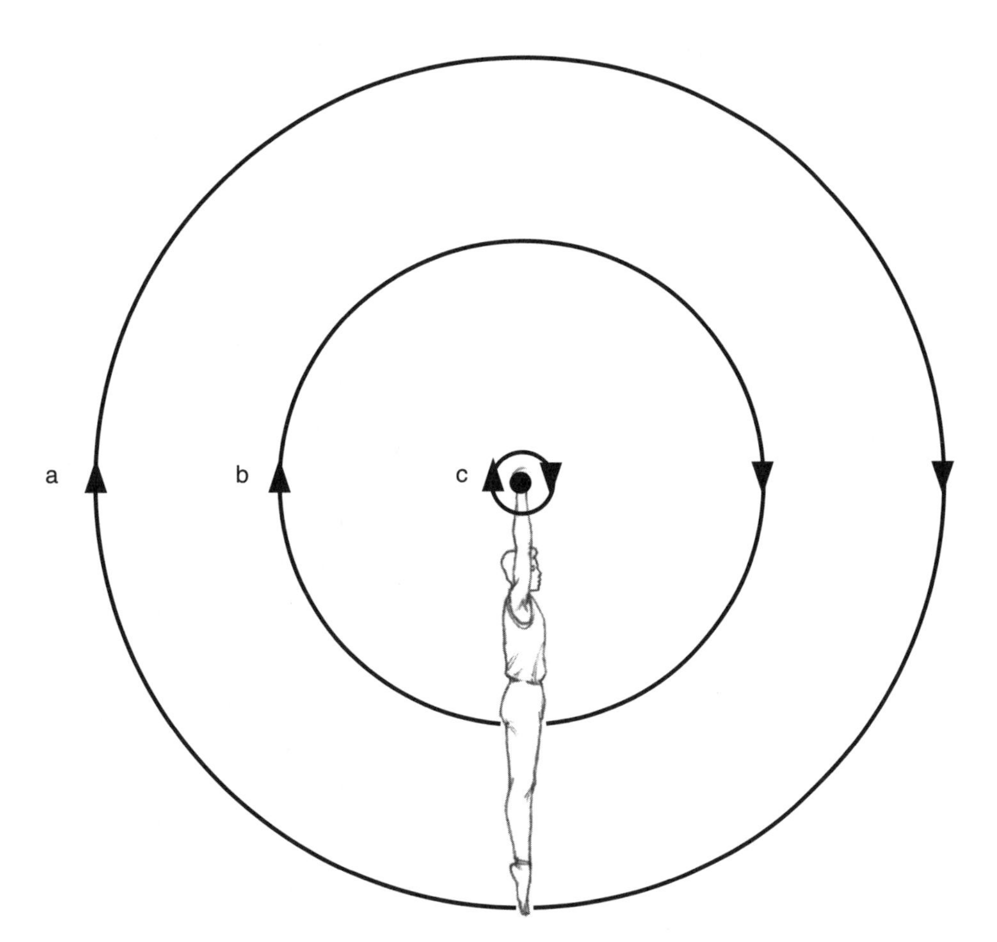

FIGURE 4.20 The gymnast's feet *(a)* travel faster than his hips *(b)*, and his hips travel faster than his hands *(c)*.

rotates (which is the gymnast's angular velocity) and how far the gymnast's feet are from the axis of rotation. (The words "how far" refer to their radius, or distance, from the axis of rotation.) The axis of rotation is the high bar itself.

Let's take what we've learned from this high bar example and apply this information to a golf scenario. Imagine that you're getting a lesson from your club pro and the pro tells you, "If you want to drive the ball farther you'll need to produce more club-head speed." Now you can think to yourself, "Okay, for more club-head speed I can increase the angular velocity of my club by swinging it faster through its arc, or I can hold higher up on the grip of the club. This puts the club head farther from my axis of rotation. By gripping higher on the club I increase the distance (or radius) from my axis of rotation to the club head. I can even use a longer club. Any of these changes will increase the speed of the club head" (see figure 4.21). Long drive competitions in which athletes swing extremely long clubs at phenomenal angular velocities produce drives of more than 400 yd. Long drive competitions don't require extreme accuracy, but they certainly demonstrate what's required to hit a golf ball a long way.

You can apply the reasoning that you used during your golf lesson to any sport that uses bats, rackets, and clubs. To hit a ball as hard as possible, the batter should hold higher up toward the end of the grip to increase the radius and then swing as fast as he can. Swinging the bat through its arc as quickly as possible maximizes its angular velocity. Maximum angular velocity combined with the optimal radius gives the greatest speed to the striking portion of a club, bat, or racket.

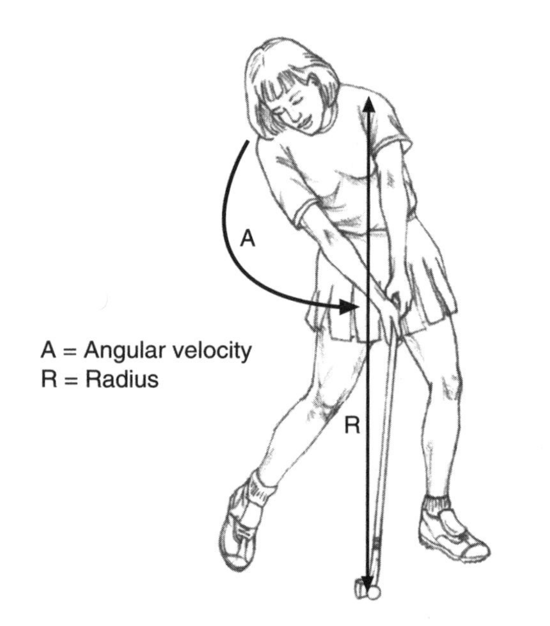

FIGURE 4.21 In a golf swing, angular velocity (A) multiplied by the radius (R) determine the speed of the club head.

Inertia and Centripetal Force

Whenever there is rotation, there is always an interplay between inertia and centripetal force. Inertia and centripetal force are present in the spin of a volleyball spike serve, a spiral pass by a football quarterback, and even a gentle putting stroke by a golfer.

Rotation is a battle between inertia and centripetal force. The inertia of an object when it is moving is expressed in its desire to travel in a straight line. Changing straight-line motion into circular motion requires a **centripetal force.** This is a force that pulls (or pushes) an object toward the axis of rotation in order to make it follow a curved, or circular, pathway. When an athlete swings a baseball bat, he applies a centripetal force to make sure that the bat follows the arc of the swing. The inward pull of centripetal force produced by the athlete battles the inertial desire of the bat to travel in a straight line. The athlete will certainly feel that the bat is trying to pull outward away from his grip. This outward pull is often called a centrifugal force, which as we will see is really inertia under another name.

Inertia and Centripetal Force in the Hammer Throw

To understand how inertia and centripetal force are interrelated, let's have an Olympic hammer thrower spin around and throw the hammer. Like anything else that has mass, a hammer has no wish to move to begin with, and second, no desire to follow a circular path. The inertia of the hammer makes it resist movement. A competitive hammer for men weighs 16 lb (7.26 kg) and for females 8 lb 12 oz (4 kg), so these implements have plenty of mass and consequently a lot of inertia. If an athlete applies sufficient

Technology Gets Involved With the Hammer Throw

The men's hammer that's thrown in track and field weighs 16 lb and measures 4 ft from the handle to the farthest point on the surface of the ball. Until the rules were changed, technicians realized that if they used extremely dense tungsten, they could put almost all the hammer's 16 lb in the distal portion of the ball. This shifted the hammer's center of gravity farther away from the thrower than a hammer made of less dense material. It also made the ball about the size of a baseball, which reduced its air resistance during flight. The handle was then made of extremely light titanium so there was virtually no weight in close to the thrower. Unfortunately, there was an unexpected problem—hammers with 16 lb concentrated in a sphere as small as a baseball buried themselves so deep in the turf that athletes and officials had a hard time pulling them out! Rules now outlaw these types of hammers.

muscular force to get the heavy ball of a hammer to follow a circular pathway, then its inertia will be expressed by its desire to travel in a straight line and not around a circular pathway.

As the athlete performs spins in the throwing ring, the ball of the hammer reluctantly travels around a big circle at the end of its wire. Most of the mass of a hammer is in the ball. There is very little mass in the wire and the handle (see figure 4.22). To keep the hammer on its circular path, the athlete must continuously fight the hammer's desire to get away and travel in a straight line. To do this he constantly pulls inward, applying a centripetal force to change the hammer's path from one that is straight to one that is circular. This pull travels from his body, down his arms, along the wire, and out to the hammer ball.

The amount of centripetal force that hammer throwers produce depends on what occurs. For example, if a thrower doubles his angular velocity by spinning around twice as fast, then he must increase his inward pull on the hammer ball fourfold. Why? Because the hammer goes around twice as fast and tries to follow a straight-line pathway at every instant with twice the force. In other words, an increase in an athlete's rate of spin squares the demand for centripetal force. Double the rate of spin (i.e., angular velocity) and the athlete has to pull inward (i.e., increase the centripetal force) four times as much. Triple the rate of spin and centripetal force must be increased nine times as much.

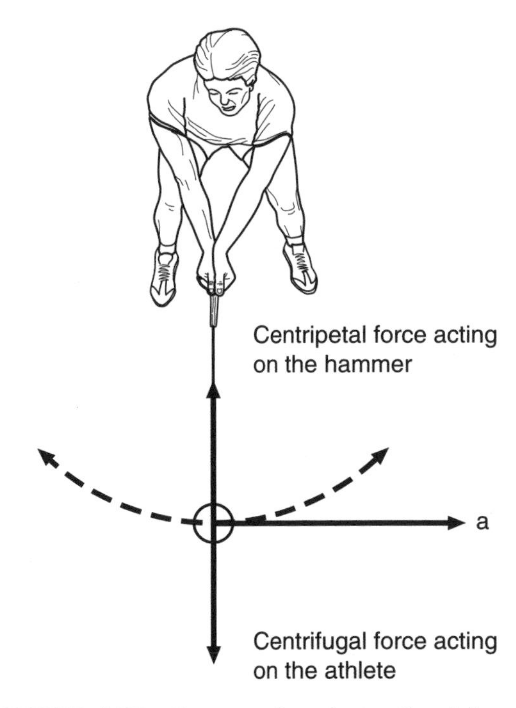

FIGURE 4.22 Centripetal and centrifugal force in the hammer throw. The hammer wants to travel in the direction of arrow *a*.

What happens if there is no change in the thrower's angular velocity (i.e., rate of spin) but instead he spins around with a hammer twice as heavy? In this case, the athlete must double his centripetal force because the extra mass of the hammer tries to travel off in a straight line at every instant with twice the force. Athletes who spin around with a more massive hammer or who attempt to spin faster (i.e., increase their angular velocity) must be prepared to

FIGURE 4.23 As the hammer travels faster, the thrower increases centripetal force by leaning away (arrow *a*) from the pull of the hammer (arrow *b*).

increase their centripetal force by flexing their legs, leaning backward, and pushing against the earth. Leaning backward and pushing with the legs against the earth causes the earth to push against them and this counteracts the extra pull of the hammer (see figure 4.23). Failure to carry out these actions can throw athletes off balance, often with disastrous effects.

Centrifugal Force

Centrifugal force is a commonly used term, yet it is described by physicists as a phony or fictitious force. It is best to consider it as "inertia in disguise." Using the hammer throw as an example, the following description will explain why.

When athletes make a hammer follow a circular pathway, they apply a centripetal force by pulling inward on the hammer at every instant the hammer goes around. As the athletes pull inward on the hammer, they experience an outward pull that we call **centrifugal force** (see figure 4.22). *Centrifugal* means "pulling outward from the axis of rotation." Since the athletes provide the axis of rotation around which the hammer travels, they certainly feel that the hammer is trying to drag them outward. In reality, this outward pull is the hammer wanting to travel along a straight line, or more precisely, to fly off at a tangent to the circle that it is being forced to follow. The more massive the hammer and the faster the athlete makes it moves, the greater its resistance to following a circular pathway and the greater its desire to fly off in a straight line.

Immediately upon release, the hammer gets its chance to quit going around in a big circle and instead fly away along a straight-line pathway. And if it were not for gravity and air resistance, it would do just that, following the trajectory that the athlete gave it at the instant of release. Throwers no longer apply a centripetal force to the hammer once they've released it. The distance the hammer travels through the air depends on the velocity and trajectory of the hammer ball at the instant of release—not on centripetal force and certainly not on centrifugal force.

Keep in mind that the battle between inertia and centripetal force occurs in all sport skills in which there's rotation, not just in hammer throwing. For example, when speedskaters glide around the oval and lean into each curve, the athletes are obeying the same principles as the hammer throwers. By leaning into the curve, speedskaters push outward at an angle against the ice (i.e., against the earth). The earth pushes back (equal and opposite) and provides the inward push of a centripetal force. The more massive the skaters, the faster they travel; the tighter the curve, the more they must lean and push outward against the earth in order to get the earth to push them inward.

The fact that a speedskater must push outward to get around the curve indicates that some of the athlete's leg thrust is used pushing outward while the remainder is used for propulsion forward along the ice. Once the athlete is out of the curve and into the straightaway, then all of the athlete's force can be used in driving toward the finish line. The same situation faces a track sprinter rounding the curve in a 200-m sprint. Compared with running on a straightaway where all the sprinter's effort can be spent driving toward the tape, running a curve makes the athlete spend some precious energy thrusting outward to get around the curve. Proof of this division of effort shows up in comparing the records for 220 yd on a straightaway and 200 m on a curve. (Keep in mind that 220 yd is 1.28 yd longer than 200 m.) In 1966, Tommie Smith ran the

220 yd on a straightaway in 19.5 sec. In 1979, Pietro Mennea of Italy set a world record for the 200 m around a curve at 19.72 sec. This record held up until 1996, when the USA's incredible Michael Johnson ran 19.66 sec. He lowered his record to 19.32 sec less than two months later at the Atlanta Games. Can you imagine what times he'd run on a straightaway?

Is there any way to help a sprinter negotiate a curve without expending precious energy pushing outward? Yes. Bank the curves of the track. A banked track pushes the athlete inward the same way that the drum of a washing machine holds clothes to a circular pathway during the spin cycle. On tight indoor tracks, high banking allows the athletes to run flat out without fearing that their inertia will cause them to fly off the track and into the spectators. Outdoor tracks are not normally banked, so the tight inside lanes can be difficult to negotiate. This is particularly the case for heavier sprinters who have more inertia and must push outward more vigorously to get their extra body mass around the curve.

Rotary Inertia

Rotary inertia can be thought of as rotary resistance or rotary persistence. In physics texts it's commonly described as "moment of inertia." All four terms mean the same thing. **Rotary inertia** is the tendency of all objects or all athletes to initially resist rotation and then to want to continue rotating once they have had the turning effect of torque applied against them and there's a centripetal force to keep them following a circular pathway. Rotary inertia occurs in every situation in which athletes rotate, spin, or twist and in every situation in which bats, clubs, and other implements are swung. In short, rotary inertia exists in all sporting situations where angular (i.e., rotary) motion occurs.

Rotary inertia is the rotary equivalent of linear inertia and so is related to Newton's first law of inertia. Massive linemen have great linear inertia. Their mass doesn't want to move, and if forced to move, their mass wants to travel in a straight line. If these linemen got on a merry-go-round (like those in your local playground), it would require a lot of effort (i.e., torque) to get the merry-go-round spinning. But once under way, the merry-go-round (with the added mass of the lineman) will want to continue spinning. If the teammates of the linemen got on as well, they would increase the merry-go-round's rotary inertia even further. The merry-go-round with all its riders would act like a giant flywheel—similar to the ones you see on old steam engines driving threshing machines at agricultural fairs. With all these heavyweight riders, it would take a lot of torque to get the merry-go-round spinning, but once spinning it would want to keep spinning.

The following two important factors determine how much inertia a rotating object will have:

- The mass of the object. The more massive an object, the more resistance it puts up against

Traveling Around a Banked Track at High Speed

Athletes who race indoors are familiar with tracks that are banked on the curves. The banked track allows athletes to run flat out without having to reduce their speed as they negotiate the curves. Texas Motor Speedway in Fort Worth, Texas, is banked at 24 degrees. The steep banking allows CART racecars to hurtle continuously around the track at full throttle. Drivers complained of dizziness, vertigo, and other problems with their vision and hearing after traveling at speeds in excess of 200 mph around the Speedway. These physiological difficulties can be related to a battle that occurs between centripetal force and inertia. Centripetal force keeps the race car and driver following the curve of the track whereas inertia wants the race car and driver to travel in a straight line. The outward inertial pull on the drivers as they rounded the track was estimated at 5 g, or five times their normal body weight.

being rotated. In addition, the more mass the greater the persistence the object has in wanting to continue rotating once rotation is established. A heavy (i.e., more massive) baseball bat is more difficult to swing than a light one. It resists being accelerated through the swing more than the lighter bat. Once a batter has applied sufficient turning effect, or torque, to get a bat moving, a heavy bat wants to continue the swing more than a lighter one. The heavier the bat, the stronger the athlete must be to get it moving and to control and stop it once this motion has been initiated.

- The radial distribution of mass. The phrase "radial distribution of mass" refers to how the mass of an object is distributed (i.e., positioned) relative to the axis around which it's spinning. Two extremes in the distribution of mass are whether the object's mass is far from the axis of rotation or whether it is close to the axis of rotation.

Here's an example of radial distribution of mass in a sport situation. Imagine that you are given two golf clubs. We'll call them club A and club B. The two clubs are alike in length and shape, and on a scale they weigh exactly the same. Club A is like any other golf club that you would buy in a sports store. But as soon as you pick up club B you sense that it differs tremendously from club A. Except for a couple of ounces of weight in the shaft, all of its weight (i.e., mass) has been concentrated in the club head. Club B will therefore have more rotary inertia than club A because almost all of its mass is way out in the club head. Compared with club A, you'll find that a swing with club B will be more difficult to initiate. It will also be more difficult to stop once the swing is under way.

Let's take club B, with most of its weight in the head, and start repositioning its mass by shifting it up the shaft of the club toward the grip. We'll leave the club head the same shape but hollow it out so that it has hardly any mass. As we progressively move the mass of the club from the head toward the grip, its rotary inertia (i.e., rotary resistance) will be gradually reduced so it becomes easier to swing and likewise easier to maneuver during the swing. Finally, when almost all of its mass is in the grip, the club will feel like a fencing foil, which has most of its mass in the handle. You'll feel that you can maneuver this golf club quite easily, but you'll also know that a club designed in this fashion will never hit a golf ball very far.

How Differences in Distribution of Mass Affect Rotary Inertia

We have just seen that the rotary inertia of any object, whether it is a golf club or an athlete, depends on how much mass it has (the more mass, the more rotary inertia) and how its mass is positioned relative to its axis of rotation (the farther the mass is from the axis of rotation, the greater the rotary inertia). How does the positioning of the mass of an object relative to its axis of rotation affect its resistance to being made to rotate and, once rotating, being made to stop rotating? The answer to this question illustrates the huge difference between the inertia of an object traveling in a straight line and an object that is spinning. Objects traveling in a straight line increase their inertia in direct proportion to their mass. More mass equals more inertia. But with a rotating object, any time its mass is moved in closer to the axis, or conversely farther away from the axis, it causes a dramatic change in the object's resistance to rotation. Here's an example of how this phenomenon works. Imagine that you have a ball with every particle of its mass concentrated at its center. (Of course, it's impossible for all the mass of a ball to be at its center, but for this example let's imagine that one exists!) In figure 4.24a we will discount gravity and air resistance, and the ball is on a string and rotating around an axis at 1 rev/sec. The distance of the ball from the axis is 2 units. In figure 4.24b, this distance is suddenly reduced to half of the original. This reduction in radius causes the rotary inertia of the ball to be reduced fourfold (i.e., to a quarter of its original). Because of this huge reduction in rotary inertia, the ball finds it easier to rotate and it speeds up (i.e., increases its angular velocity) proportionally from 1 rev to 4 rev/sec. In figure 4.24c, the distance of the ball from the axis has been doubled from 2 units to 4. The rotary inertia of the ball is now increased fourfold. The ball finds it more difficult to rotate and slows to 1/4 rev/sec. This example tells us that the rotary

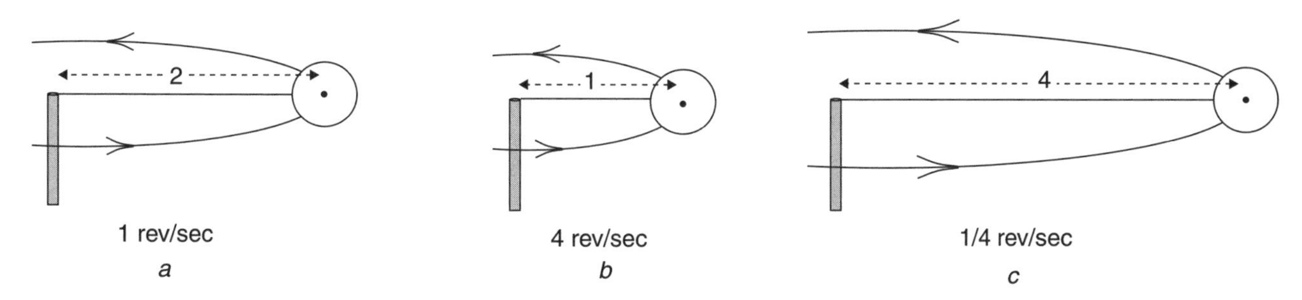

FIGURE 4.24 Reducing the radius from 2 units *(a)* to 1 unit *(b)* reduces rotary inertia fourfold. Doubling the radius from 2 units *(a)* to 4 units *(c)* increases rotary inertia fourfold. (Air resistance and gravity have been discounted.)

inertia of a spinning object is proportional to the square of the radius. Halve the radius and the rotary inertia is reduced to one quarter of the original ($1/2 \times 1/2$). Double the radius and the rotary inertia is increased to four times the original (2×2). What's the effect of changing the mass of the rotating object? The answer is that it is directly proportional. Doubling the mass of the object doubles its rotary inertia. Halving it will halve its rotary inertia.

In sport there's no situation in which every little bit of an object or an athlete is shifted the same distance toward or away from its axis. You cannot move every particle of a baseball bat to one end because there would be nothing left to make up the remaining part of the bat. Nor can you shift every particle of the bat halfway along its length because that leaves nothing at the ends. It's the same with athletes.

But there *is* a big difference between athletes and inanimate objects such as baseball bats! Athletes can change their shape at will. They can tuck their bodies up tight, or they can extend them. By carrying out these maneuvers, they pull their body mass in close to their axis of rotation, or they push their body mass out from their axis of rotation as far as possible. Let's see how this occurs in diving.

Manipulating Rotary Inertia in Diving

When springboard and tower divers somersault in the air and move from an extended body position to a tuck, they flex their torsos, legs, and arms. Some parts of their bodies (i.e., their arms and legs) shift a large distance toward their transverse axis (hip to hip). Other parts, such as their heads, move a short distance. Nevertheless, there is a dramatic difference in rotary inertia between an extended and tucked body position. A diver's legs and arms are relatively heavy and have a lot of body mass. Moving them a large distance toward the diver's axis greatly reduces the resistance of the diver's body against rotation.

Divers in flight rotate slowly when they assume an extended body position. If they pull their bodies into a tight tuck, they rotate much faster. The more body mass they pull toward their axis of rotation (which always passes through their center of gravity) the faster they spin. This means that a lean body coupled with great flexibility plays an important role in determining how much faster divers spin when they pull into a tuck. Huge, muscular athletes have a tough time pulling themselves in as tight as lean athletes. Their excess body mass gets in the way. That's one reason today's great Chinese divers look positively skinny! With little to no excess body mass, and being phenomenally flexible and strong, they are able to pull themselves into the tightest tuck possible. The technique they use for multiple somersaults is similar to that used by trapeze artists. Trapeze artists perform quadruple somersaults by gripping their shins and pulling their knees up as high as possible toward their shoulders. The more compressed and compact they are, the faster they spin (see figure 4.25, a and b).

Rotary Inertia in Sprinting

Athletes vary their rotary inertia in many events, not just in sports like diving. For example, in sprinting 100 m, a sprinter extends his legs to the rear of his body and flexes them when they come forward for the next stride. Likewise, when he drives his legs forward, the thigh is elevated and the leg flexed at the knee. Flexing at the knee

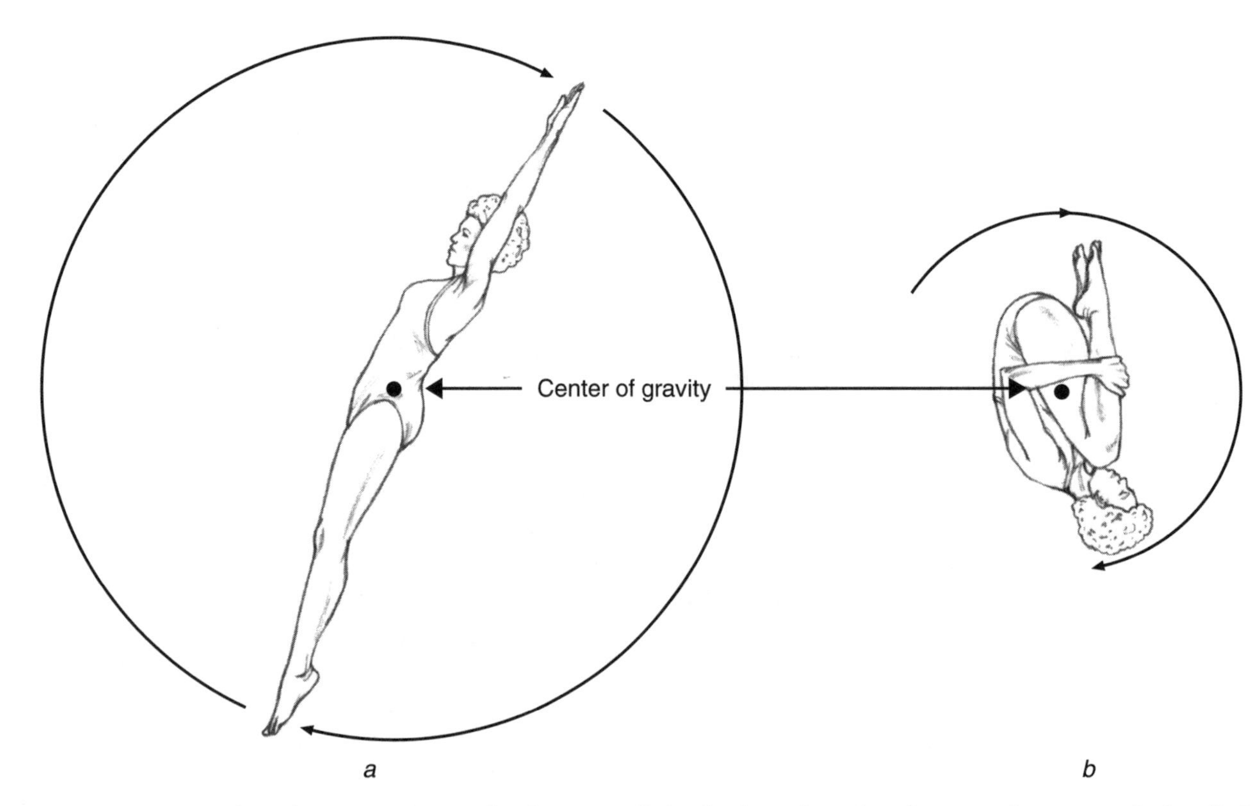

FIGURE 4.25 Angular velocity increases as the diver *(a)* pulls her body in close *(b)* to her axis of rotation (which in flight passes through her center of gravity).

brings the mass of the sprinter's leg closer to the hip joint (the axis around which the leg rotates). This reduces the leg's rotary inertia. If the rotary inertia of the leg is reduced, it makes the task of moving the leg forward much easier for the muscles involved (see figure 4.26).

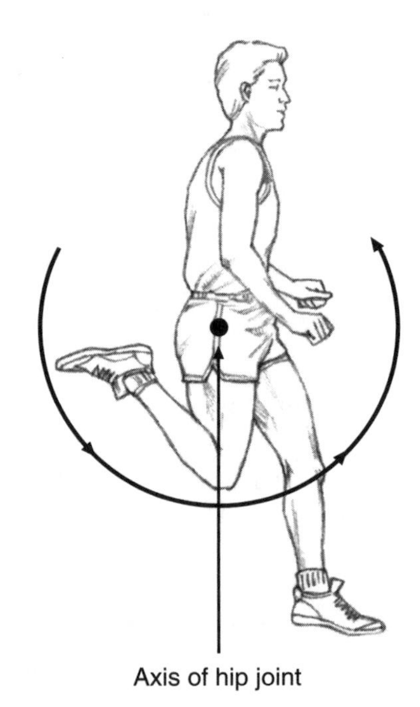

FIGURE 4.26 Rotary inertia of the sprinter's leg is reduced when it is flexed.

Angular Momentum

Angular momentum applies to mass that is rotating, spinning, or turning. More precisely, **angular momentum** describes the quantity of motion that a rotating athlete or an object possesses. In sport it's often important for athletes to generate as much angular momentum as possible, whether it's to their own bodies, to an opponent, or to a bat or a club. On other occasions, athletes must reduce angular momentum to minimal values. To help you understand how angular momentum is used in sport, let's review linear momentum, and then we'll look at angular momentum.

In chapter 3 we used a football lineman charging straight ahead as an example of linear momentum. The more massive the lineman and the faster he moves, the more momentum he produces. We can apply the same concept to

Olympic Pole Vault Differs From Dutch Canal Vaulting

One of the rules in pole vault says that a vaulter cannot shift the upper handhold. Why? Because in the old days, vaulters were monkey-climbing the pole! If you visit Holland, you'll see a competition in which climbing the pole still occurs. The Dutch, for fun, pole-vault for distance across water-filled canals. Climbing the pole gives them more distance and a dry landing on the other side of the canal. If they climb the pole too soon or too high, they can stall and drop in the water, or even fall back toward the takeoff. Spectators love the drama and the inevitable dunkings! When they climb the pole, the vaulter's rotation slows. A musician's metronome does the same thing. Raise the weight and the metronome clicks at a slower tempo. Keep raising the weight and the metronome will stop altogether. The same principle applies to Dutch canal vaulting.

objects that rotate or to rotating objects (such as bats being swung) that meet other objects (such as baseballs being pitched).

In baseball, a pitch travels in a predominantly linear manner to meet a bat moving in an arc. The bat being swung has angular momentum. The ball has linear momentum, and if it spins, it has some angular momentum too. Even if a pitcher hurls a fastball at 100 mph, the ball will still not possess much momentum because it doesn't weigh much and so doesn't have much mass. The bat swinging around its arc at a slower speed is much more massive than the ball and so has more momentum. At the instant the bat and ball collide, the bat is traveling in a linear manner one way and the ball is traveling in a linear manner the other way. At impact, mass × velocity (bat) overwhelms mass × velocity (ball). The bat slows down and the ball changes direction—it is driven backward.

Suppose a pitcher could fire the ball over the plate at 1,000 mph. Even though the mass of the ball has not changed, its velocity has increased tremendously, and as a result its momentum has increased too. In this imaginary situation, the ball could possess more momentum at impact than the bat. The bat will be driven back, or if it's made of wood rather than aluminum, it could snap off at the handle.

What can a coach do to compete against a pitch traveling at 1,000 mph? Start scouting for a batter who can swing a bat at phenomenal speed. If this athlete can react fast enough to handle a 1,000-mph pitch, the bat will win instead of the ball. In all likelihood the ball will disappear out of the park and land in the next city!

In this fantasy scenario, the swing of the bat gives us some idea of the components that make up angular momentum:

- Mass (i.e., how massive the object is)
- How the mass is positioned (i.e., distributed) relative to the axis the object is spinning around
- Rate of rotation, or swing (i.e., its angular velocity)

In reality, when bats (or clubs) are swung, there's always a trade-off between the mass of the bat, its length and mass (weight) distribution, and the rate at which the bat is swung. In baseball, no slugger on record has used a 42-in bat, which is the limit the rules will allow. Batters also tend to choose bats weighing between 32 and 34 oz, although there are no legal restrictions on using heavier ones. The reason? A huge, long, heavy bat with most of its mass in the barrel (i.e., hitting end) demands tremendous power from the batter and inevitably takes longer to accelerate than a light bat. The speed of a pitch gives a batter only a fraction of a second to react. So batters opt for a lighter bat that they can swing quickly. Perhaps they also know that the angular velocity of the bat (i.e., its rate of swing) is more important than how massive the bat is in determining how far a baseball will travel.

Generating as much angular momentum as possible is just as important in sports in which athletes rotate as in sports in which bats and clubs are swung. In diving, particularly in those dives that require numerous twists and somersaults, it is essential that the diver generates both linear and angular momentum at takeoff. The diver uses linear momentum to get high above the board as well as sufficiently far enough away from the board to be safe. At the same time the diver will initiate rotation at takeoff and make sure that her body is extended and spread out. By extending and spinning, the diver has produced a considerable amount of angular momentum. She then uses this angular momentum to perform all the somersaults and twists that occur later in the dive. Without a large amount of angular momentum initiated at takeoff, dives that contain a high number of twists and somersaults would not be possible.

Increasing Angular Momentum

The batting example we looked at earlier indicates that it's possible to increase angular momentum in three ways:

1. Increase the mass of whatever is rotated. Athletes can choose a heavier bat to swing. But if athletes are rotating, they must suddenly (and magically!) gain weight and spin at the same rate to increase angular momentum. Obviously, this is not possible.
2. Shift as much mass as far from the axis of rotation as possible. If athletes are rotating, they must extend their bodies. If they are swinging a bat, it must be long and with most of its mass at the barrel end.
3. Increase the angular velocity of whatever is rotated. Athletes can increase their rate of spin. Batters can swing a bat faster and so give it more angular velocity.

Keep in mind that you always need to find a balance between mass, distribution of mass, and angular velocity. In striking and hitting skills, a huge increase in mass placed a long way from the axis of rotation is like putting a superheavy boot on the kicking foot of a field goal kicker. If the athlete doesn't possess the strength to swing his leg, the extra mass is worthless. The right amount of mass combined with a long leg and tremendous angular velocity is what's required.

Using Angular Momentum at Takeoff

The rules of high jump demand that athletes jump from one foot and cross a bar. A two-foot takeoff is not allowed. If an elite high jumper pushed down on the earth with one leg and did nothing else, the athlete would not go very high. What a high jumper adds to the thrust of the jumping leg is a strong upward swing by the arms and the free leg. By performing these actions, the high jumper generates angular momentum that is then transferred to his body as a whole. The upward swing of the arms and free leg adds to the thrust of the leg that is pushing down on the earth. More push down at the earth means that the earth in reaction pushes more against the athlete. The result is that he gets higher in the air (see figure 4.27).

When gymnasts perform back somersaults on the floor, they use a similar technique to a high jumper. They swing their arms upward to assist in getting off the ground. For greatest effect, their arms are extended and swung upward with tremendous velocity.

Elite high jumpers and gymnasts always make sure that their takeoff actions occur while in contact with the earth. Figure skaters do the same. Great skaters push against the ice because the ice is part of the earth, and if they push down, the earth pushes them up. The more they push, the more the earth pushes back against them.

FIGURE 4.27 Momentum transfer in a high jump takeoff.

It's this reciprocal response from the earth that gets the skaters up in the air for their triples and quadruples.

If athletes try the same actions while in the air and not in contact with the earth, a totally different effect occurs. We'll discuss this phenomenon in the following sections.

Conservation of Angular Momentum

If an athlete or an object such as a wheel is spun, both athlete and wheel will continue to spin forever providing there is nothing to stop their rotation. More precisely, if the turning effect of torque is applied to an athlete or a wheel, both would spin continuously if not for the fact that ultimately torque is applied by something else or someone else to stop them spinning. The something or someone can be friction with a supporting surface that the athlete is spinning on, it can be the resistance applied by air or water, and it can be the resistance (i.e., torque) applied by another athlete. The principle of **conservation of angular momentum** is the rotary equivalent of Newton's first law of inertia, which tells us that mass on the move wants to continue moving in a straight line and will do so in the absence of oppositional forces. In a rotary sense, spinning objects will continue spinning as long as there is nothing to stop them.

In sport there are many instances when the resistance opposing the spin of an athlete is so small as to be negligible. Think of gymnasts rotating in the air as they dismount from a high bar routine. Consider high jumpers and long jumpers in flight or divers and trampolinists rotating in flight. In all these cases, air resistance is so small that it can be discounted. In these situations we say that their angular momentum is conserved.

Conservation of Angular Momentum of Athletes in Flight

The word *conserve* means to stay the same, or to be maintained. This applies to the amount of angular momentum an athlete possesses during a particular phase of a skill. For example, the angular momentum an athlete generates at takeoff in high jump, long jump, and diving remains the same while the athlete is in flight. Why? Because an athlete cannot push against the air to increase or decrease angular momentum, and during flight the air has a negligible effect in reducing angular momentum. The only force that's working on the athlete is the earth's gravitational force, which pulls at her center of gravity. This force doesn't have any effect on the athlete's angular momentum, although gravity certainly increases her linear momentum as she accelerates toward the earth. Consequently, the athlete's angular momentum doesn't increase or decrease. It stays as is. This means that the angular momentum generated at takeoff is conserved during flight. Let's look at diving again to see how the conservation of angular momentum is tied in with a diver's ability to control the rate of spin while in the air.

Controlling the Rate of Spin in Diving

When divers accelerate down from the 10-m tower, they hit the water in less than 2 sec. In flights of such short duration, the diver's angular momentum is conserved. Discounting the negligible amount of angular momentum lost pushing the air around, we can say that the amount of angular momentum generated by the diver at takeoff stays virtually the same throughout the dive.

In flight, divers, just like a gymnasts, shift from layout body positions to tight tucks. To get to the latter position, the diver uses muscular force to pull the legs and arms inward, tuck in the chin, and flex the spine. By doing this, the diver pulls his or her mass closer to the axis of rotation. When this happens, rotary inertia (i.e., the diver's resistance against rotation) is reduced and the diver spins faster (i.e., the diver's angular velocity increases). But what causes this faster spin? Where does the extra angular velocity come from? The answer is found by examining what happens when one of the items that make up angular momentum is increased or decreased.

We know that the amount of angular momentum divers possess in flight is determined by three items: their rate of spin, their mass, and finally, the distribution of their mass. In other words, how much spin, how much mass, and how extended or tucked up they are. All three of these factors combine to make up the divers' angular momentum.

If divers are in a situation (e.g., in flight) where their total angular momentum stays at a set amount (i.e., is conserved) and one factor involved in creating angular momentum is reduced (e.g., they tuck and pull their body mass inward), then another component of angular momentum must increase to keep their total angular momentum unchanged. Since divers cannot possibly change their mass (i.e., gain or lose weight) while in flight, when they tuck and pull themselves in toward their axis of rotation, their angular velocity must increase. Pull your body inward and you spin faster. Spread your body out and you spin slower.

Controlling the Rate of Spin in Figure Skating

Another example of athletes reducing rotary inertia by pulling themselves in tight to increase the rate of rotation occurs when figure skaters perform a multiple twisting skill in the air, such as a quadruple toe loop or a triple Axel. In these skills the skaters complete several rotations (i.e., twists) around their long axis (i.e., from head to feet) during their flight. During the short time that they are in flight their angular momentum is conserved.

A skater drives up into an Axel with one leg forward and the other back. The arms are extended sideways. As a result, his body mass is spread out relative to the long axis of his body. (This is similar to divers being extended at takeoff from the board.) In a spread-out position, the skater's rotary inertia (resistance) is considerable. Consequently, his angular velocity (i.e., rate of spin) around the long axis is minimal. But in flight the skater pulls his arms and legs inward, which greatly reduces their resistance against rotation. He now spins with tremendous angular velocity. When he spreads his arms and legs out again on landing, rotary inertia is again increased and the rate of spin reduced.

The principles just discussed apply even when a skater performs high-speed spins while in contact with the ice. The rate of spin is increased as the limbs are pulled inward. The difference between this and a skill performed in the air is that a skater's blades pressing and turning on the ice generate a little more resistance than the air does when the skater is in flight. So a skater loses some angular momentum because the skates experience friction with the ice.

Making Use of Angular Momentum During Flight

When tower or springboard divers are in flight, there is no large mass for them to push against. Any muscular action that they perform while in the air causes an equal and opposite reaction to occur elsewhere in their bodies. All divers, gymnasts, and other high-flying athletes experience this phenomenon. For example, imagine a diver stepping off the tower and dropping toward the water in an upright position. In this position, he is not rotating. As the diver drops, imagine the coach shouting for him to raise his extended legs 90 degrees from perpendicular (i.e., pointing directly downward) to horizontal. The muscles that rotate his legs forward and upward around the hip joint pull equally at both origin and insertion (i.e., at either end of the muscle) and so simultaneously pull down on his trunk. As a result, during the time frame that the diver's legs rotate upward, his trunk must rotate downward. Do his legs and trunk rotate in equal-size arcs? No, because they do not have the same rotary inertia. The rotary inertia of the diver's trunk and upper body is approximately three times that of his legs. So his trunk and upper body resist rotation three times more than his legs. When the diver's legs move upward 90 degrees to a horizontal position, the trunk and upper body, which have three times more rotary resistance, move downward in an arc a third of the size, or approximately 30 degrees (see figure 4.28).

The difference between the movement of the diver's legs compared with that of his trunk and upper body may not seem like an equal and opposite reaction, yet it is. In our example, the action is the 90-degree arc moved in a counterclockwise direction by the diver's legs. The reaction is the 30-degree arc moved in a clockwise direction by his trunk and upper body. This reaction is equal because the diver's trunk and upper body have three times the rotary inertia of the legs. It is opposite because the movement of the trunk and upper body is in the opposing direction to that of the legs.

We've seen that rotary inertia depends not only on how much mass is involved but also

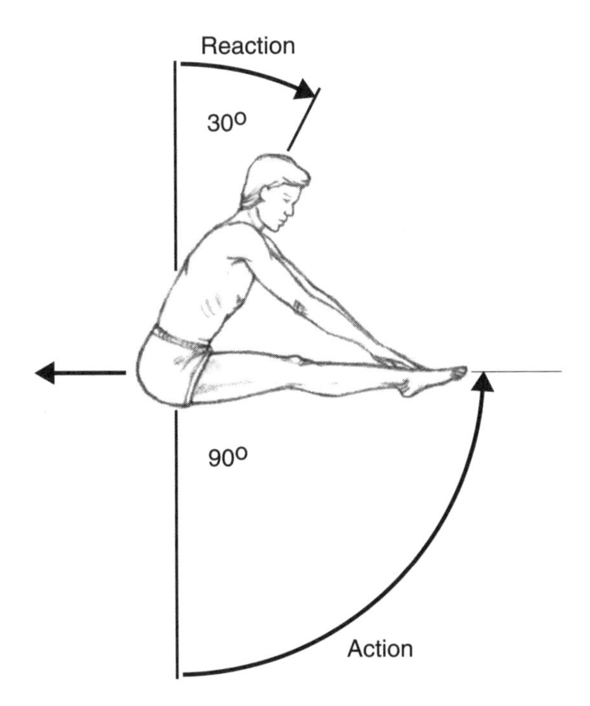

FIGURE 4.28 When the diver's extended legs are raised 90 degrees in a counterclockwise direction, the upper body reacts by moving 30 degrees in a clockwise direction.

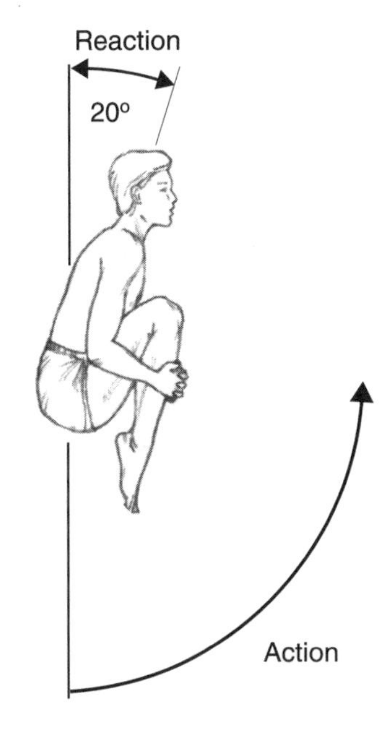

FIGURE 4.29 When the diver's flexed legs are raised, they have less rotary inertia than when they are raised in an extended position. The upper-body response is less.

on how it is distributed relative to its axis. In the previous example, the diver keeps his legs extended throughout their 90-degree movement. If he flexes at the knees and lifts the thighs, then the rotary inertia of the legs is reduced. The reaction of the trunk and upper body is also reduced. The upper body and trunk flex approximately 20 degrees (see figure 4.29).

Figure 4.30 shows that another reaction will occur in the diver's movements. Notice that as the diver's extended legs rotate counterclockwise, the trunk and upper body rotate clockwise. In the illustration, these body parts move toward the right as you look at them. In the air, body mass moving to the right is counterbalanced by body mass moving to the left. In our example the diver's buttocks and hips react (equal and opposite) by moving toward the left.

This phenomenon of equal and opposite reactions occurring in the air is visible in many sports. A flop high jumper always arches to clear the bar. Figure 4.31 shows a high jumper arching the upper body down toward the pit in a counterclockwise direction. The legs respond (equal and opposite) by moving in a clockwise direction. Although upper body and legs are moving in opposing directions, both are moving down

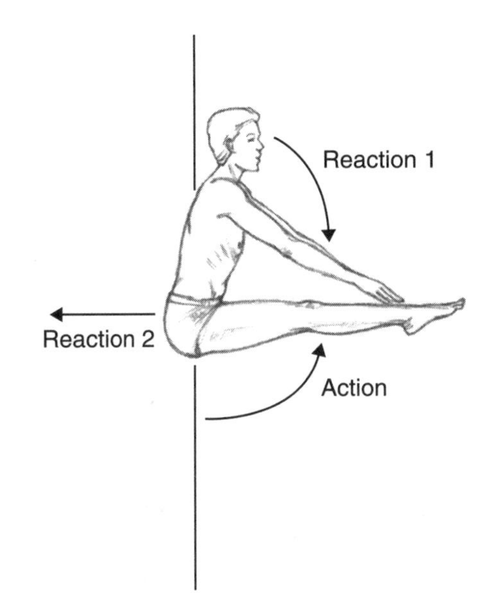

FIGURE 4.30 In the air, when the lower and upper body flex forward, the hips react by shifting backward.

toward the pit. So the hips react by moving upward. Correctly timed, a high jumper uses this action when passing over the bar. The athlete's hips and buttocks move upward, and this helps him clear the bar.

Another example of this phenomenon occurs when a volleyball player jumps to spike a ball (see figure 4.32). As the upper body flexes

FIGURE 4.31 The high jumper's hips lift upward when the upper and lower body flex downward.
© Claus Andersen

backward in a counterclockwise direction, the lower body reacts by moving in a clockwise direction. In figure 4.32, you will see that even though the upper and lower body are rotating in opposing directions, both move to the left. An equal and opposite reaction must balance this movement. The hips and abdomen shift in the opposing direction. Do you see how this replicates the action of the high jumper?

In events such as the long jump, high jump, figure skating, diving, and gymnastics, the athletes are in the air for a short time. During flights of short duration we know that their total angular momentum remains constant. If athletes introduce more angular momentum during flight by suddenly rotating the arms or legs, another part of the body (or the body as a whole) must give up some angular momentum to keep the total constant. In other words, if a long jumper gives herself 10 units of angular momentum at takeoff she cannot increase this 10 units to 12 while in flight. If the long jumper introduces 2 units of angular momentum by rotating her arms and legs, then 2 units must disappear somewhere else in her body to keep the total constant at 10.

Interestingly enough, the principle of introducing angular momentum with the arms or legs in order to get rid of angular momentum elsewhere in the body gives athletes some control over their movements while in the air. Let's first look at what occurs in ski jumping, then we'll take a closer look at the long jump, where elite jumpers put this principle to good use.

In ski jumping, a jumper who mistakenly gives himself too much forward rotation at takeoff knows that unless something is done about the problem, he's likely to land on his face! So the ski jumper will rotate his arms in the same direction as the unwanted forward rotation. The angular momentum generated by the arms helps put the brakes on the forward rotation of his body. If performed vigorously enough, the arm action can rotate the skier's body backward into a more favorable position. If the skier rotates his arms backward, his body will rotate farther forward, which in the situation we've described is the last thing he wants to do.

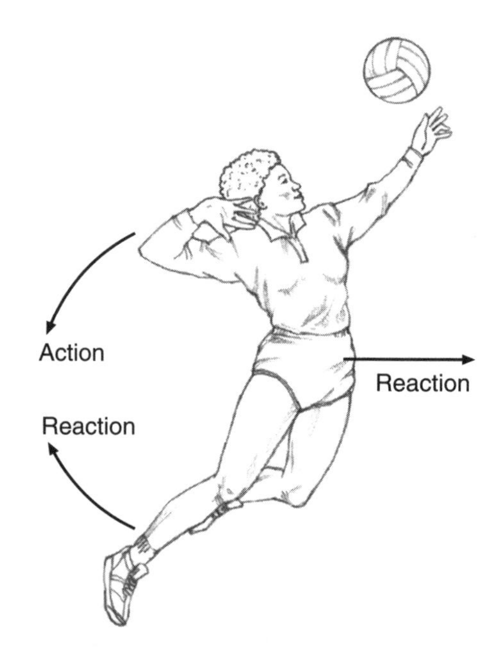

FIGURE 4.32 Action and reaction in a volleyball spike.

Controlling Forward Rotation in the Long Jump

All long jumpers rotate forward at takeoff. Even world-class jumpers face this problem. It's impossible to avoid because the takeoff foot pushing back at the board causes the athlete's body to rotate in the same direction as the direction of push of the takeoff foot. A long jumper taking off from left to right will push toward the left with the takeoff foot. In this case, the jumper will have given his body clockwise rotation. Worse still, his body continues to rotate forward throughout his flight unless he does something about it. His legs and feet will rotate down toward the sand and the distance jumped will be greatly reduced. Novice jumpers who hold a bunched-up position in flight get themselves in trouble because a tightly flexed position means that they also spin very quickly. As a result, a novice's body and feet quickly rotate down toward the sand, resulting in a poor distance (see figure 4.33).

To counteract unwanted forward rotation initiated at takeoff, an expert long jumper rotates the arms and legs in the same direction (i.e., forward) while in the air. Elite jumpers who have perfected the cycling action of a "hitch kick" can stop their bodies from rotating forward and make them rotate in the opposing direction (see figure 4.34). This change in rotation helps them achieve a good body position for landing and so increases the distance jumped.

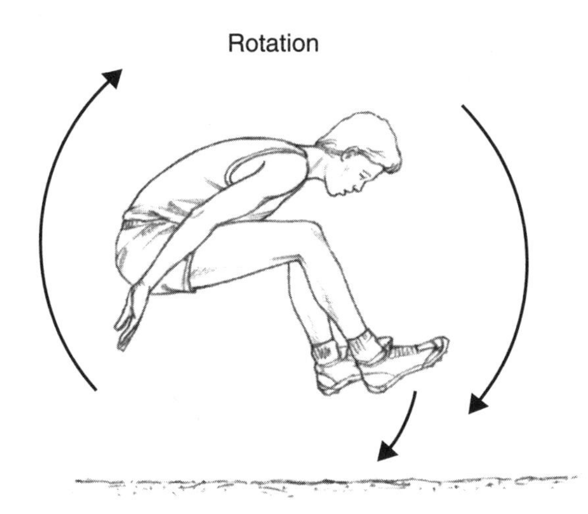

FIGURE 4.33 In the air, a bunched position increases the long jumper's angular velocity and the feet hit the sand earlier than necessary.

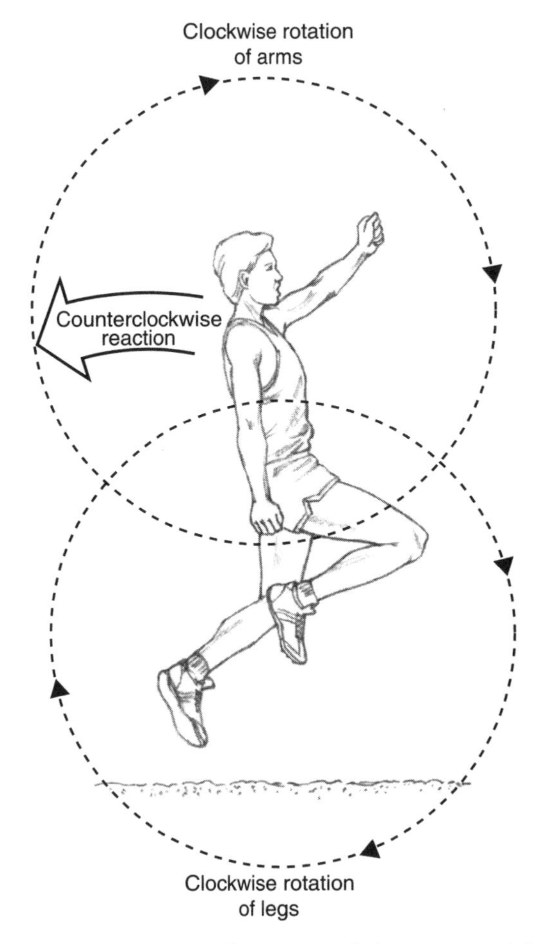

FIGURE 4.34 Forward rotation of the arms and legs causes backward rotation of the long jumper's body.

Just how much reaction an athlete gets from rotating the arms and legs depends on how much angular momentum the athlete's body has been given at takeoff and how much angular momentum is introduced by the rotary motion of the arms and the legs. Vigorous rotary actions with arms and legs extended produce the most angular momentum. Look for expert long jumpers bringing their arms and legs forward in a flexed position and then rotating them backward in an extended position. This has maximum effect on counteracting the forward rotation of the athletes' bodies.

A good place to see a desperate use of angular momentum is at your local swimming pool. Youngsters run and jump off the springboard. They don't expect the board to flip their feet upward to the rear, and in horror they find themselves rotating in the air and heading for a belly flop. It's at this point that they introduce a wild flailing of the arms. These youngsters, uneducated in mechanics, are introducing angular momentum in the correct direction. They rotate their arms, and frequently their legs, in the same direction as the unwanted rotation they received from the springboard. Without knowing it, they replicate in rough form the precise movements of an elite long jumper.

Transferring Angular Momentum Between Somersaults and Twists

Divers, gymnasts, and ski aerialists frequently combine somersaults with twists. In these complex skills, athletes simultaneously somersault around their transverse axis (hip to hip) and twist around their long axis (feet to head). The most remarkable technique used to combine somersaults with twists is one in which athletes begin by somersaulting with no twist apparent. Let's look at diving to examine the mechanics of these actions.

At takeoff, divers push against the board to help get their body rotating and so generate angular momentum. As with long jumpers, the amount of angular momentum produced at takeoff remains virtually the same throughout the diver's flight down to the water.

In flight, the diver begins by rotating around the somersaulting axis (i.e., transverse, or hip to hip). Then the diver performs a series of body actions that borrow some angular momentum from the somersault to put into the twist. The diver now somersaults and twists. After somersaulting and twisting for the required number of revolutions, a second series of actions is performed. These remove the twist and the diver somersaults without twisting.

There is one dominant method for borrowing, or stealing, angular momentum from the somersault and placing it in the twist. This is called the body tilt technique.

Body Tilt Technique

The body tilt technique of twisting requires that the diver's body be tilted away from its somersault axis so some angular momentum goes into the twist axis. This is best understood by imagining the diver as a solid block somersaulting around the transverse (somersault) axis (see figure 4.35, a-c). The diver's angular momentum while in flight is set at takeoff, so in figure 4.35a, a specific amount of angular momentum is given to the block around its transverse axis. Suppose that during the somersaulting rotation, the block is made to tip over sideways so that it lies horizontally (see figure 4.35b). Rotation continues, but the block now rotates around its longitudinal axis. In effect, the block's long axis has become its somersault axis.

Now let's tilt the block over just a few degrees (shown in figure 4.35c) rather than all the way to horizontal. The block will twist and somersault at the same time. Why? Because the angle of tilt forces some angular momentum into the twist axis (i.e., the block's long axis) but some still remains in the original somersault axis. In addition, the rotary inertia (i.e., resistance) of the block around its twist axis is less than that around the somersault axis. The block twists faster than it somersaults. If the block is brought back to its

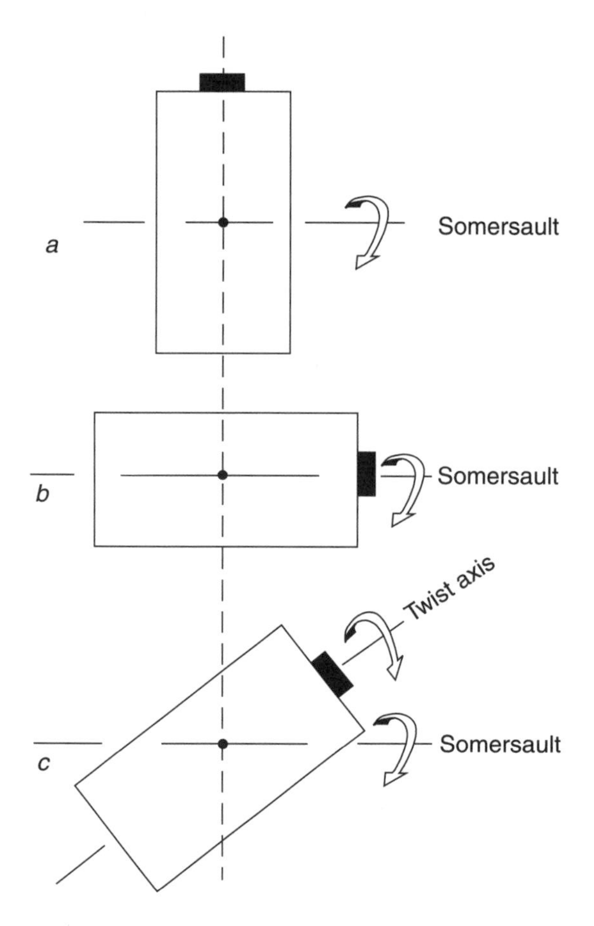

FIGURE 4.35 Mechanics of the body tilt technique of twisting.

Divers Make Splashless Entries

At one time, a feet-first entry was used in a dive from the 10 m tower. Divers found if they flattened their feet on entry rather than pointed their toes, there was little or no splash and the water simply bubbled at the surface. Similar experiments with headfirst entries led to divers clasping their hands and having the palms flattened and facing toward the water. As with the flat feet entry, this technique produced a low-pressure area that sucked the water downward behind the diver's hands and produced very little splash. A problem with this technique was that the impact with the water could injure a young diver's wrists. To counteract this problem, young divers wear wrist supports in training and occasionally in competition.

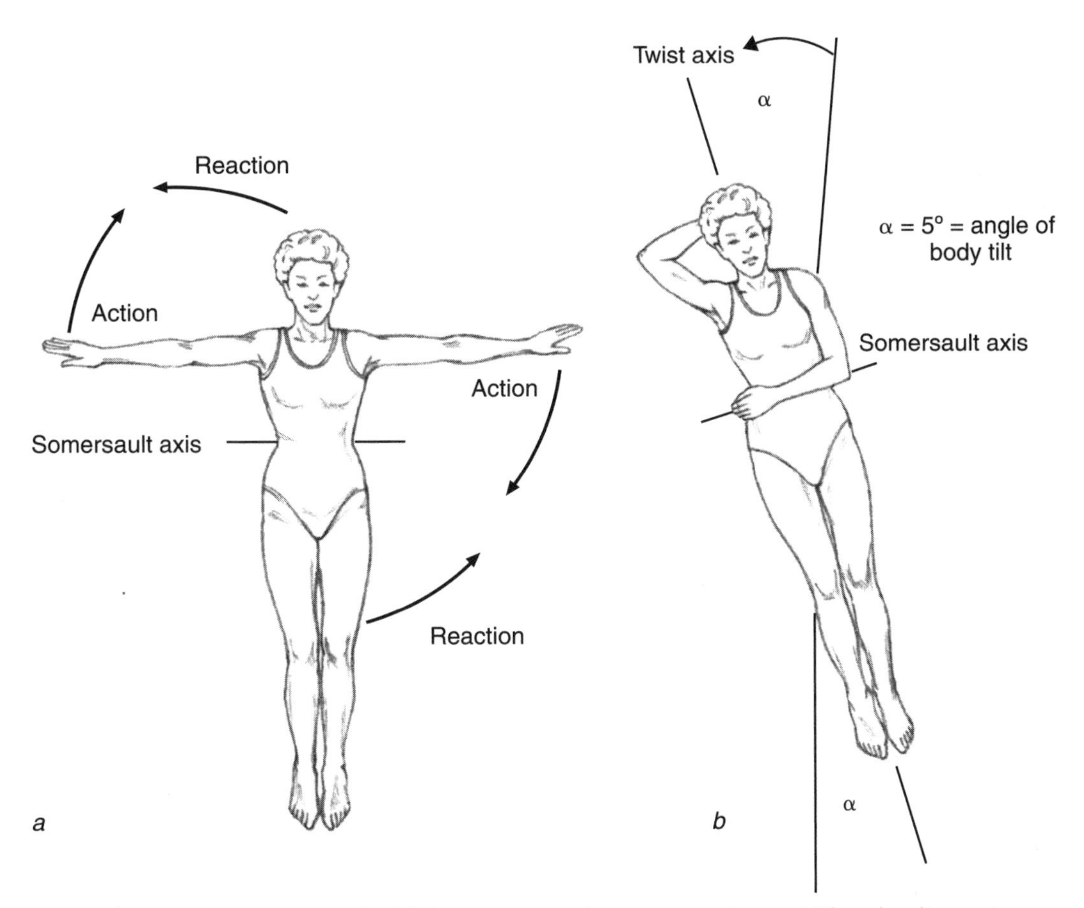

FIGURE 4.36 A diver with arms outstretched *(a)* rotates around the somersault axis. When the diver raises one arm and lowers the other *(b)*, her body tilts. This causes some somersault momentum to be transferred to the twist axis.

original position, the process is reversed. The twist disappears and all angular momentum goes back into the somersaulting axis.

Initiating Body Tilt

The principles of the body tilt technique indicate that if divers start a dive by somersaulting, they will somersault and twist if they tilt over slightly while in flight. But how do divers tilt themselves over while in the air? Here's how it occurs. When divers take off from the board, they extends their arms sideways (figure 4.36a). In flight, they swing one arm downward and the other simultaneously upward (figure 4.36b). Notice that even though one arm goes up and the other down, both arms rotate clockwise. The angular momentum generated by these clockwise arm actions causes the diver's body to react by rotating in a counterclockwise direction. A double arm swing of 90 degrees (one arm swung up 90 degrees and the other swung down 90 degrees) causes the body to tilt approximately 5 degrees from the vertical (see figure 4.36b). Five degrees is sufficient for a diver to initiate a twist. Swinging the arms back reestablishes the original body position and eliminates the twist.

A common sequence for the combination of a somersault and twist using body tilt follows: The athlete begins the dive by initiating a somersault at takeoff. In flight, the diver raises one arm and lowers the other. A diver's body tilts over, which causes twists and somersaults to occur at the same time. When the diver returns the arms to their original positions, the twist is eliminated. The somersault action continues and is reduced to a minimum when the diver's body is fully extended for entry into the water.

Divers are also aware that the more body tilt they can establish, the more twists they can perform. So what can be done to exaggerate the body tilt? After a half twist around the long axis is complete, the arm that was lowered is then raised and wrapped to the rear of the head. The other arm, which had been raised, is lowered and pulled in tight across the waist. This repetitive arm action tilts the diver over farther, causing twists to occur at an incredible rate and making dives with triple twisting somersaults possible.

Cat Twist Technique

This twisting technique is called the cat twist because a cat, like many other animals, performs the mechanics of this technique naturally, without training or coaching. Experimentation has shown that if a cat is held inverted in a static position 2 to 3 ft from the ground and allowed to drop, the animal can initiate a twist in the air so that it can land safely on all fours. Figure 4.37, a through e, shows a cat as it falls. Notice the piked angle at the cat's midsection in figure 4.37a. This important angle allows the cat to twist the upper body against the rotary inertia of the lower body and vice versa. The first move

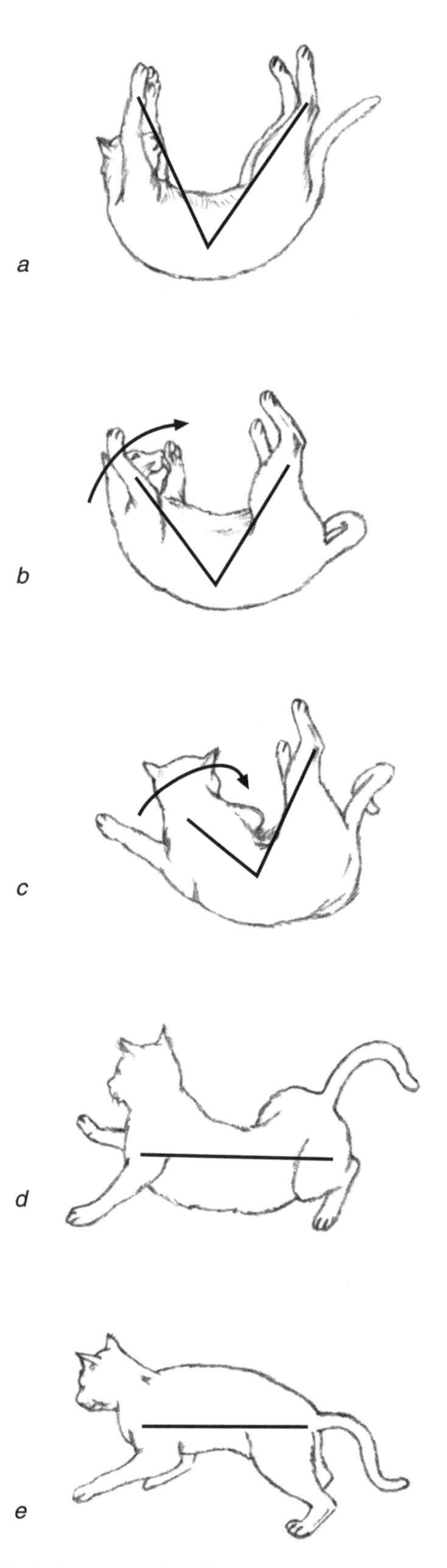

FIGURE 4.37 Cat twist. The cat reduces the rotary inertia of the upper body and twists the upper body against the larger rotary inertia of the lower body. Notice the angle of the body in *(a)* to *(c)*.

the cat makes is to establish visual contact with the ground. The cat pulls in the forelegs and turns the upper body toward the ground (see figure 4.37, b-c). By pulling in the forelegs, the cat reduces the rotary inertia of the upper body in relation to its lower body, which has more rotary inertia because the rear legs are fully extended. Consequently the upper body, with its smaller rotary inertia, can be rotated through a large angle with minimal reaction in the opposing direction from the lower body. Once the cat sees the ground, the forelegs are extended in preparation for landing (see figure 4.37d). This increases the rotary inertia of the upper body. The rear legs are pulled inward, reducing the rotary inertia of the lower body and allowing the lower body to rotate against the greater rotary inertia of the upper body. Once fully aligned, the cat is ready to land safely (see figure 4.37e).

The cat twist differs from the body tilt technique because it can be performed without borrowing, or stealing, angular momentum from a somersault. In other words, an athlete does not have to be somersaulting to perform the cat twist action.

The cat twist always requires the athlete's body to have an angle, or to be bent at the waist in some way. It doesn't matter what direction. Flexion at the waist can be forward, backward, or sideways—they all work successfully. A 90-degree angle at the waist is preferable (i.e., with the upper body piked forward), but the cat twist can be performed with far less than 90 degrees. An elementary skill on the trampoline that employs the cat twist technique is the swivel hips.

The piked position used in the swivel hips also occurs in diving. Figure 4.38a shows a diver in a piked position with a right angle existing at the hips. In this position the mass of the legs is a

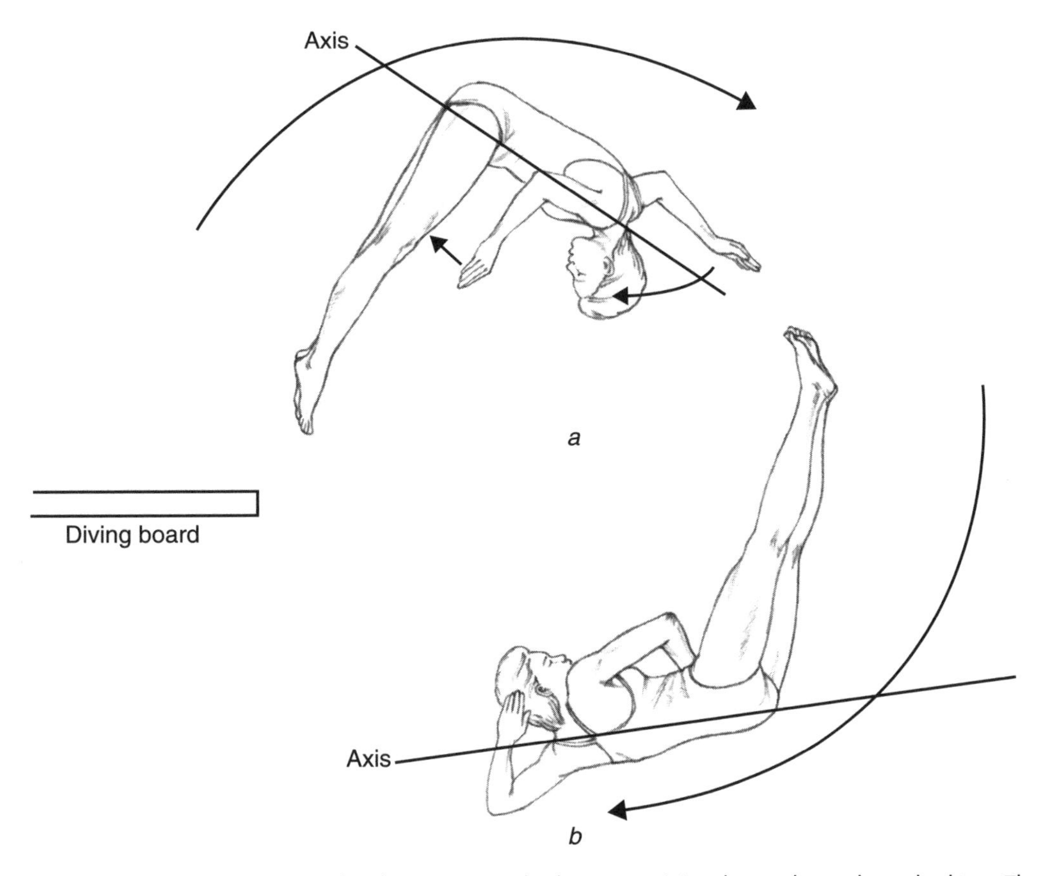

FIGURE 4.38 Diver using a cat twist. The diver is in a piked position *(a)* with a right angle at the hips. The diver twists *(b)* the upper body against the greater rotary inertia of the legs.

Adapted, by permission, from M. Adrian and J. Cooper, 1989, *Biomechanics of human movement* (Times Mirror Higher Education Group), 505. Reproduced with permission of The McGraw-Hill Companies.

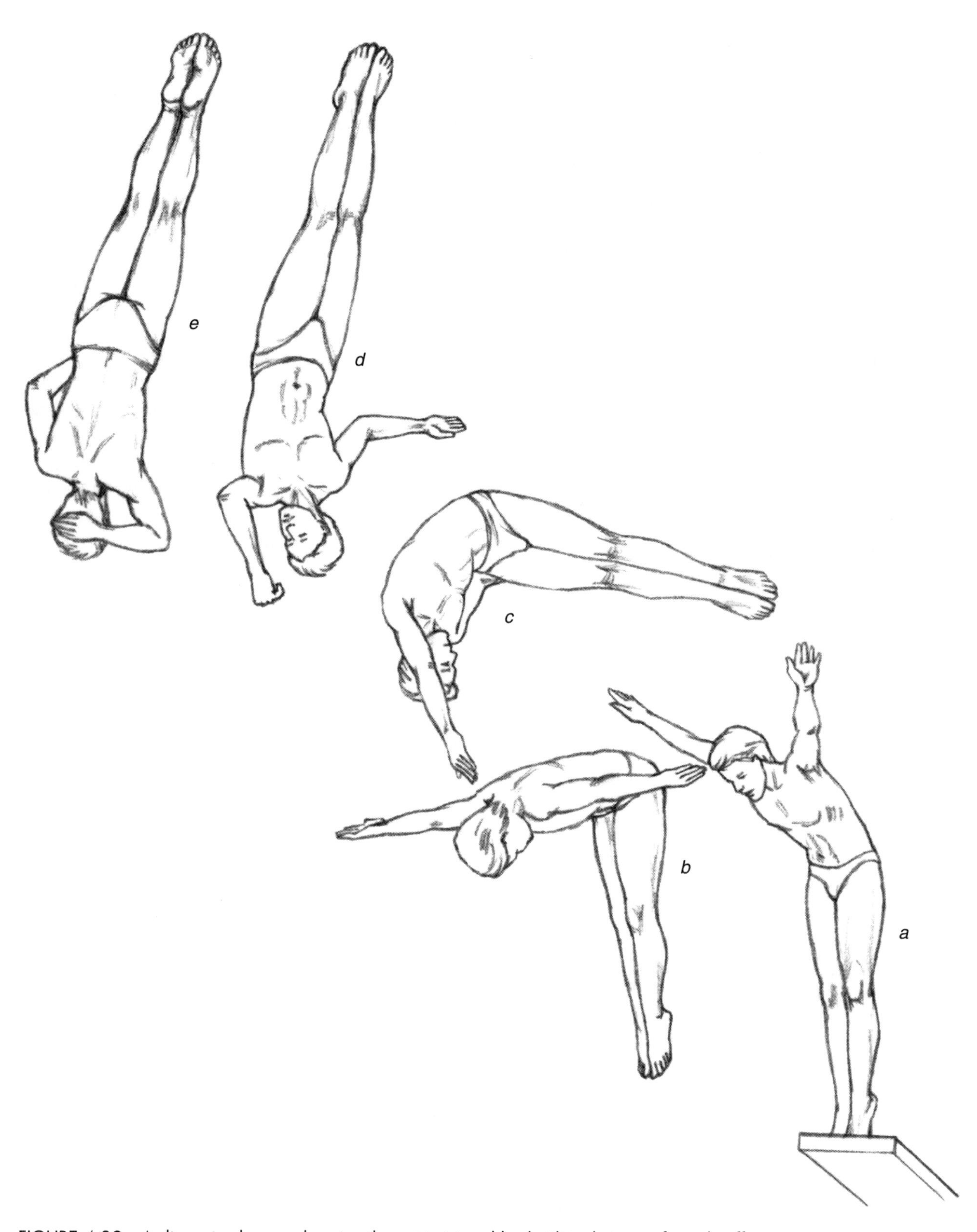

FIGURE 4.39 A diver simultaneously using the cat twist and body tilt technique after takeoff.

long way out relative to the long axis, which extends from the athlete's hips to the head. Figure 4.38b shows the athlete twisting the upper body against the greater rotary inertia of the legs. This is the basic principle of the cat twist. A twist is achieved by turning one part of the body that has reduced rotary inertia against another part that has much larger rotary inertia.

The cat twist is frequently used at the start of forward dives that contain somersaults and twists. In these dives the athlete combines aspects of both the cat twist and the body tilt technique. Figure 4.39, a through e, shows a diver taking off for a dive that will contain both somersaults and twists. Notice in figure 4.39a the diver's arms are extended outward above the shoulders. Rotation around the transverse somersault axis is initiated at takeoff. Figure 4.39b shows the 90-degree pike required by the cat twist and the raising of one arm and the lowering of the other using the principles of the body tilt. Once the upper body is twisted against the inertia of the lower body, the angle at the waist is removed (see figure 4.39, d-e) to allow the twist to continue around the long axis of the diver's body.

Divers and particularly ski aerialists use other twisting and somersaulting techniques. Unfortunately a discussion of these techniques is beyond the scope of this text. If you coach or intend to coach gymnastics, diving, figure skating, ski aerials, trampoline, or any sport in which your athlete twists and turns in the air, be sure you understand the mechanics of the event and always use proven safety techniques.

SUMMARY

- All rotational, turning, spinning, and swinging motions are forms of angular motion. Angular motion implies that an object or athlete rotates around an axis.
- Motion in an athlete's body is predominantly rotational. Muscles pull on bones, and bones rotate at the joints.
- Levers are simple machines that transmit mechanical energy. A lever incorporates a rigid object that rocks or rotates around an axis, or fulcrum. Force is applied at one position on the lever, and a resistance applies its own force at another.
- The two most important functions of a lever system are magnification of force or the magnification of speed and distance. Both cannot occur at the same time.
- There are three classes of levers. In a first-class lever, the axis is positioned between the force and the resistance. First-class levers can be made to magnify either force or speed and distance. In a second-class lever, the resistance is positioned between the axis and the force. Second-class levers magnify force at the expense of speed and distance. In a third-class lever, the force is positioned between the axis and the resistance. Third-class levers magnify speed and distance at the expense of force.
- Third-class levers predominate in the human body. Most muscles in the human body apply great force in order to move light resistances over large distances at great speed.
- Levers produce a turning effect called torque. Torque is increased by magnifying the applied force and/or the distance from the axis of rotation that force is applied.
- Angular velocity is synonymous with rate of spin. It refers to the angle, degrees, or revolutions completed in a particular time frame (e.g., 1 sec) in a specific direction (e.g., clockwise or counterclockwise).
- All objects that rotate or swing have an inward pulling or pushing force, called a centripetal force, that acts toward the axis of rotation. Centripetal force counteracts the inertial desire of objects to travel in a straight line.

- When athletes or objects rotate, a centripetal force will exist. In response they will experience the outward pull of what is commonly called centrifugal force. Centrifugal force is caused by the inertial desire of whatever is rotating to travel in a straight line and not in a circle. Centrifugal force is called a fictitious or phony force.
- The inertia of all objects makes them resist rotation. Once forced to rotate, however, an object's inertia is expressed by wanting to continue rotating.
- Rotary inertia varies according to the mass of a spinning object and the way its mass is distributed. The greater the distance that mass is spread out from its axis of rotation, the greater the rotary inertia. The more compressed that mass is relative to its axis of rotation, the greater the reduction of rotary inertia.
- Angular momentum is the rotary equivalent of linear momentum: It describes the quantity of rotary momentum. The angular momentum of an object is determined by the product of its mass, its angular velocity, and the distribution of its mass.
- In flights of short duration (e.g., diving, long jump, gymnastics, and high jump), the amount of angular momentum generated by an athlete at takeoff remains the same for the duration of the flight. This indicates that the athlete's angular momentum is conserved.
- When an athlete's angular momentum is conserved, the athlete's rate of spin (angular velocity) increases or decreases in relation to changes in the distribution of the athlete's mass. For example, the tighter the tuck, the greater the angular velocity.
- In flight, the angular velocity of an athlete's body increases or decreases in relation to the introduction of angular momentum by other body parts. During flight, clockwise rotation of the athlete's body as a whole can be counteracted by introducing clockwise rotation in the athlete's arms and legs. The same principle applies to counterclockwise rotation.
- Divers use combinations of the body tilt, cat twist, and other techniques to twist and somersault.
- The body tilt technique of twisting requires athletes to be somersaulting before initiating the twist. This technique uses specific arm actions to tilt the body out of the somersault axis. The action transfers angular momentum from the somersault axis to the twist axis. Athletes then twist and somersault simultaneously. When the body tilt is removed, the twist is eliminated and the somersault continues.
- The cat twist technique can be initiated in flight with zero angular momentum. No somersault need be initiated beforehand. This technique requires athletes to be flexed at the waist. The athlete twists by rotating one part of the body that has reduced rotary inertia against another part that has larger rotary inertia.

KEY TERMS

angular momentum
angular velocity
centrifugal force
centripetal force
conservation of angular momentum
first-class lever
force arm
frontal axis
fulcrum
lever
longitudinal axis
resistance
resistance arm
rotary inertia
second-class lever
third-class lever
torque
transverse axis

REVIEW QUESTIONS

1. Two athletes perform strict biceps curls with 30-lb dumbbells. One athlete has much longer forearms. Why does the biceps curl demand more muscular force from the athlete with the longer forearms?
2. Swing a baseball bat forward and backward around your body, noting how much effort you put into starting, stopping, and maneuvering the bat during the swing. Now hold the bat at the hitting end (i.e., the barrel end) and repeat the process. Why is it easier to start, stop, and maneuver the bat when held at the hitting end?
3. Perform a vertical jump and reach, swinging your arms upward as you take off. Now perform your jump without an arm swing but with both arms extended above your head. Explain the mechanical reasons that account for the better jump performance when you swing your arms upward as you jump.
4. What components make up an athlete's angular momentum? Why is angular momentum conserved when divers, high jumpers, trampolinists, long jumpers, and gymnasts are in flight?
5. What equal and opposite reaction occurs when a high jumper simultaneously lowers her legs and torso toward the pit as she arches over the bar?
6. What twisting technique mentioned in the text does not require the athlete to be somersaulting before initiating the twist?

PRACTICAL ACTIVITIES

1. **Torque.** Close a door by pushing with the tip of your index finger at the handle. You will find this easy to accomplish. Open the door again. This time close the door by pushing with the tip of the your index finger near the door hinges. You will find this action extremely difficult. Why? Explain the differences that occur, using in your explanation the components that make up torque.
2. **Angular momentum.** Draw the design for a spinning top that you are going to build for a spinning-top competition. Here are the rules for the competition to determine which top will spin the longest.
   a. The timing of the your top's spin begins the instant that it leaves your hand and stops the instant that any part of it's perimeter hits the supporting surface.
   b. The top can only be spun once using the thumb and finger of one hand. No machine crank or string is allowed to increase the top's angular velocity.
   c. The top can be any size and weight.
   d. The top can be made of any material.
   e. The top can be spun on any surface.

   In your design, you should take into consideration the mechanical principles relating to torque, rotary inertia, mass, distribution of mass, stability, and friction. Add explanations to your diagram.
3. **Rotary inertia.** Using your favored hand and employing wrist action only, rotate the following items from the left to the right horizontally through a 180-degree range of motion. Second, with each of the items "write" a large figure eight in the air in front of you as fast as possible. (a) A golf driver gripped at the upper extremity of the grip, (b) the same golf driver held at the opposite end (i.e., at the base of the handle just

above the head of the club), (c) a baseball bat gripped at the upper extremity of the handle, and (d) the same baseball bat gripped at the lower extremity of the barrel. Rank these items for difficulty in moving through the required pathways, and relate this difficulty to their rotary inertia. Explain why the rotary inertia of these items increases when you grip the items one way and decreases when you hold them in the opposing manner. In your explanation, show that you understand what the rotary inertia of an object depends on.

4. **Inertia and centripetal force.** Standing outdoors and well clear of other people, swing a bucket half full of water back and forth by your side and then vigorously swing it in a full circle. The water stays in the bucket. Using the mechanical terms of centripetal force and inertia, explain why the water stays in the bucket.
5. **Action and reaction relative to rotary inertia.** Sit in a swivel chair with your feet off the ground. Hang one arm (your right) by the side of your body. Stretch your left arm out horizontally to the side. Swing your left arm vigorously from the side to the front so that its range of movement is an arc of 90 degrees. Estimate how many degrees your body moves in the opposing direction. Write a detailed explanation showing that you understand why the response of your body is much less than 90 degrees.
6. **Centripetal force, inertia, and frictional forces.** Jan Ullrich, a great German cyclist and rival of American cyclist Lance Armstrong, is heavier than Lance and is also much taller. Because of Jan's size, he has to ride a bike that is bigger and has the seat set much higher than that of Lance Armstrong. During the final time trial of the 2003 Tour de France, Jan fell on the tight curve of the slick road surface. Using stick figures, draw both a tall, heavy cyclist going through a curve on a flat surface and then a shorter, lighter cyclist. Include arrows to indicate the forces acting into the curve and the forces acting out of the curve, and underneath write an explanation that tells why a taller, heavier cyclist going through a curve can overwhelm the frictional surfaces between the bike's tires and the road surface.

5

Don't Be a Pushover

When you finish reading this chapter, you should be able to explain

- the importance of balance and stability in sport skills,
- how athletes make use of linear and rotary stability,
- the mechanical principles that determine different levels of linear and rotary stability,
- why some sport skills require minimal stability, and
- why the rotary stability of a spinning object is proportional to its rotary resistance.

This chapter discusses the importance of balance and stability in the performance of sport skills. You'll see why some sport skills require athletes to maximize their stability while other skills require athletes to temporarily reduce their stability to minimum levels. You'll also see how stability is related to mass and inertia, and in particular how the preservation of an athlete's stability is always a battle of torques. The turning effect of one torque that disrupts an athlete's stability is battled by the turning effect of torque that is applied by the athlete's muscles to regain stability.

Many athletes naturally sense how to move and seem to know instinctively what they should do to maximize their stability. Unfortunately not all athletes are gifted in this way. Young athletes—particularly those struggling to learn a new skill—assume poor body positions, which reduces the quality of their performances. They don't plant their feet properly when throwing, striking, or hitting, and they find that the reaction forces resulting from their actions cause them to stumble or "fly" in the opposite direction. These athletes don't assume efficient stances when checked, blocked, or challenged by an opponent. When they lose their balance, they don't make the best maneuvers to quickly regain control. If they want to move suddenly, they are unable to do so because their stance simply doesn't allow them to get off the mark quickly. All of these examples are errors involving the amount of stability that's required at a particular instance in the performance of a sport skill. Most of these errors are easy to correct providing you teach the mechanical principles relating to balance and stability. You'll read about these principles in this chapter.

Balance and Stability

Balance implies coordination and control. Athletes with great balance are able to neutralize those forces that would otherwise disrupt their performances. When a world champion gymnast works flawlessly through a routine on the beam she successfully counteracts the forces that would cause her to fall off. Compare the gymnast with NFL running backs. These athletes twist and turn and keep driving for the end zone even though they are bombarded with repeated tackles from their opponents. In spite of the great differences in their sports, world class gymnasts and elite running backs both demonstrate superior control and balance.

Athletes must maintain their balance in skills where there is little apparent movement (like a handstand) and in skills that are highly dynamic (like the running back driving for a touchdown). The "enemy" they fight can be any external force. Gravity, friction, air resistance, and forces applied against them by opponents can all destroy their performance.

Stability specifically relates to how much resistance athletes "put up" against having their balance disturbed. The more stable an athlete, the more resistance the athlete puts up against any disruptive forces. An athlete can be in a balanced position and be as stable as the Rock of Gibraltar. At the other extreme, an athlete can be in a balanced position yet be highly unstable. A giant sumo wrestling champion squatting low with both hands on the ground is obviously in a more stable position than a ballerina balancing on the tips of her toes. A child can produce enough force to push the ballerina off balance, but it's unlikely that the same force will do anything but bring a smile to the sumo wrestler's face. Both ballerina and sumo wrestler are balanced, yet the sumo wrestler has far more stability than the ballerina.

Linear Stability

Athletes can put up a certain amount of resistance against being moved in a particular direction. The same athletes can possess a certain amount of resistance against being stopped or having their direction changed once on the move. Both situations are types of **linear stability.** For example, the sumo wrestler can resist being forced by his opponent in a particular direction. Then when he charges across the ring to slam into his opponent, his opponent needs to apply force to stop or change his direction.

If a 300-lb sumo wrestler bulks up to a massive 400 lb, then he will need more strength to overcome his own inertia and get himself moving. It will also require more force from an opponent to move the more massive sumo wrestler from a static and stable position or to change his direction once he is charging across the ring. Consequently, the more massive the sumo wrestler becomes, the more inertia he has and the more stability he has too. So linear stability is directly related to mass. Sumo wrestlers are well aware of this fact. It's for this reason that they lift weights to get more powerful and stuff themselves with food to increase body weight.

Whether athletes actually want to have great linear stability or not depends on the demands of the skill. In a rowing race, athletes want to row the shortest distance between the start and the finish. To do this they try to counteract any forces that might shift them off course. A heavyweight rowing eight powering toward the finish has tremendous linear stability. Collectively, the eight rowers and their shell form a huge mass. The long narrow shape of their shell coupled with their rowing actions propels them at high speed in a straight line. The opposition trying to push them off their straight-line course or slow them down is predominantly wind, waves, and friction generated by the shell and the rowers moving through the water (and also through the air).

Surfers, figure skaters, and slalom skiers differ from a rowing eight because these athletes must contend with sudden directional changes. They will also want a certain amount of linear stability, but they don't want to keep going in a straight line when it's necessary to make sudden tight turns. It's for this reason that athletes who want to shift and turn quickly avoid tipping the scales like a sumo wrestler. Imagine a slalom skier trying to maneuver 400 lb of body weight though a series of tight turns! It's no different for squash players, badminton players, or goalkeepers in soccer. Tremendous body mass means great inertia, and great inertia needs huge amounts of physical power to get going. Once on the move, great inertia implies straight-line movement. So too much mass means too much stability, which can be a liability in skills where high-speed movements, first one direction and then the other, are required.

Linear Stability and Friction

Friction is of assistance when athletes need traction to get moving. It can also increase the stability of an athlete in a static position. Consider a massive wrestler in the Olympic heavyweight division lying in a defensive position on the wrestling mat. Lying flat and spread-eagle, the athlete is very difficult to move. The wrestler's huge mass is pulled down tight onto the mat by gravity. If before the competition the athlete increased his body weight to the absolute limit allowed by the rules, then the additional pressure from his extra body mass would press him down onto the mat even more, increasing his friction that much more too. You can see why an opponent would have to be phenomenally powerful to pull, push, or roll this athlete out of his defensive position.

Compare this scenario with what is required of a speedskater. Speedskaters are strong, muscular athletes, but they are all considerably lighter than heavyweight Olympic wrestlers, and the frictional differences between a wrestling mat and slick ice are obvious. The slightest backward thrust from the skater causes the blades of the skates to glide over the ice. The static stability of the speedskater in this situation is much less than that of the wrestler.

Regardless of the surface type, a heavy, massive athlete presses down onto a supporting surface more than a lighter athlete. In a sport like football, the pressure generated by the mass of a huge lineman in contact with the earth produces more friction between his cleats and the turf than one of the lighter running backs on his team. This gives the lineman better traction than a running back. (Of course, the lineman has considerably more mass and inertia than a running back, so although he may have more traction he certainly doesn't have a running back's maneuverability.) An opponent must overcome the lineman's friction with the turf if he wants to block him out of a play.

Rotary Stability

Rotary stability is the resistance of an athlete or an object against being tilted, tipped over, upended, or spun around in a circle. But, if the turning effect of a torque is sufficient to set an athlete or an object (e.g., a discus) spinning, rotary stability then describes the ability of the athlete and the object to keep spinning and resist whatever would slow down their rate of spin. We'll look at the principles that help an athlete avoid falling or being tipped over, upended, or spun around. You'll notice that these extremely important principles are universal in sport.

The amount of effort required to maintain balance varies from sport to sport. An elite sprinter is normally not thinking about balance and stability during a 100-m race, and a top-class basketball player doesn't expect to fall over when bringing the ball up-court. However, in other sport skills athletes must actively work to maintain balance. For example, in judo, athletes battle to maintain their balance and try to destabilize their opponents as they attack and counterattack. In weightlifting, athletes struggle to control immense barbells held at arms' length above their heads, and in sprint cycling, you'll often see competitors balancing almost motionless as they try to outmaneuver each other at the start of a sprint race. All these athletes are maintaining various levels of rotary stability. Excellent rotary stability makes athletes better able to resist the destabilizing and turning effect of a torque applied against them. The more stable they are, the more torque that's necessary to upset their balance. Let's see how this battle of torque is played out.

A destabilizing torque that upsets an athlete's balance can come from any external source. It can be generated by gravity, air resistance, an opponent, or any combination of forces. The axis around which athletes rotate when a torque is applied against them can be any place on their body, an opponent's body, or some external object. If a gymnast loses her balance on the beam, the axis of rotation will be where her body contacts the beam. It can be the ball of one foot during a pirouette or her hands during a handstand. In a hip throw in judo, the point of rotation is where the athlete's body contacts the opponent's hip. If a sprinter stumbles during a lunge for the tape, the point of rotation is likely to be where his spikes contact the track.

When a gymnast performs a difficult one-handed handstand, she is balanced but not very stable. She needs to fight to maintain her balance. If she allows her body to shift even the slightest distance out of a balanced position, a turning effect (or torque) occurs as she starts to rotate. The earth pulls at her center of gravity. The axis of rotation will be where her hand contacts the floor or the apparatus. The farther she shifts out of a balanced position, the greater the turning effect. In a one-handed handstand, a gymnast must use the strength in her supporting hand and forearm (plus other muscles in her body) to counter a shift out of a balanced position. She does this by producing an opposing torque. If the turning effect of this opposing torque is powerful enough, it will rotate the gymnast in the opposite direction until a state of balance is regained. But if she allows her center of gravity to shift too great a distance from its correct position (which is directly above the supporting hand), then she is not likely to have the strength to pull herself back to a balanced position. Gravity then wins this "battle of torque" and the gymnast collapses out of the handstand. Figure 5.1 shows a gymnast who has allowed her center of gravity to shift too far to the left. The greater this distance (d), the greater the torque produced by gravity.

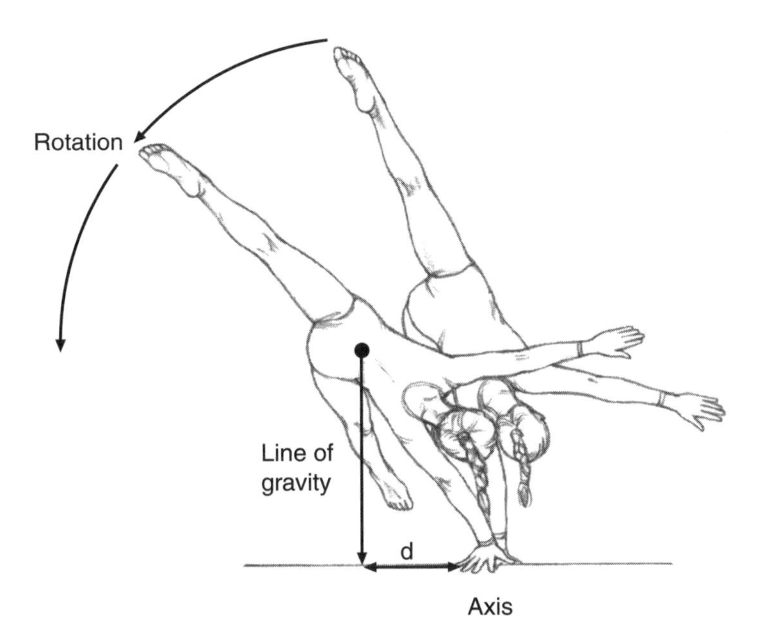

FIGURE 5.1 When the center of gravity is no longer above the supporting base, gravity applies a destabilizing torque. The greater the distance (d) of the line of gravity from the axis, the greater the torque produced by gravity.

FIGURE 5.2 The attacking wrestler in contact with the mat has his center of gravity well outside of his base. In this totally unstable situation, he has put himself at risk in order to throw his opponent.

Compare a gymnast balancing in a one-handed handstand with either of the two Greco-Roman wrestlers shown in figure 5.2. It is easy to see that neither of the two wrestlers has any stability at all. The defender is being thrown and is totally out of contact with the mat. The attacker has only one foot solidly in contact with the mat, and his center of gravity and line of gravity are both well outside of his supporting base. In this highly unstable position, he has gambled everything to throw his opponent, and his gamble has paid off!

Figure 5.3, another example from wrestling, shows how rotary stability is a war of one torque versus another. The torque applied by the attacker (i.e., force × force arm) competes against the torque (i.e., resistance × resistance arm) generated in the opposing direction by the defending wrestler. In this situation the defending wrestler's body weight acts as the resistance and his hand is the axis of rotation. The attacker tries to increase the turning effect he's applying to his opponent by increasing both the force he's applying and the length of the force arm he's using. The defending wrestler obviously cannot increase his rotary stability by gaining body weight during the wrestling match. But what he tries to do is keep his resistance arm as long as possible. Figure 5.3 shows that the defending wrestler could lengthen his resistance arm even more by shifting his hand on the mat farther away from his opponent. (Don't forget that the mechanical word *arm* in this context refers to the perpendicular distance from his center of mass to the axis of rotation and not to the length of the wrestler's arm.)

Let's stay with the sport of wrestling and imagine a heavyweight wrestler lying facedown and spread-eagle on the mat. Every time the opponent grabs hold of an arm or leg to turn him over, the wrestler immediately shifts his body so that the attacker never gets into a position that gives him good leverage. This situation would be like using a crowbar to raise a rock and finding that the rock keeps moving and never lets you get into a position where you can use the crowbar to your advantage.

Compare the heavyweight wrestler's situation with that of a gymnast balancing on one foot on the beam (see figure 5.4). The wrestler does everything possible to maximize his rotary stability. He is massive, his center of gravity cannot be positioned any lower to the ground, and he constantly shifts his mass and spreads his arms and legs to counteract the torque applied by his attacker. The gymnast's center of gravity is high above the beam, and then to make matters worse, she's balancing on one foot! Both the gymnast and the wrestler are successfully maintaining their balance. But the gymnast has minimal rotary stability because it takes very little torque to destabilize her. The wrestler, on the other hand, constantly repositions himself to maintain a tremendous amount of rotary stability.

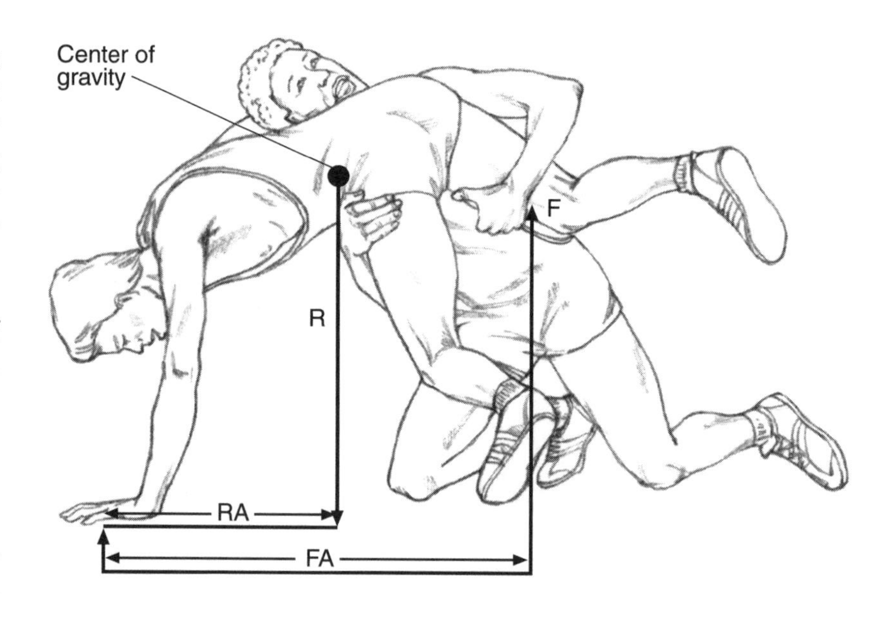

FIGURE 5.3 Wrestling is a battle of torque. The attacker's force (F) × force arm (FA) competes against the defender's resistance (R) × resistance arm (RA).

Factors Determining Rotary Stability

The conditions that give the gymnast minimal stability and the wrestler a considerable amount of stability are clues to the mechanical principles that determine different levels of rotary stability. These principles occur in every sport skill. Let's take a look at each of them one at a time to see how they affect an athlete's performance.

Athletes Increase Their Stability When They Increase the Size of Their Base of Support

The bigger an athlete's base of support, the greater the athlete's stability. A **base of support** commonly refers to the area on the ground enclosed by the points of contact with the athlete's body. However, the base of support is not always below the athlete. Anything that provides resistance against forces exerted by the athlete can become a base. So if a student in an aerobics class leans against the wall in a calf-stretching exercise, her base includes the wall and the ground. A gymnast hanging from one of the uneven bars has a bar for a base, and it happens to be overhead.

What do we mean by *area* when we talk about the base of support? Imagine a gymnast facing along the length of the beam and being on one foot. In this situation, she uses the area of her foot as a base of support. If she places her other foot on the beam, her base of support now stretches from one foot to the other. The gymnast

FIGURE 5.4 The gymnast's supporting base is the area covered by her foot on the beam. The line of gravity falls within the area of the base.

Increasing the Stability of the Leaning Tower of Pisa

The 180-ft-tall Leaning Tower of Pisa in Italy was reopened to tourists in December 2001 after being closed for almost 12 years. Tourists can now climb the 300 stairs to the top. Because it sat on a marshy sand and clay riverbed, the Tower began to lean from the time it was constructed in 1173. With each additional inch of lean, the Tower became more unstable. At one point it was estimated to be more than 14 ft out of line. In 1993, engineers decided to strengthen the sides of the Tower and temporarily attached heavy lead ingots (650 tons) and steel cables opposing its direction of lean while work was done to improve the ground supporting the Tower. In 1995, another 250 tons of lead ingots were added. In June 2001, the Tower had been straightened up sufficiently to be declared safe. It still has its characteristic lean and is expected to remain stable and attract tourists for many years to come.

is now more stable than when on one foot alone. As you can guess, with two feet on the beam, her stability is better forward and backward than it is from the side. It's more difficult to shove the gymnast off the beam if you push her from the front or the back than if you push her from the side. In figure 5.3 the defending wrestler's base of support stretches from the single hand on the mat to where he contacts his opponent. Given that his opponent is trying to turn him over onto his shoulders for a pin, it's obvious that an opponent can hardly be considered part of a stable base of support!

You can see from the previous examples that stability is directly related to the size of the supporting base. Make your base as big as possible and you are more stable. A gymnast in a one-handed handstand has a base that is solely the area of one hand. If the gymnast moved to a headstand, the base now becomes the triangular shape that runs from the head to the hands and back again (see figure 5.5). When a wrestler lies facedown and flat on the mat and spreads out his legs and arms as wide as possible, he covers a huge area that runs from his fingertips to his feet. As far as the size of his base is concerned, the wrestler has made himself maximally stable. If you compete in the sport of wrestling you'll make use of this position when you're defending and also when you're attacking. If you're attacking and you want to hold your opponent in a particular position and stop him from moving, you maximize your stability just as you would do if you were defending. Maximizing your stability makes it as difficult as possible for your opponent to mount a counterattack.

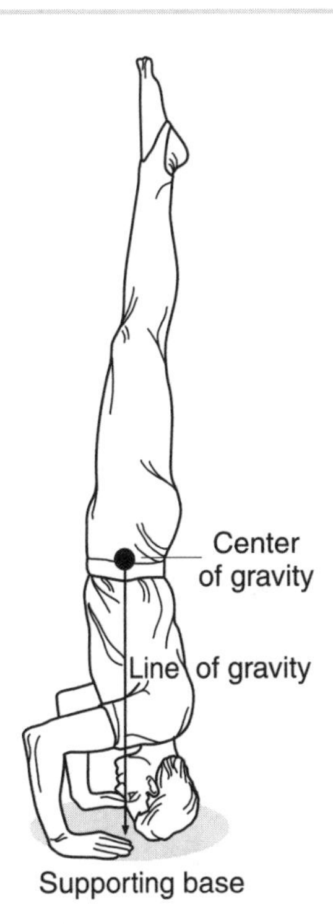

FIGURE 5.5 In a headstand, the supporting base is the triangular area from the head to the hands.

Athletes Increase Their Stability When Their Line of Gravity Is Centralized Within Their Base of Support

An athlete's balance is maintained as long as a vertical line passing through the athlete's center of gravity falls inside the perimeter of the base of support. The closer to the center of the base that this **line of gravity** (i.e., a perpendicular line from the athlete's center of gravity) falls, the more stable the athlete becomes. Conversely, the closer to the edge of the base that the line of gravity falls, the more unstable the athlete becomes. The larger the base, the easier it is for an athlete to make sure that this vertical line falls well within the base.

Unlike inanimate objects, living beings can maneuver and shift position and in this way keep their line of gravity within the perimeter of their supporting base. If you are balancing above a very small base and you allow your center of gravity to shift the slightest distance, then you've given gravity or an opponent a chance to apply a torque that you may not be able to counteract. That's one reason a gymnast's one-handed handstand is such a phenomenal feat requiring tremendous strength and control. The base of support is solely the area covered by the gymnast's supporting hand. The gymnast's center of gravity cannot be allowed to shift from being directly above the supporting hand. So the gymnast's muscles are put to work to maintain this position.

In total contrast to the one-handed handstand is the wrestler's spread-eagle defensive position. Assuming that the wrestler's center of gravity is close to his navel, the line of his center of gravity would have to be shifted several feet in any direction before it gets anywhere near the perimeter of his base. It's no wonder that this position is used so often in wrestling.

You must not think that athletes always want to have their line of gravity within their base. When you walk or run, you take a series of steps forward. With each leg movement your center of gravity shifts outside of your base and you shift from a stable to an unstable position. When you put your foot down at the end of the step, your center of gravity moves back in between your feet and you are in a stable position again (see figure 5.6). When athletes sprint, they go from one unstable position to another as they drive themselves along the track. One moment their line of gravity is outside their supporting base, the next instant it is back inside again. This happens all the way down the track. The same situation occurs when hockey players skate along the ice.

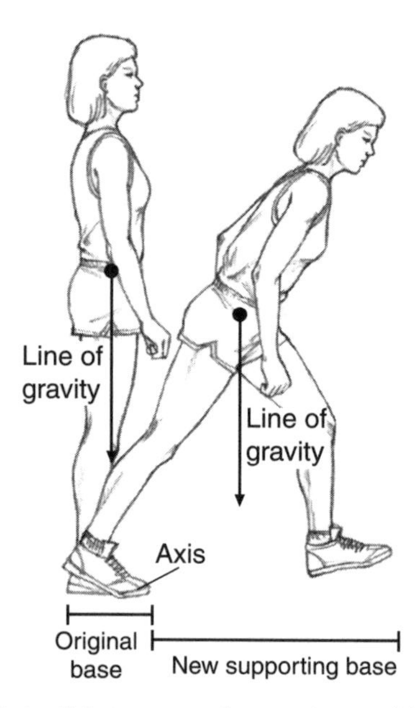

FIGURE 5.6 Taking a step forward, an athlete becomes unstable, restabilizing when both feet again contact the ground.

When runners, speedskaters, cyclists, and slalom skiers go around a curve, they all lean into the turn. Here you have an example of dynamic balance (i.e., balance while on the move). If they came to a sudden stop in the middle of the curve, they would fall. As they lean into the curve their line of gravity no longer falls within their supporting base. The faster they move and the tighter the curve, the more they lean inward and the farther their center of gravity (and their line of gravity) shifts into the curve. The correct amount of lean balances the forces acting into the curve with those acting in the opposing direction (see figure 5.7).

Athletes Increase Their Stability When They Lower Their Center of Gravity

Athletes with a center of gravity that is elevated high off their base of support tend to be less stable than athletes whose center of gravity is lower down. It's for this reason that great running backs who can twist and turn and suddenly cut one way and then the other tend to be shorter than other football players. They run low to the ground and in this way they are more stable

FIGURE 5.7 An athlete rounding a curve is in a state of dynamic balance. Forces acting into the curve counteract those acting outward.

and maintain their balance better than taller athletes.

To understand how the degree of stability relates to the height of an athlete's center of gravity, let's have an athlete using the same size base first stand erect and then crouch down. In both positions we'll tip the the same angle. 5.8 that for the s the line of the a gravity shifts be the supporting base when the athlete is standing erect. When the athlete is crouching down, the line of gravity is still within the base. This means that if other stability variables remain the same, the lower the athlete's center of gravity, the more stable the athlete becomes.

The principle of lowering the center of gravity to increase stability is one of the reasons athletes crouch or lie flat on the mat when they are defending in combatives. Athletes in wrestling and judo not only lower their center of gravity but also widen their base of support. This increases their stability twofold, making it considerably more difficult for an opponent to destabilize them. Ski jumpers know that they lose style points if they stagger on landing. Lowering their center of gravity by using a telemark landing (see figure 3.4, page 42) in which the legs are flexed helps them maintain

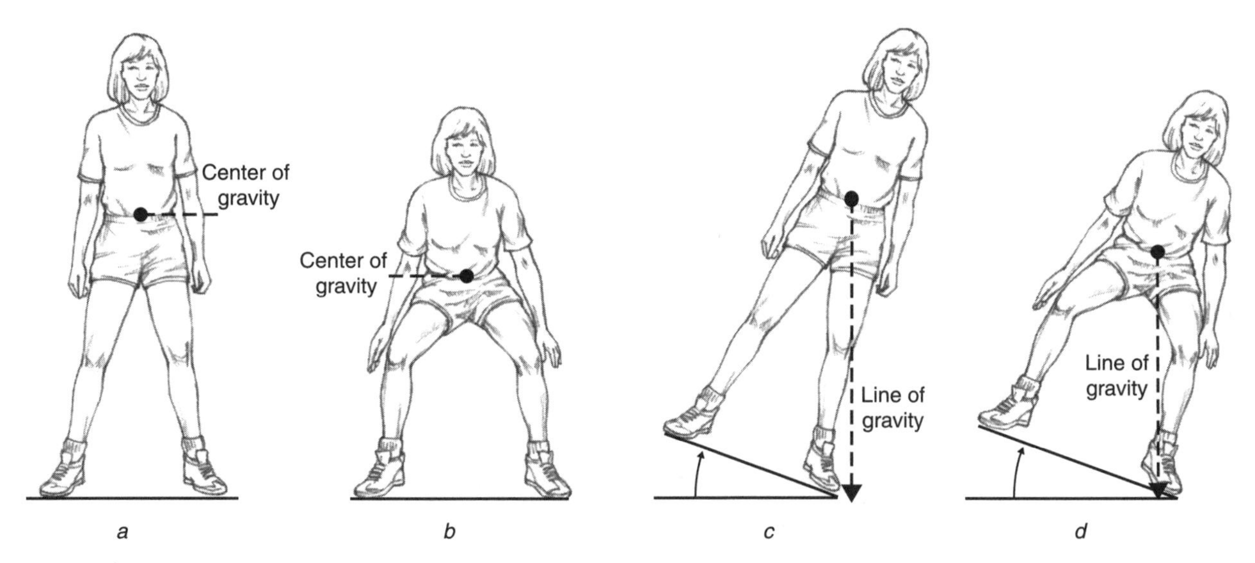

FIGURE 5.8 When all else is equal, the higher the center of gravity, the less stable is an object or an athlete. An athlete stands erect *(a)* and crouches *(b)*. For the same angle of sideways tilt, the line of the athlete's center of gravity shifts beyond the edge of the supporting base when the athlete is standing erect *(c)* but remains within the base when the athlete is crouching or squatting *(d)*.

The Sinking of the Vasa

Today's ships are carefully designed so that they are stable enough to withstand the force of waves and wind. Not so the *Vasa*. In the 17th century King Gustav Adolphus of Sweden ordered the building of the *Vasa,* which was to be the biggest battle galleon ever built. It had two gun decks with a total of 64 heavy bronze cannons. This differed from the common practice where there was either a single row of cannons, or if there were two rows, the upper row was made lighter in weight than the lower. In addition to huge masts, enormous sails, and cannons that alone weighed 100 tons, the ship was covered in elaborate ornamentations that decorated the whole upper section, including the high stern castle. The ballast meant to stabilize the ship proved woefully inadequate. Prior to its maiden voyage, the *Vasa* was given a stability test: 30 men ran back and forth in unison across the deck while it was moored. The men had to stop after only three runs because the ship was ready to capsize. True to this prediction, on its maiden voyage there was a sudden squall, the ship listed heavily to port, and water poured in the open gun ports. The *Vasa* sank in moments within one nautical mile of the shore. The *Vasa* was raised in 1959 and is now a popular tourist attraction in Sweden.

their balance. Kayakers who are tall and who have most of their weight in their upper bodies have a high center of gravity. They are less stable than shorter athletes or athletes who have most of their weight in the area of their hips and seats. (Taller kayakers can counteract this problem by using kayaks that have a wider beam, which widens their base of support.)

Weightlifting gives us many great examples of reduction in stability when the center of gravity is elevated. The more weight a weightlifter hoists and the longer the athlete's arms, the higher the combined center of gravity of both athlete and barbell. The line of this common center of gravity must stay centralized above a relatively long but narrow base formed by the lifter's feet. In addition, the barbell and the athlete's extended arms have a tendency to rotate at the axes of the shoulder joints. If an athlete allows the barbell to shift the slightest distance out of line, he must immediately reposition his base and also fight with his shoulder muscles to bring the barbell back in line so it's again centralized above his base. Figure 5.9 shows the final position in the jerk phase of a clean and jerk. Stability in this position is better forward and backward because the base is large in that direction. But the base is narrow side to side. Even small movements by the athlete in a sideways direction can shift the common line of gravity of athlete and barbell outside the periphery of the base. From the final leg split position of the jerk, the weightlifter must now bring his feet together, stand erect, and

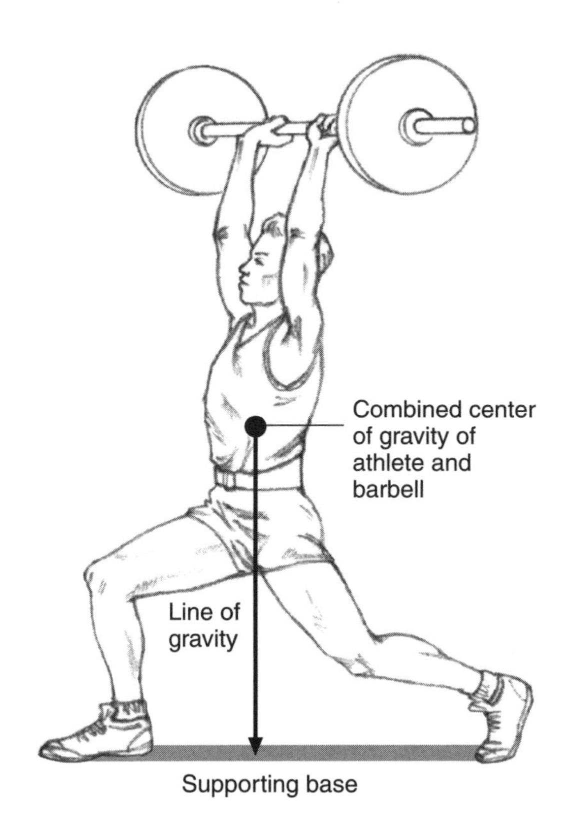

FIGURE 5.9 In a clean and jerk, the common center of gravity of both the weightlifter and bar must be centralized above the supporting base. The barbell also must be positioned directly above the shoulder joints.

control the barbell for 3 sec. It's easy to appreciate the difficulty of controlling a heavy barbell in this position. The slightest shift of the barbell out of line and control can be lost and the lift declared a failure.

Athletes Increase Their Stability by Increasing Their Body Mass

This principle simply says that if all other factors relating to stability remain unchanged, a heavier and more massive athlete is more stable than one with less body mass. This is why weight divisions exist in combative sports. What hope would a featherweight wrestler have trying to lift and rotate an athlete in the superheavyweight division?

The value of body mass in combatives is duplicated in football where huge linemen are given the duty of maintaining their position in the face of whatever is thrown against them. The heavier they are, the more force (and torque) it takes to throw them off balance and to knock them out of position. It's for this reason that they weigh close to or in excess of 300 lb. There are no weight divisions in football as there are in Olympic wrestling.

In all sport skills, heavier athletes who get out of control and lose their balance must exert more muscular force to regain their balance than lighter athletes. If heavier athletes don't have the muscular strength to control their actions, then their extra body mass becomes a great disadvantage and very much a liability. In judo, athletes always try to make use of their opponent's body weight and, if their opponents are on the move, they try to make full use of their opponents' momentum. It's inefficient and exhausting to try to halt an opponent's push or thrust and then drive the athlete in the opposing direction. Better to use the opponent's movement and have him rotate around an axis (such as your hip or leg), and then add your force to that of gravity as the opponent rotates toward the mat. Smaller sumo wrestlers who try to overcome heavier opponents also attempt to carry out these maneuvers. It's obviously a good idea to shift out of the way when an athlete weighing 400 lb is charging at you. If you're a lighter opponent you could try to destabilize him (by quickly stepping aside and simultaneously tripping him) as his huge mass rushes by. You then add your own force to that of gravity to drive him to the floor—or more satisfying, to heave him amongst the spectators!

Athletes Increase Their Stability When They Extend Their Base in the Direction of an Oncoming Force

Irrespective of the type of base, stability is increased if an athlete's supporting base is enlarged in the direction in which force is being received or applied. For example, a force can come from an opponent who is trying to block or tackle you, or you can apply a force in a particular direction when you are throwing, hitting (e.g., in baseball), or lunging (as in fencing). When a running back wants to keep his balance and continue running when hit by an opponent, he must take into consideration not only the force of the hit but also the direction it's coming from. To maintain stability and stay upright, the running back must widen his base in the direction of the applied force. If the hit is coming from the front, he widens his base from front to back. If the hit comes from the side, he widens his base in that direction. Naturally, he is going to lean into the hit as well. The mechanical principles of leaning into a hit are explained later in this chapter.

The second application of this principle applies to situations in which athletes apply force in a particular direction. If you watch throwers, pitchers, and hitters, you'll notice that they widen their base in the direction they are applying force. Why do this? Because it gives them a good stable base, and it allows them to apply force over a considerable distance without losing balance. If they didn't do this, they'd be thrown in the opposite direction. It doesn't matter whether you're blasting a home run, hurling a fast ball, or just having fun in a pickup softball game, the same principles apply (see figure 5.10).

The actual size of the base an athlete should use depends on how much force is applied. Imagine tossing a superlight table tennis ball a few yards. You could do this without any trouble while balancing on one foot. The table tennis ball has very little weight (i.e., mass) and in

FIGURE 5.10 An athlete's supporting base is lengthened in the direction in which force is applied (i.e., in the direction of the pitch).
© Terry Wild Studio

this situation has hardly any velocity. Now try throwing a huge medicine ball as far as possible while balancing on one foot. You'll find that as the medicine ball goes in one direction you get pushed in the opposite direction.

Let's now compare catching a table tennis ball that is lobbed gently toward you with catching a heavy object (like the medicine ball) traveling at a much higher velocity. You could easily catch the table tennis ball while balancing on one foot. Its momentum would be minimal, and you'd also be able to maintain your balance on one foot without any trouble. The medicine ball presents a different situation. If you want to maintain your stability and not have the medicine ball knock you over, you must assume a wide stance with both feet well planted on the ground. Reach toward the ball and extend your stance in the direction from which the medicine ball is approaching. The more massive the ball and the greater its velocity, the more stable your stance must be.

Athletes Should Shift Their Line of Gravity Toward an Oncoming Force

The principle of shifting your line of gravity (and your center of gravity) toward an oncoming force is directly related to the principle that states you should widen your base toward an oncoming force. Running backs purposely lean into tackles, hockey players lean into opposing players who are trying to body-check them, and wrestlers lean into their opponents when they grapple.

What is interesting about this principle is that athletes are temporarily destabilizing themselves by moving their line of gravity closer to the perimeter of their supporting base. This position requires the attacking athletes to apply considerable force (and momentum) to drive the defenders back beyond the rear perimeter of their supporting base. If the attackers are successful, then the defenders are totally destabilized. But the time taken by the attackers in thrusting the defenders from the forward perimeter of their base to beyond the rear perimeter is frequently sufficient for the defenders to spin away from the attack and to turn defense into attack.

The principle of shifting your line of gravity toward an oncoming force applies not only in tackles and checks but also in catching a heavy medicine ball. You widen your base and shift your center of gravity toward the oncoming medicine ball. In this way the momentum of the medicine ball pushes you back into a stable position. Equally important, you give yourself plenty of time to apply force to slow down the medicine ball. Do you recognize that "lots of time to apply force" is an expression of impulse used for stopping (i.e., a small force applied over a large time frame)? It's the small force applied over a long time frame rather than a huge force over a short time frame that allows you to bring the heavy medicine ball comfortably and painlessly to a stop.

An important difference exists when you are an athlete applying force. In this case you lengthen your base in the direction that you are applying force. But you position your center of

FIGURE 5.11 In the javelin throw, the athlete's center of gravity begins to the rear of the supporting base and ends in front of the thrower's base.

gravity to the rear of your base—and in many cases, temporarily outside the rear of your supporting base. All good javelin throwers start their throws from a position where they are leaning backward with their center of gravity outside the rear of their base of support. Then in the follow-through they finish with their center of gravity beyond the front edge of their base. In this way they apply great force over the longest possible distance and the longest possible time frame (see figure 5.11).

In combatives like wrestling and judo, athletes must be aware of the danger of shifting their center of gravity too close to the perimeter of their base. Imagine that you're in a judo competition. You're pushing and leaning into your opponent, and the opponent is doing the same thing to you. Your center of gravity is close to the front edge of your base. Suddenly your opponent stops pushing and instead pulls hard. Now the force of your own push has suddenly been increased by the addition of your opponent's pull. You find that your center of gravity is suddenly outside your base and that you're totally unstable. You're about to be thrown!

If you sense that a push is about to change into a pull, immediately shift your center of gravity in the opposing direction and widen your supporting base in the same direction. Attack-counterattack sequences involving the positioning of the athletes' center of gravity are the very essence of judo. One instant you can be stable, and then if you are not careful the next instant you're rotating toward the floor around an axis set up by a leg sweep from your opponent.

How All Principles of Rotary Stability Are Interrelated

Remember that all the factors that control rotary stability are closely interrelated. It's no good if an athlete makes one maneuver to increase stability if all the other factors controlling stability aren't satisfied as well. For example, if an athlete widens her base of support, she must also move it in the direction of an applied force. A soccer player can widen her base toward a tackle coming from the right, but if her line of gravity stays close to the left edge of her base, she will easily be knocked over and beaten by the tackle. An athlete in the superheavyweight division is more stable than an athlete in a lower weight division, but extra body weight is of little use if this superheavyweight uses a narrow base and stands with his center of gravity high off the ground and close to the edge of his base. Likewise, an athlete who lowers her center of gravity improves her stability, but this action is only a valuable maneuver if her line of gravity is kept well within her base. All the principles of stability are related, and each depends very much on the others. Just obeying one principle of stability isn't good enough.

Skills That Require Minimal Stability

We've been talking at length about maximizing stability. Are there skills where athletes purposely minimize their stability? Yes. Examples occur during the explosive acceleration of basketball or hockey players during a fast break and swimmers and sprinters at the start of their races. In a sprint start, sprinters aim to get out of the blocks as fast as possible. On the "Set" command, they shift their line of gravity forward so that it is very close to their hand positions on the track (see figure 5.12). This highly unstable position satisfies two requirements. It extends the sprinters' legs into a powerful thrusting stance. Second, prior to the start, it moves the athletes as far as possible in the desired direction—toward the finish line.

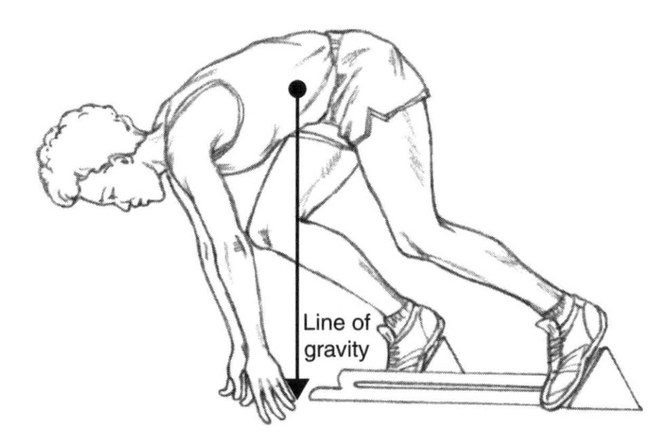

FIGURE 5.12 In a sprint start, the athlete's line of gravity is shifted close to the forward edge of the supporting base.

In sport skills where sudden direction changes occur, athletes want to be able to shift quickly in any direction. The set position in a sprint start is excellent for sudden and fast movement toward the finish in a 100-m race. But you'd be highly amused if you saw tennis players using a sprinter's set position when they're waiting to receive a serve. The set position might be good for a sudden move in one direction, but it's useless for moving quickly in other directions. Volleyball players receiving a serve and soccer goalkeepers defending the goal all need to be quick off the mark no matter what direction. They cannot guarantee the direction, velocity, and spin of the ball, and they certainly don't want to commit themselves too early. Consequently, these athletes will use a fairly small base with their line of gravity centralized. In this way it takes no more than a split second to shift in the direction they want to move. They can react quickly and move fast in any direction.

Rotary Stability When Twisting, Turning, or Spinning

When objects or athletes rotate, swing, or turn, their rotary stability depends on their rotary resistance (i.e., rotary inertia). The more rotary resistance, the greater the stability. From chapter 4 you'll remember that rotary resistance is made greater by increasing the mass of the object and by moving the mass as far away as possible from the axis of rotation. How does this type of rotary stability show up in sport? Here is an example: The men's discus at 4 lb 6 oz (2 kg) is twice as massive as the women's discus and it's over 2 in.

A Tightrope Walker's Pole Increases Stability

Circus tightrope walkers like the Great Wallendas and Philippe Petit of France have walked across cables strung between tall buildings. Why are these daredevil performers more stable when they use a long pole, and why are they even more stable when the pole is really long, curved downward, and weighted at either end? A pole that curves downward lowers the center of gravity of performer and pole combined. If weights are added at the ends of the pole, it lowers the center of gravity even farther. In addition, the longer the weighted pole, the greater its resistance against rotation. This helps stop the performer from tipping sideways and falling off the wire. Without a long, curved, weighted pole, tightrope walkers would have a tough time maintaining balance!

larger in diameter. If all other factors are equal, the men's discus (by virtue of its greater mass and mass distribution) will battle the destabilizing effect of air currents better than the women's discus.

Similar to a discus, a football will remain more stable in flight when it's given spin. The more spin, the greater the angular momentum of the football and the greater its stability. In this situation you're increasing angular momentum simply by increasing the ball's angular velocity, or spin. You're not changing its mass or the distribution of its mass. When a quarterback rifles a pass to a receiver the ball simulates the flight characteristics of a bullet. The spin around the ball's long axis gives the ball gyroscopic stability (i.e., stability resulting from spin). This helps the ball resist destabilizing forces produced by air currents and air resistance. Without the spin, the ball will tumble and flutter in the air. It's flight will not be "true," and it will not travel as far. The same thing happens when a discus is thrown without spin.

SUMMARY

- For athletes, balance indicates control of their movements. An athlete with good balance is able to counteract those forces that would otherwise disrupt his balance.
- Stability refers to an object or athlete's resistance to having his balance disturbed. There are degrees of stability. An athlete can be highly stable or minimally stable.
- There are two types of stability, linear stability and rotary stability.
- Linear stability can exist when an athlete or an object is at rest or moving in a straight line. At rest, linear stability is proportional to an athlete's mass and the frictional forces that exist between the athlete and any supporting surfaces. When an athlete is moving, linear stability is directly related to mass and inertia. The more massive the athlete, the greater the inertia and the greater the linear stability.
- The rotary stability of athletes or objects is their resistance against being tipped over, or upended, and their resistance against being stopped or slowed down once they are rotating.
- Rotary stability is a battle of torques. The turning effect of a torque that disturbs an athlete's balance must be countered by the turning effect of a torque that the athlete applies to regain balance.
- Several principles relate to an athlete's stability. The athlete should satisfy as many of these principles as possible to ensure that stability is maximized: (a) Stability is increased when an athlete's area of supporting base is made larger; (b) stability is increased when the athlete's line of gravity falls within the boundaries of the athlete's supporting base, especially when the athlete's line of gravity is centralized within the supporting base; (c) stability is increased when an athlete lowers her center of gravity; (d) stability is proportional to an athlete's mass; the greater the athlete's mass, the greater the athlete's stability; (e) in order to maintain stability after impact, an athlete's supporting base should be extended toward an oncoming force; (f) to maintain stability after impact, the athlete's center of gravity and line of gravity should be shifted toward the oncoming force; (g) to maintain stability in the application of force, the supporting base should be extended in the direction that force will be applied; (h) for the greatest application of force in a particular direction, the center of gravity and the line of gravity should be initially positioned as far as possible in the opposing direction above an extended base, and to complete the application of force, the center of gravity and the line of gravity should extend well ahead of the supporting base.
- Some sport skills require minimal stability. Sprinters and swimmers in a set position shift their line of gravity toward the edge of their base in the direction they will race.

Athletes who need to move quickly in any direction keep their base small and centralize their line of gravity.

- In a situation where objects or athletes are rotating, rotary stability is proportional to angular momentum. Rotary stability is increased by increasing mass, increasing angular velocity, and increasing the distribution of mass away from the axis of rotation.

KEY TERMS

balance
base of support
line of gravity
linear stability
rotary stability
stability

REVIEW QUESTIONS

1. What is linear stability and why is the linear stability of an object or an athlete on the move dependent on the magnitude of its mass and not on its velocity?
2. Why is rotary stability a battle of one torque versus another? As an example in your answer, use an athlete balancing on one foot on the beam.
3. Using the example of a defending wrestler lying on a mat, explain why stability increases when the wrestler extends his base of support in the direction of force (and torque) applied by an attacker.
4. Skills in which an athlete must change direction suddenly or be ready to thrust in any direction require the athlete to assume a position that does not have maximal stability. Why is this true?
5. When a discus is rotating, its stability depends not only on its angular velocity but also on its rotary resistance (rotary inertia). Why?
6. An Olympic weightlifter stands erect with his feet close together (ankles touching) and with a 400 lb barbell at arm's length above his head. Why is it difficult for the weightlifter in this situation to control the barbell?

PRACTICAL ACTIVITIES

1. **Static stability: Exceeding the line of gravity.** Take a block of wood 4 in. × 3 in. × 2 in. and place it on its 3-in. side on a small plank of wood. Slowly raise the plank of wood from one end until the block tips over. Get a partner to measure with a protractor the angle of the plank at the instant the block tips over. (You may need to put sandpaper on the incline so that the block doesn't slide.) Lay the block on all the remaining sides to determine maximum angles achieved before tipping over. How does the angle of the plank relate to the choice of base used for the block?
2. **Maximum stability.** You will need a partner for this activity. Assume the most stable position you can think of, and maneuver in opposition to the efforts of your partner, who tries to destroy your stability and to move, lift, or roll you over. Switch your roles of attacker and defender and repeat the activity. Record what you felt was your most stable position and what you tried to do in response to the destabilizing actions of your

attacker. Take into consideration the mechanical principles relating to torque, levers, area of supporting base, and position of your center of gravity.

3. **Line of gravity relative to the supporting base.** Stand with your heels and back against a wall. With your legs kept as straight as possible, bend down to touch your toes. Do you tumble forward? Compare the performances of males and females in this activity. Are females more successful than males? Now perform the same movements away from the wall. Record what happens in terms of your center of gravity and the line of your center of gravity relative to your base of support. Why are there differences between males and females?
4. **Dynamic stability: Stability on the move.** Run a tight figure eight at increasing speed on a flat surface. Why is it difficult to run the figure-eight course at high speed? Have a partner note the changes in your body position as you speed around your course. Switch and repeat the exercise with you performing the observations. Draw stick figures in your explanations, and use arrows to indicate the forces acting into the curve and those acting out of the curve as you (or your partner) maintain dynamic stability. Explain why it's difficult to run the figure-eight course at high speed.

6

Going With the Flow

© Sport the Library

When you finish reading this chapter, you should be able to explain

- how the fluid forces of hydrostatic pressure, buoyancy, drag, and lift affect athletes and objects as they move through air and water;
- how pressure, temperature, and the nature of air and water affect the way fluids act;
- how surface drag, form drag, and wave drag affect the movements of athletes and objects through air and water;
- how competitors in high-velocity sports counteract or make use of drag and lift; and
- how the Magnus effect and drag forces affect the flight path of a spinning ball.

We could just as easily call this chapter "Going Against the Flow" because it deals with how the forces of air and water help or hinder an athlete. In the following pages you'll read about swimmers using the resistance of water to maximize their propulsion, and you'll understand why one athlete can be a floater and another a sinker. You'll learn why competitive cyclists use aerobars, disc wheels, and slick racing suits and why discus throwers love to compete in stadiums where they can launch the discus into a headwind. You'll also read how a pitcher uses spin to produce a curveball and why golf balls travel farther and faster when they're covered in dimples rather than smooth. This chapter is about making the most of the flow!

Hundreds of sports are affected by the forces produced by air and water. They range from the underwater sport of scuba diving to speed events like downhill skiing and auto racing, and from these land-based events to high-flying sports like hot air ballooning and skydiving. You'll also find many other sports in which the forces exerted by air and water play little part in the outcome of the competition. Wrestlers don't worry about air resistance when they're pinning an opponent, and gymnasts don't move along a beam at speeds that require them to be streamlined. The same can be said about basketball. Basketball players would have to play on an outdoor court in a gale before they'd worry about the effect of wind on the quality of their performance.

Although air is obviously not as thick and dense as water, both air and water act like fluids. They flow around and easily mold to the shape of an athlete or an object like a bicycle or a javelin. This characteristic means that air and water exert similar forces. These forces are hydrostatic pressure, buoyancy, drag, and lift. Let's look at each of these forces to see how they affect athletic performance.

Hydrostatic Pressure

Hydrostatic pressure is the force exerted by a fluid such as air or water in supporting its own weight. To understand the characteristics of hydrostatic pressure, let's look at our own atmosphere. Most of us don't think of the atmosphere around us as pressing down on our bodies, but it does. In fact, gravity pulls the earth's atmosphere onto the surface of the earth, and since we move on the surface of the earth, the atmosphere presses on us, too. An easy way to picture this concept is to think of the atmosphere as blankets layered on you when you're lying in bed. Each blanket holds up those above and presses down on those below. Imagine the surface of the earth is a bed with you resting on it. At sea level you have the most blankets of atmosphere on top of you. At high altitudes, only one or two blankets are lying on you instead of many. In other words, the greatest atmospheric pressure exists at sea level, and it lessens as you increase altitude. At sea level, the weight of the atmosphere is 14.7 lb on every square inch of the earth's surface. In the metric system, 14.7 psi (pounds per square inch) is just over 100 kPa (kilopascals). Close to 14.7 psi presses on every square inch of you, too! So if you measured the area of the top of your head, you could work out the weight of the atmosphere pressing down on your head. For adults, this can be more than 1,000 lb. As humans, we've evolved so that we can put up with this kind of pressure. You don't feel it, but we'd all be in big trouble if this pressure was ever removed.

At sea level, the air we breathe is 21% oxygen and 78% nitrogen with the remaining 1% made up of other gases. If you go from sea level to higher altitudes and compare the constituents of a volume of air, the air keeps to its ratio of 21% oxygen and 78% nitrogen. But there's less oxygen and less nitrogen because at higher altitudes the air is at a lower pressure and so less oxygen and less nitrogen are squeezed into the volume you are considering. Climbers on Mount Everest find that as they pass 26,000 ft on their way to the peak at 29,028 ft (8,848 m), there is a third less oxygen available in the air compared with what existed in a similar volume of air at sea level. This is why the majority of Everest climbers carry additional supplies of bottled oxygen. Even at lower altitudes around 5,000 ft (1,524 m), it is important to become acclimatized to air that is at a lower pressure and that gives you less oxygen each time you inhale. Tourists who casually decide to climb Mt. Kilimanjaro (19,340 ft, or 5,895 m) as part of a trip through Tanzania will find they breathe more rapidly as their bodies attempt to get the oxygen they need and frequently suffer from altitude sickness as they climb to the summit.

Water is more than 800 times as dense as air, and a cubic foot (12 in. × 12 in. × 12 in.) of salt water with its additional mineral content weighs approximately 64 lb, which is approximately 1.6 lb more than a cubic foot of fresh water. Because water is virtually incompressible, the pressure it exerts increases regularly with depth. Scuba divers (using their self-contained underwater breathing apparatus) learn that for every 33 ft (10 m) they go down in the ocean, the pressure exerted by seawater increases by the equivalent of one atmosphere (i.e., 14.7 psi). So at 33 ft, you have 14.7 psi, which is the weight of the atmosphere resting on the ocean, plus another 14.7 psi as a result of the pressure exerted by the water. This gives a total of two atmospheres. At a depth of 66 ft, pressure increases to 44.1 psi (or three atmospheres), and at 99 ft it's exerting a pressure of 58.8 lb (or four atmospheres) on every square inch of a diver's body (see figure 6.1). Most of a diver's body can put up with this pressure, but not a diver's air cavities. The sinuses, lungs, and inner ear all have air cavities, and pressure in these areas must be balanced with whatever pressure is exerted by the surrounding water.

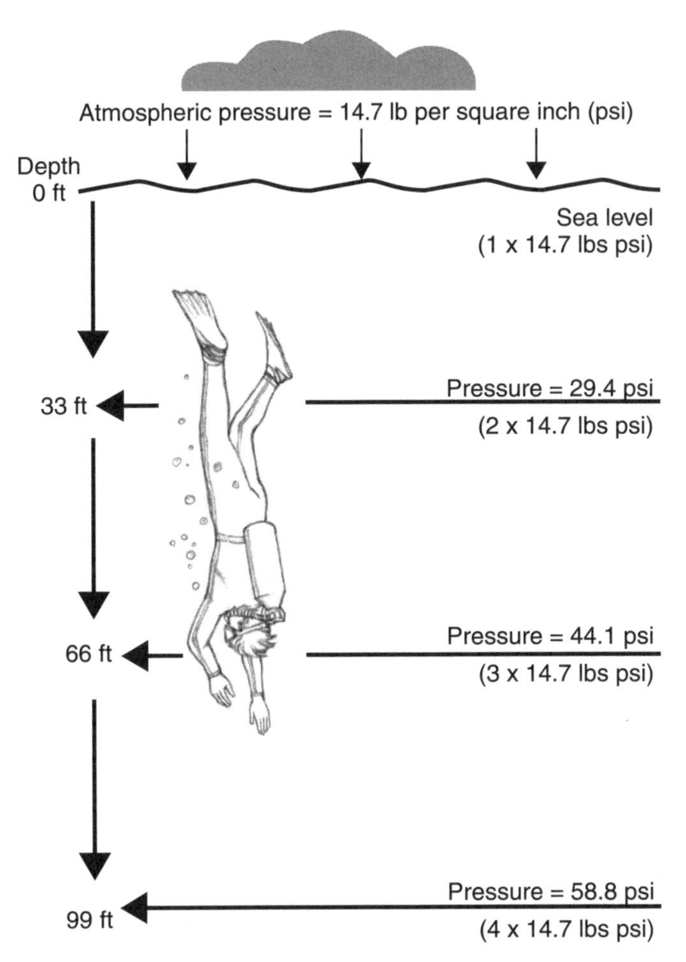

FIGURE 6.1 Pressure increases by one atmosphere (14.7 psi) for every 33 ft of depth.

When scuba divers descend in the ocean, they take with them compressed air in tanks and it passes through regulators designed to provide a balance in pressure between the air in the divers' lungs and air cavities and the pressure exerted by the water on their bodies. Air that is inhaled through the regulator at a depth of 33 ft is reduced by the pressure existing at that depth to half the volume it would have on the surface of the ocean. At a depth of 66 ft the air is reduced to one third of its surface volume, and at 99 ft to one quarter of its surface volume. Ascending to the surface, the reverse occurs and the compressed air expands as water pressure is reduced. One lungful of compressed air inhaled at a depth of 33 ft, if it's held in the lungs, will expand to twice its size at the surface. At 66 ft, one lungful of air be-comes three lungfuls at the surface, and at 99 ft a single lungful expands to four times its size at the surface. A scuba diver who fails to check his gauges and who empties his tank at 99 ft, and then in a panic holds his breath all the way to the surface can expect the air in his lungs to expand four times its volume by the time he gets to the surface. There's no way the lungs can handle this expansion. The unfortunate diver who forgets the cardinal rule of diving, "Never hold your breath," is likely to suffer an embolism—expanding air ruptures the lungs and enters the blood stream. Scuba divers avoid this dangerous situation by breathing normally on the way to the surface and by always having sufficient air to avoid rush ascents.

The increased pressure that a diver experiences with depth also means that a tank of air that lasts an hour on the surface will last approximately 30 min at 33 ft, 20 min at 66 ft, and only 15 min at 99 ft. Not only is the diver's "bottom time" reduced the deeper the diver goes, but breathing air at such high pressures also allows high concentrations of nitrogen to get dissolved into the bloodstream where it can affect the diver's central nervous system. Jacques-Yves Cousteau referred to the effects of nitrogen on a diver as "raptures of the deep." It is also known as nitrogen narcosis, a condition in which high concentrations of nitrogen make the diver act as though he is inebriated. Other symptoms include loss of memory, slowed response time, faulty judgment, and disorientation.

Nitrogen concentrations can also cause problems for a scuba diver on the way to the surface. If the diver rises too rapidly, the nitrogen that was dissolved in his blood under the higher pressure that existed in deep water changes from a liquid state back into tiny bubbles in the same way that a bottle of carbonated water fizzes when the cap is taken off. These nitrogen gas bubbles must be expired slowly through the lungs. But with the diver's rush to the surface, there is insufficient time for the gas to be expired through the lungs; instead the bubbles can become lodged in the joints, muscles, and other tissues of the body. This produces a condition descriptively and painfully known as "the bends" because on the surface the diver can be crippled and bent over. Other symptoms of the bends, or more accurately *decompression sickness,* include disorientation, rashes, itching, dizziness, nausea, and pain in the joints. Divers normally avoid decompression sickness by having enough air in reserve to ascend slowly and by making timed decompression stops at various levels during the ascent, thereby providing sufficient time for the nitrogen to be safely transported to the lungs for elimination.

Scuba diving is a much safer sport than it used to be. Advances in gas mixtures, equipment design, and the availability of excellent training all combine to make scuba diving a highly enjoyable sport. Two of the golden rules for scuba diving are (1) always scuba dive with a partner and (2) understand the basic laws of physics as they apply to diving. Following these rules will provide the scuba enthusiast with many years of safe, pleasurable diving.

Buoyancy

Buoyancy, the tendency of an object to float or rise when submerged in a fluid, is one of the few forces that lift upward and fight gravity. Buoyancy is related to hydrostatic pressure. When an athlete or an object is fully immersed in water, the water presses on the athlete from all directions. Because pressure in a fluid increases with depth, there's more pressure pushing on the athlete from below than there is from the sides or from above. The result is a **buoyant force,** a more powerful upward force pushing on the athlete from below.

Buoyancy doesn't apply only to water. It exists in the air as well. Hot air ballooning is based on the fact that a volume of hot air weighs less than an equivalent amount of surrounding cooler air. When you sit in the stadium seats and watch a television blimp circling overhead, you are looking at an object filled with a gas (helium) that is lighter than air. The blimp rises in the air because a buoyant force is acting on it. It remains circling the stadium at a set altitude when buoyancy and other upward-lifting forces acting on the blimp are balanced against the downward pull of gravity.

Because numerous sports like swimming, rowing, kayaking, and yachting are affected by the buoyant force of water, let's look at how it works. A simple law known as **Archimedes' principle** will tell you how strong the upward lift of buoyancy is going to be. The buoyant force acting on an athlete is equal to the weight of the water that the athlete's body takes the place of, or pushes out of the way, once the athlete is immersed in the water. When this buoyant force is greater than the force of gravity pulling the athlete down, the athlete is pushed to the surface (see figure 6.2). This means that an athlete will float if he weighs 200 lb and the water he displaces weighs more than 200 lb. Because bone and muscle are more dense and weigh more than the same volume of body fat, an athlete with big bones, big muscles, and very little fat has a lot of mass compressed into the space his body occupies. So an athlete with a body type that is predominantly bone and muscle is likely to weigh more than the water he displaces. Consequently, gravity pulls down more than the buoyant force pushes up . . . and the athlete sinks. These types of athletes were once featured in televised superstar competitions. Huge, muscular athletes who were best at land sports turned the swimming competition into a submarine race because most of them were "sinkers." The buoyant force acting on their bodies lost out to gravity. But if these superstars had left the fresh water of the swimming pool and competed in the ocean, there was a better chance they'd be floaters. Because of its additional salt content, the water the athletes displace in the ocean weighs more than the fresh water they displaced in the pool. If the amount of displaced ocean water weighs more than they do, then they'll float.

The percent body fat that athletes possess relative to bone and muscle plays an important role in distance swimming, particularly in colder ocean waters. It not only assists in buoyancy but also keeps the athletes warm. Coupled with a grease coating that swimmers spread over their bodies, body fat helps resist the loss of body heat and the onset of hypothermia. Athletes with minimal body fat experience hypothermia more quickly than those who have a high proportion of body fat.

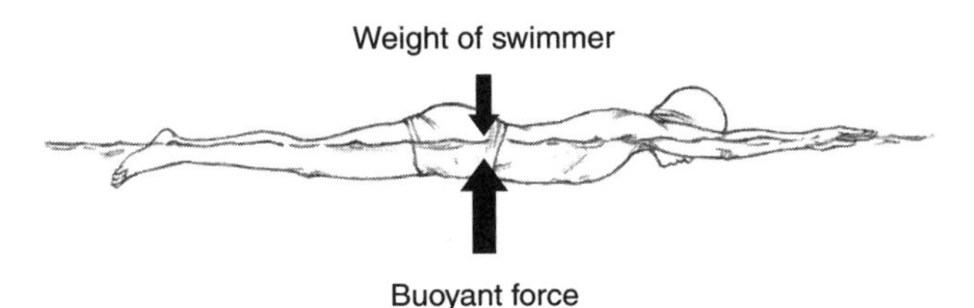

FIGURE 6.2 When the buoyant force is greater than gravity's downward pull (i.e., the athlete's weight), the athlete will float.

If an athlete with a low percentage of body fat puts on a wet suit, she will have an easier time staying afloat. A wet suit increases the athlete's space in the water with little increase in weight (because the wet suit weighs very little). The thicker the wet suit, the greater the space the athlete takes up in the water (i.e., the greater the athlete's **volume**) and the more water she displaces. The buoyant force pushing up is increased with little change in the athlete's body weight.

If an athlete wearing a wet suit descends well below the surface of the ocean, the increased pressure with depth tends to compress the suit. The deeper the athlete goes, the more compression. In this situation the athlete can become a sinker. Scuba divers wear lead hip weights to counter the additional buoyancy of the wet suit when they're close to the surface. As they descend, the increased pressure compresses the wet suit. The scuba divers must now balance the loss of buoyancy that occurs with increasing depth by pumping air from their tanks into what is called a buoyancy vest. The expanded buoyancy vest increases the space they take up in the water, and so the buoyant force is increased as well. Too much air in the buoyancy vest and the scuba divers will rise to the surface. On the other hand, if they want to remain at a particular depth in the water, they pump just enough air into the vest to balance the upward push of the buoyant force against the downward pull of gravity.

All of us can use our lungs as a mini-buoyancy vest. If we take a deep breath and expand our rib cages, we take up more space in the water. The buoyant force is increased without any change in body weight. For many athletes (but not all) this action is sufficient to hold them at the surface. Exhaling reverses the situation and the athletes start to sink. This phenomenon can be tested easily in a swimming pool.

The buoyant force acting on an athlete varies according to the density and temperature of

Breath-Hold and Free Diving Competitions

Breath-hold and free diving competitions are extreme sports in which athletes compete to achieve the greatest time, depth, or horizontal distance on a single breath without using scuba equipment. There are six categories, and the greatest publicity has been given to a category called "No Limits." Competing in this category in 2005, Herbert Nitsch of Austria achieved a depth of 172 meters (564 feet) by holding onto a weighted sled which rocketed him down a cable toward the ocean floor. He was then pulled back to the surface by a lift bag which he inflated after achieving his record depth. By comparison, sperm whales, the world's deepest diving mammals, have been recorded at depths in excess of 7000 feet.

the water. The warmer the water, the less dense it becomes. So it's easier to float in a cold, salty ocean than in warm, fresh water. Cold ocean water is more dense and weighs more than an equal volume of fresh water. In cold ocean water there is a greater likelihood that the amount of water displaced by an athlete will weigh more than the athlete does.

Center of Buoyancy

The place where the buoyant force concentrates its upward push on an athlete's body or on any object (such as a boat) is called the **center of buoyancy.** An athlete's torso and upper body contain the lungs and collectively take up plenty of space in the water. Compared with the athlete's legs, the upper body takes up more space and generally weighs less than the water it displaces. Consequently, the upper body is pushed upward more than the legs. The result is that the center of buoyancy for most athletes is not the same place as the center of gravity but a little closer toward the upper body, generally positioned just below the rib cage. If you could somehow hold the water in the manner and shape that it was displaced by an athlete, you would find that the athlete's center of buoyancy is the center of gravity of the displaced water.

Floating Position

The position an athlete assumes when floating is determined by the fact that the center of buoyancy is usually closer to the upper body than the athlete's center of gravity is. Normally the buoyant force concentrates its upward push just below the rib cage, whereas gravity pulls downward at approximately waist level. These two forces, one acting up and the other down, cause rotation. The athlete's legs drop downward while the chest lifts up. Rotation ceases when the center of gravity is positioned directly below the center of buoyancy. The chest is up and the lower body and legs hang beneath. Figure 6.3a shows gravity pulling down and buoyancy pushing up on a swimmer. The alignment of the pull of gravity and the buoyant force acting on the swimmer are shown in figure 6.3b.

The tendency of an athlete's body to rotate when immersed in water is magnified when the athlete's legs are muscular and lean. A swimmer who has muscular legs with little fat must counter the legs' tendency to sink by lowering the head in the water as a counterweight and by using an efficient leg kick so that a streamlined body position is maintained. Figure 6.4a shows an inefficient swimming position in which the athlete's body is angled downward with the legs low in the water. When the athlete's body is parallel to the direction of movement, the resistance from the water is reduced and the energy expenditure for propulsion is also reduced (see figure 6.4b).

Lean, muscular athletes such as those in triathlon competitions frequently find that they are negatively buoyant in fresh water and positively buoyant in the salty environment of ocean water. The extra buoyancy of the ocean water helps the athletes maintain an efficient swimming posi-

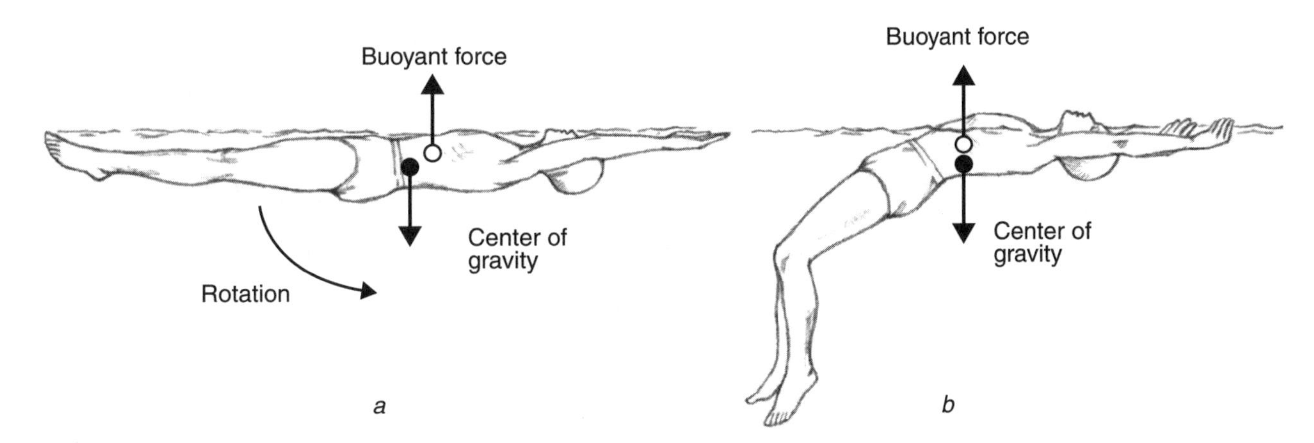

FIGURE 6.3 Torque caused by the forces of gravity and buoyancy *(a)* makes an athlete rotate to a floating position where the center of buoyancy is directly above the center of gravity *(b)*.

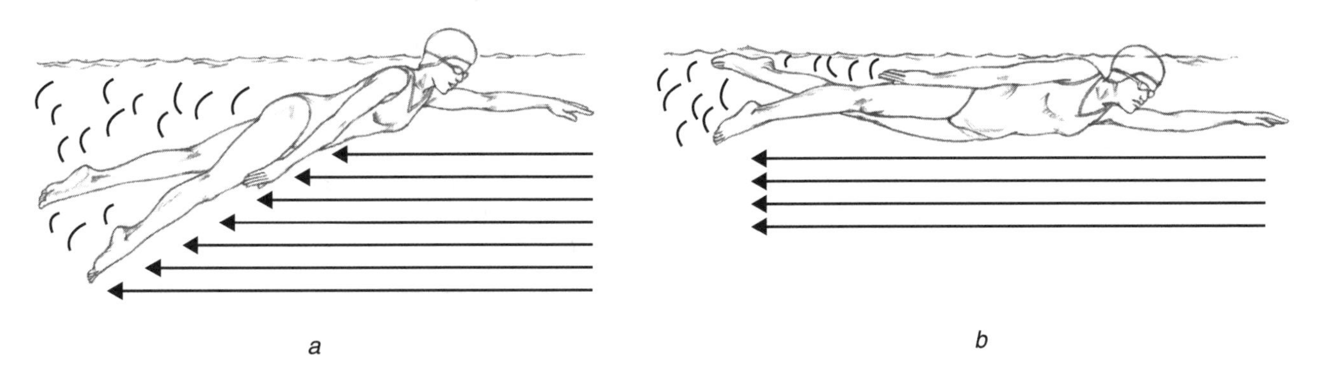

FIGURE 6.4 In swimming, considerable drag results from a low leg position *(a)*. A high leg position *(b)* reduces drag.

tion. If the rules allow the use of a wet suit, then the wet suit not only keeps the athletes warm but also improves their buoyancy. This is particularly beneficial when the swim portion of the triathlon competition is in a cold fresh-water lake. The increased buoyancy from wearing the wet suit pushes the athletes' bodies (particularly their legs) into a streamlined position. As a result their swimming stroke is more efficient and they consume less energy.

When it comes to the stability of boats, it's important that a boat's center of gravity be positioned well below its center of buoyancy. The keel on a sailboat counteracts the rotating effect of the waves and the wind blowing in the sails. A kayaker who has more mass in the upper body than in the hips and legs can raise the center of gravity (of the kayak plus its occupant) to a position where it is close to the center of buoyancy, or worse, to where it is positioned above the center of buoyancy. In this situation it takes little force to set up the turning effect of a torque sufficient to flip the kayaker upside down. Hanging below the kayak and fully immersed in water, the kayaker now becomes the equivalent of the keel on a sailboat. To spin the kayak 180 degrees and raise himself out of the water, the immersed kayaker must flex his body upward toward the kayak so that his center of gravity is as close as possible to the center of buoyancy. Once in this position the kayaker uses the paddle and a sharp hip thrust to spin himself and the kayak so that he is lifted out of the water. This action, commonly called an Eskimo roll, is much easier to perform in agile river kayaks than in larger touring ocean kayaks.

Drag

In most cases, **drag** is "a drag" because it's a collection of fluid forces that tend to oppose the actions an athlete is trying to perform. Drag pushes, pulls, and tugs on an athlete. If an athlete runs to the left, drag forces act to the right. Whatever the direction, drag forces act in opposition. It's easy to guess what will increase these drag forces. If athletes sprint faster, the air will push and pull at them more. If they sprint down the beach and then into the water as occurs at the start of triathlons, the drag on their bodies is greater in the water than in the air.

Drag varies according to the type of fluid (water or air), whether the fluid is warm or cold, and how dense and viscous (sticky and clinging) it is. Drag also depends on the size, shape, and surface texture of the athlete and likewise on the size, shape, and surfaces of the equipment the athlete is using. A giant sprinter running in a full-length fur coat that grabs at the wind will generate considerably more drag than a smaller athlete who wears a slick, polished bodysuit.

Let's look at the characteristics of drag. We'll start by examining the velocity of the athlete and the fluid as they pass and rub against each other. When you consider drag, remember that it doesn't matter what or who is doing the moving—it can be the athlete, or the fluid, or both. If the fluid is moving, then drag occurs even when an athlete is standing still. As you'll see, what's important is the relative motion (i.e., how fast the fluid and athlete are passing each other).

Relative Motion

Elite swimmers frequently practice in special observation swim tanks called flumes. The coach views their strokes through a side window. In a flume, a turbine drives water past the swimmer's body. The speed of the turbine can be adjusted to drive the water at various velocities toward the swimmer. Because the water is moving, the athlete must then adjust his swimming rate to stay in the same spot. In a swimming pool the reverse occurs. The water is stationary, and swimmers travel through and past the water. In a flowing river you can have a situation where the water moves and so do the swimmers. These situations (swimming in a flume, the pool, and a river) give you the three possible variations in what is called **relative motion,** which is the motion of one object relative to another.

In our three examples, one item is a swimmer and the other is the water. In these situations, if the relative movement of swimmer and water past each other is the same, then the drag forces produced by this movement are the same also. In other words, drag forces can be the same no matter who or what is doing the moving.

Let's look further at this concept of relative motion. Figure 6.5 shows three variations in relative motion. In figure 6.5a, an athlete is sprinting at 20 mph into a headwind of 5 mph. The athlete moves one direction at 20 mph and

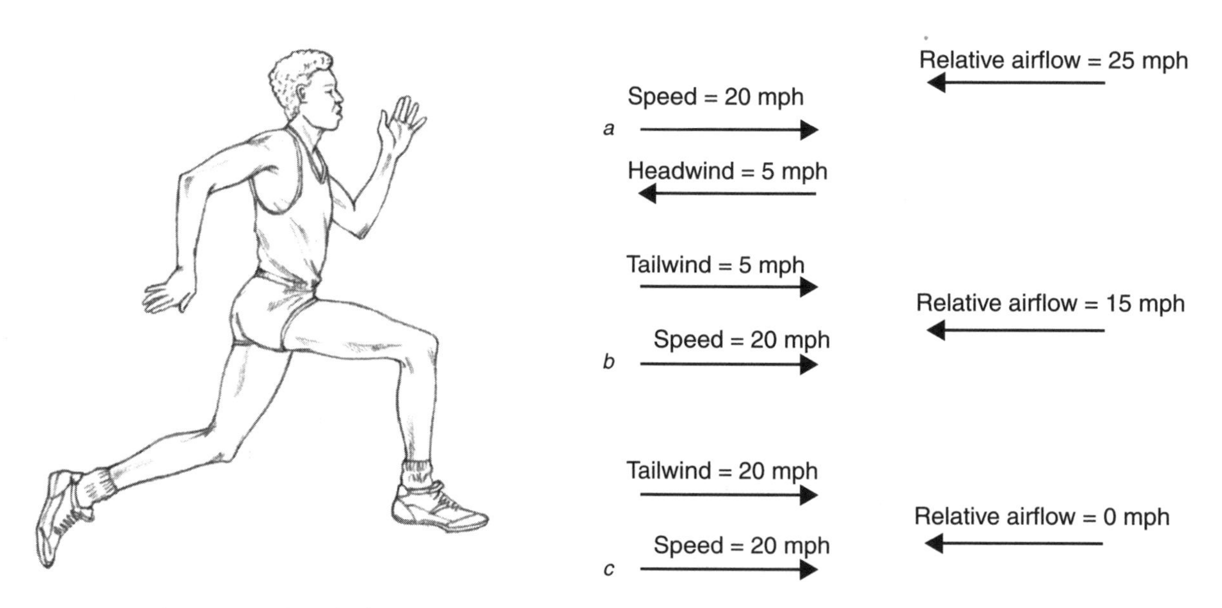

FIGURE 6.5 Three examples of relative motion in sprinting.

airflow is in the opposite direction at 5 mph. The relative velocity of airflow past the athlete in this situation is 25 mph. In figure 6.5b the athlete sprints at 20 mph but with a tailwind of 5 mph. The relative velocity of airflow past the athlete is now 15 mph. In figure 6.5c the athlete is again sprinting at 20 mph but with a tailwind of 20 mph. The relative velocity of airflow past the athlete in this situation is zero, because both athlete and airflow are traveling in the same direction at the same velocity. In the three examples we've just described, the frictional forces acting on the athlete occur maximally in figure 6.5a and minimally in figure 6.5c.

The fact that similar drag characteristics can occur whether the athlete moves or the fluid moves is the principle employed not only in flumes but also in wind tunnels. Wind tunnels are used to assess drag forces that occur when objects move through the air. Just as a coach checks a swimmer's technique by watching through the observation window of a flume, a technician can use wind tunnels to perfect the aerodynamic qualities of planes, race cars, speedskaters, ski jumpers, cyclists, or any object affected by the forces exerted by airflow. For example, when air is driven past sprint cyclists in a wind tunnel, assessments can be made of the drag forces that occur with different bike designs, various types of competitive clothing, and different body positions used by the cyclist.

Wind tunnels give valuable data on how to reduce the drag forces generated by airflow over a particular object, along with information on the shapes and body positions that provide optimal flight characteristics. This information is then used in such sports as gliding and ski jumping and in throwing events like the javelin and discus. It is common knowledge among throwers that a discus travels farther when it is spun at release and launched (like a glider) at an appropriate angle into a headwind. Wind tunnel tests can determine the optimal trajectory and spin of the discus relative to wind velocity and wind direction. From wind tunnel tests using a spinning discus and from practical experimentation by throwers, it has been found that a headwind of 15 to 20 mph can add 20 ft or more to the distance thrown, as compared with throwing in still air. Consequently, athletes try to adjust the flight of the discus according to the conditions that they meet. Discus throwers quickly learn which venues have winds that blow regularly from a favorable direction. They then try to compete at that venue as often as possible. Currently there are no illegal wind speeds for setting records in discus, javelin, hammer, and shot put as there are in sprint races and long jump.

Characteristics of Air and Water

Air is obviously not as thick, dense, or sticky as water, but both air and water vary in density and

How Does an Airplane Wing Produce Lift?

In level flight, four forces act on an airplane: (1) Thrust, which pushes the airplane along; (2) drag forces caused by the resistance of the air, which oppose thrust; (3) gravity, which pulls the airplane toward the earth's surface; and (4) lift generated primarily by the airplane's wings, which opposes gravity. To produce lift, air flows past the leading edges of the airplane's wings and continues past the trailing edges. The wings are given an "angle of attack," which basically means that the leading edges are tilted upward relative to the air flow. Air passing over the upper surfaces of the wings leaves a low pressure area to the rear of the leading edges. Atmospheric pressure drives air into this low pressure area and accelerates it, making it follow the downward tilt of the upper surface of the wing. Air passing along the underside of the wing is deflected downward. For an airplane to remain in level flight, the weight of the mass of air directed toward the surface of the earth must equal the weight of the airplane itself. For an airplane to climb, the weight of the mass of the air directed toward the surface of the earth must be greater than that of the airplane.

in consistency. Pressure variations, temperature changes, and differences in what air and water contain (like water droplets in the air and salt in water) all change the way these fluids act and the way they affect sport performances.

Athletes compete when it's hot or cold, from sea level to high altitudes, and at times when humidity (i.e., the percentage of water vapor in the air) is high. Variations in these conditions affect how fast athletes move, how far baseballs and badminton shuttles fly, and how much movement can be put on a curveball or a knuckleball. Because there are so many variations and combinations of air temperature, pressure, and humidity, here's a summary of what happens with variations in each.

- Air temperature. When air temperature increases, its density decreases and its resistance against a moving object decreases.
- Barometric pressure. If barometric pressure (i.e., air pressure) decreases (as you go from sea level to higher altitudes), the density of air decreases and its resistance against a moving object decreases.
- Humidity. When humidity increases, the density of the air decreases and its resistance against a moving object decreases.

The statement concerning humidity may seem contrary to what you'd expect. In explaining the characteristics of humidity, sport enthusiast and physicist Peter Brancazio (1984) writes:

> You may find this surprising, since the air often feels "heavy" on humid days—but this feeling arises from body physiology (perspiration evaporates more slowly when the air is damp) and not from the actual density of the air. In damp air, oxygen and nitrogen molecules are replaced by lighter-weight water molecules. Thus damp air weighs less than dry air at the same temperature and pressure. Changes in air density due to humidity are normally not very large, . . . for example at 86 degrees Fahrenheit and 80% humidity the air is about 1% less dense than dry air at the same temperature and pressure. Nevertheless, moving objects are subject to less drag at higher humidity. (p. 325)

Of course, the drag-reducing effect of humidity doesn't apply when humidity switches to precipitation. In a downpour, water is no longer in a vapor state and a baseball thrown by a pitcher has to push its way through the rain. In this situation the ball must contend with the considerable resistance of falling rain.

These atmospheric characteristics tell us that an athlete or an object (e.g., a baseball or a golf ball) will travel faster and farther in warm conditions than in cold, faster and farther at high altitudes than at sea level, and faster and farther on a humid day than on a dry day. It's for this reason that batters hit farther and pitchers hurl fastballs faster in Denver, where the ballpark is well above sea level. On the other hand, pitchers hurling curveballs and knuckleballs need thick air that grabs at the seams of the ball and helps move the ball around. What's good for a fastball doesn't produce the fluttering deception of a knuckleball.

Just as a fastball travels faster at high altitudes, so will a sprint cyclist. Some of the fastest times recorded by competitive cyclists, from the 200-m sprint to the 1-hr time trial, have been set at high-altitude venues such as Colorado Springs; Bogotá, Colombia; and Mexico City. In the 1-hr time trial, what is gained from cycling at high altitude can be partially offset by the reduced oxygen available for the athlete. To counteract this problem, prior conditioning at high altitude is a necessity.

The benefits provided at high altitudes for a fastball pitcher, a home run hitter, and a sprint cyclist can cause difficulties for badminton players. A badminton shuttle hit with the same force and trajectory at high altitudes as at sea level will travel farther because there is less resistance to its flight. To counteract this problem, badminton shuttles weigh less than those used at sea level. This allows the thinner air at high altitude to reduce their velocity.

The consistency of water differs just as it does for air. Water varies in density depending on its chemical makeup. An athlete attempting to swim in the Dead Sea situated on the Jordan–Israel border will be shocked at the difference between the water there and the water in a swimming pool. The Dead Sea is incredibly thick and dense because of its high salt content. This characteristic dramatically increases its drag. But there is a benefit! Even the leanest and most

Changes in Atmospheric Conditions Affect Flight of Badminton Shuttlecocks

Top quality badminton shuttlecocks use feathers from geese and only specific wing feathers are used. Thick white feathers are preferred, which are measured for their shape and weight. Feathers from the left wing are used on the same shuttlecock and likewise with those from the right wing. Sixteen feathers are mechanically driven into a single cork head to guarantee accuracy of angle and length. The shuttlecocks are then given a speed rating according to their weight. The rating takes into consideration variations in atmospheric conditions such as barometric pressure, temperature, and humidity. Lighter-weight shuttlecocks are used at high altitude venues and heavier shuttles are used at lower altitudes and at venues that are known to have a thicker, more dense atmosphere. In an international tournament, the chief referee will test up to three different speed shuttlecocks to ensure that the speed of the shuttlecock is correct. If atmospheric conditions change during a tournament, the shuttlecock will be changed as well. Disputes can arise over this exchange because a slower shuttlecock might benefit one player while a faster shuttlecock might be advantageous to another player.

muscular athlete can lie back in the saline water of the Dead Sea and, with little to no movement, be sufficiently buoyant to read a book without fear of sinking.

Fluid Viscosity

Viscosity is a measure of a fluid's flow. You can think of **viscosity** as a fluid's stickiness, its resistance to flow, and its ability to cling to the surface of any object that passes through it. Air is obviously less sticky than water, but air differs from water in that its viscosity increases slightly with an increase in temperature, whereas the viscosity of water is reduced with an increase in temperature. The viscosity of water and air plays a big part in what is called surface drag; and with an increase in viscosity, surface drag is increased. Surface drag combines with two other types of drag, form drag and wave drag, to make up the three most important drag forces that affect the movement of an athlete or an object. Let's look at each of these types of drag.

Surface Drag

Surface drag is also called viscous drag and skin friction. These names immediately give us a clue to the characteristics of surface drag. When an athlete or an object moves through a fluid such as air or water, the fluid forms what is called a **boundary layer.** This is the layer of air or water that, as it passes by, comes in direct contact with the athlete or the object, causing **surface drag.** Because of the fluid's viscosity, it clings and drags at the surface of whatever it contacts. Figure 6.6 shows the boundary layer around a ball that is moving slowly through the air.

The amount of surface drag developed by the movement of an athlete or an object through the air (e.g., a sprint cyclist pedaling a bike) depends on how viscous or sticky the air is and how large the surface area of the athlete and the bike is that comes in contact with the air flowing by.

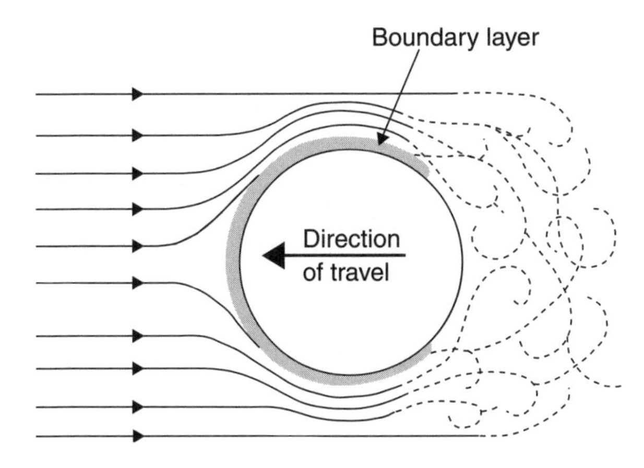

FIGURE 6.6 Boundary layer around a ball traveling through the air.

Surface drag also depends on the roughness of the surfaces of the athlete and the bike that contact the air, and finally, the relative velocity of the cyclist and the air as they pass each other. An increase in any of these factors increases surface drag.

Form Drag

To surface drag we now add its close associate—form drag, also called shape drag, profile drag, and pressure drag. As these names suggest, form drag is produced by the shape and size of an athlete or an object, like a racecar as it passes through the air. An area of high pressure develops in front where the athlete or object pushes against the air head-on. Immediately to the rear you find an area of turbulence. This trailing area, called the **wake** (like the wake to the rear of a ship), is a region of swirling low pressure and suction. With high pressure at the front and low pressure at the rear, an imbalance in pressure exists from front to rear that opposes forward motion. This is called **form drag** or, by its more descriptive title, pressure drag. Form drag will increase if the relative velocity of the athlete (or object) and the air as they meet and pass each other is increased and if the shape of the athlete or object promotes development of high pressure in front and low pressure to the rear. What kind of shapes are we talking about here? Think of an athlete pedaling as fast as possible into a strong wind sitting bolt upright on an old-fashioned bicycle. An upright body moving at high velocity would cause a large high-pressure area to develop in front of the athlete and a huge turbulent low-pressure area to develop to the rear. If the athlete wore loose, flapping clothing and strapped huge carrier bags and luggage to his bike, this would increase not only his form drag but his surface drag too.

Efforts to reduce form and surface drag as much as possible occur in many sports where athletes and their equipment move at high velocity. In these sports you see body positions and equipment that are designed to

- eliminate pushing through the air with a blunt shape that has a large cross-sectional area at right angles to the direction of fluid flow;
- eliminate lumps, bumps, projections, and rough edges and instead smooth out and polish all surfaces that contact the flow of the air; and
- eliminate the turbulent wake that occurs at the rear of the athlete or object where the low pressure occurs.

Designers working in high-velocity sports such as cycling, downhill skiing, auto racing, luge, and boat racing aim for a streamlined "teardrop" airfoil shape that also has ultrasmooth surfaces. Figure 6.7, a and b, compares the flow patterns around the blunt circular shape of a ball or a cylinder with that of a teardrop shape. Notice in figure 6.7a that a circular shape produces a turbulent wake area to its rear and in figure 6.7b that a teardrop shape causes the flow patterns to eliminate this turbulent wake. The teardrop shape in figure 6.7b could be improved further by making it more dart-shaped. Coupled with supersmooth surfaces, this reduces drag even further.

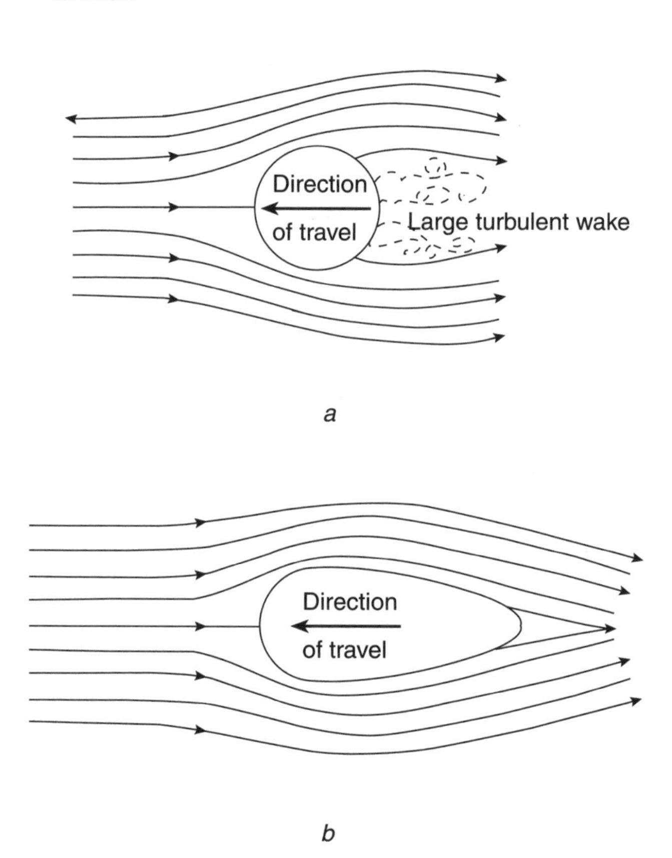

FIGURE 6.7 A teardrop shape eliminates the turbulent low-pressure wake that occurs to the rear of a circular shape (such as a ball).

Let's look closely at the sport of competitive cycling to see how drag is reduced to minimum levels. Elite athletes shed excess body weight and ride bikes made of lightweight metals and modern composite materials. This reduces nonproductive weight to a minimum. The bike is often raked (i.e., the frame is tilted downward toward the front wheel) and equipped with a larger rear wheel and extended handlebars. The combination of these features forces the rider into an aerodynamic position with the back parallel to the ground. The athlete's head is lowered and the arms are extended, with the hands together in front. This flattened cycling position reduces the frontal area (and high pressure) to a minimum and helps the cyclist simulate an airfoil shape in order to cut through the air with the hands and arms leading the way.

To reduce drag generated by surface friction, the athlete wears a super-slick, low-friction one-piece skintight suit with no wrinkles or loose flapping sections. Skintight fingerless gloves and laceless shoes, or booties, are also used. Legs, arms, and face are shaved. A teardrop aerodynamic helmet is worn. Frequently the helmet has a visor section in front to help slice through the air and a long blade shape to the rear to reduce turbulence and low pressure. A disc rear wheel is used, which eliminates any drag caused by spokes. The front wheel is made of lightweight composite materials with three large spokes that are teardrop-shaped in cross-section. An alternative design for the front wheel is using a low number (14) of high-tension blade-shaped spokes. These design features are intended to reduce the tremendous drag generated by the normal number (36) of circular spokes found on each wheel of a standard bicycle. All other components on an aerodynamic bike are teardrop-shaped. Shapes that are circular in cross-section, like the tubing on a standard bicycle, are avoided at all costs. Athletes looking for the ultimate in slipping through the air spray their bodies with silicone!

Downhill speed skiers and elite swimmers are similar to cyclists in their efforts to minimize drag forces. The full and partial bodysuits worn by swimmers not only improve buoyancy but also have surface patterns that are designed to allow water to flow over them with the minimum of turbulence. Downhill speed skiers compete to see who can reach the greatest speed as they hurtle through a "speed trap" where their speed is measured. Their crouched body positions, body-contoured poles with minimal-size baskets, super-slick uniforms, and aerodynamic helmets and boots are all designed to reduce form and surface drag to the lowest level possible.

Even with all the modifications to conquer drag, elite cyclists using equipment and body positions similar to those described previously are not as aerodynamically efficient as those who cycle in a recumbent position (i.e., inclined backward) and who cover themselves totally with lightweight fiberglass shells. Athletes enclosed in this manner have pedaled in excess of 80 mph, far faster than the 45 mph reached by the world's best sprint cyclists. Classified as HPVs (human-powered vehicles), lightweight shells are not allowed in the Tour de France or in Olympic cycling competitions. For the average person, cycling in a recumbent position inside a lightweight shell makes cycling faster and less energy consuming. But being lower down, they're difficult for motorists to see, and they are very unstable in crosswinds.

Standing in front of one of these teardrop-shaped human-powered vehicles, you'll notice how narrow and long they are. Their narrow, pointed profile reduces the cross-sectional area at the front of the vehicle and the high pressure that occurs there. The sides of the shells and fairings are smooth, polished, and slick, and all vehicles are designed with a long, tapering tail. The tapering tail fills in the low-pressure wake that occurs to the rear of the vehicle. The objective is the same as that of the speed skier, to reduce form and surface drag to a minimum (see figure 6.8).

Similar to the human-powered vehicles, luge competitors lying back on their small sleds wear form-fitting, slick uniforms and even try to reduce the drag caused by the shape of their faces. Though a luge competitor's nose is small compared with the rest of the athlete's body, it's still a projection and increases drag. To counter this problem, athletes cover their faces with plastic shields. It's worth the effort, particularly when .01 sec frequently separates one medal from the next.

FIGURE 6.8 High-speed human-powered vehicle.

Making Use of the Wake

In team pursuit races, teams of four cyclists start on opposing sides of the velodrome and attempt to gain distance on each other. The leading cyclist expends more energy carving through the air than do the following members on the team. The second cyclist "drafts" in the low-pressure wake of the first cyclist and also helps to fill in the low pressure wake of the cyclist ahead so the leading cyclist can also travel at a faster pace than when cycling alone. This is why a single cyclist in a break away in the Tour de France usually gets caught by the peleton (the large group of pursuing cyclists). Each cyclist in the peleton expends less energy individually and the whole peleton can travel at a greater speed for a longer time than a single cyclist. In a team pursuit race, the third cyclist does the same behind the second and likewise the fourth behind the third. As the team races around the oval, the leading cyclist drops to the rear and the second in line takes up the task of leading. In this way team members conserve energy by drafting behind teammates.

The benefit of cycling in this manner is that the lead cyclist travels faster and the drafting athlete can keep up the same speed as the lead cyclist while expending less energy. In essence, the drafting cyclist faces little high pressure in front and is pulled along in the wake that trails the cyclist in the lead. This technique is often called slipstreaming.

The closer the drafting cyclist follows the lead cyclist, the better. Distances less than 12 in. between the rear tire of the lead cyclist and the

Drafting Takes Cyclist Fred Rompelberg to 166.9 mph

Nothing demonstrates the advantages of drafting more clearly than comparing the top speeds attained by cyclists with and without drafting. The best time for a sprint cyclist in a velodrome with no help from a pace vehicle is just over 45 mph. For a cyclist pedaling in a recumbent position enclosed in an aerodynamic shell and riding at high altitude, the top speed is just over 80 mph. Compare these speeds with the incredible 166.9 mph achieved by Dutchman Fred Rompelberg in 1995 when he drafted in the lower-pressure suction zone behind a specially designed race car on the Bonneville Salt Flats. Rompelberg's bike was equipped with two chainwheels that stepped up the revolutions occurring at the back wheel. With this system, one revolution of the pedals moved the bike a phenomenal 35 yd (32 m). With such a gear system, Rompelberg could not accelerate the bike from a dead stop. Instead he had to be towed to 85 mph and then released from the rear of the pace vehicle.

front wheel of the drafting athlete prove most efficient. One reason that a tandem is faster than a single cyclist is that the rear rider on a tandem is drafting close behind his partner. Even though there is a greater surface area exposed to the air on a tandem, the additional power of two riders and the benefits of drafting produce faster speeds than single cyclists.

In triathlon competitions, the benefits experienced by the drafting athlete are so enormous that drafting is forbidden during the cycling part of the race. However, drafting is allowed during the swim portion of the race because it is impossible to officiate. In the Tour de France, cyclists are allowed to draft behind each other but officially they are not allowed to draft behind motorized vehicles.

How Drag Affects the Flight of Baseballs, Tennis Balls, and Golf Balls

If you examine what happens to a smooth-surfaced ball as it travels through the air and what happens to a ball with seams (like a baseball) or dimples (like a golf ball), you'll find some dramatic differences in the way that drag affects each ball. These variations play an important role in determining the flight of baseballs, tennis balls, and golf balls.

A circular and blunt object like a ball is poorly streamlined when compared with a teardrop shape; in fact, all circular shapes produce considerable drag. What's more, you can't change a ball's shape in the way that you can redesign a race car, a bobsled, or the cross-sectional tubing on a bike so they become more aerodynamic. A ball is always ball-shaped no matter how small or how large.

To understand how drag forces affect the flight of balls, let's have a smooth-surfaced ball move through the air very slowly. At this velocity the boundary layer of air that contacts the surface of the ball flows smoothly around the ball. Its flow pattern looks like laminations in a piece of plywood. This type of pattern is called **laminar flow,** or streamlined flow. Moving at such a low velocity, the ball is affected predominantly by surface drag caused by the clinging, viscous nature of the boundary layer (see figure 6.9).

If we make the ball travel faster so the velocity of airflow around the ball increases, the laminar flow starts to break up. Smooth laminar flow is now mixed with a disturbed, distorted **turbulent flow** pattern. Because the air passes the ball at a higher velocity, the air cannot follow the ball's contours the same way it did when it passed slowly. Instead, the boundary layer follows the ball's contours partway; thereafter it tears away from the ball's surface toward the rear of the ball. This causes a turbulent low-pressure wake to develop at the rear. With the increase in velocity at which air and ball pass each other, pressure also builds up where the ball hits the air head-on. So we have high pressure increasing at the front and low pressure increasing at the rear. The net result is that the ball experiences an increase in form drag (see figure 6.10).

If the ball and the air travel by each other at an even greater velocity, the place where the boundary layer breaks away from the ball's

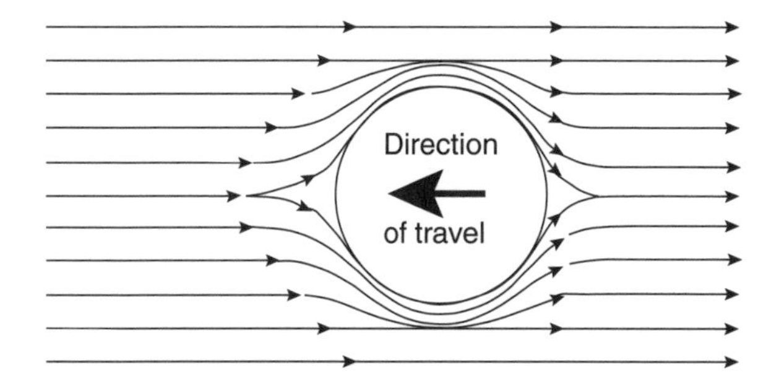

FIGURE 6.9 Laminar flow.

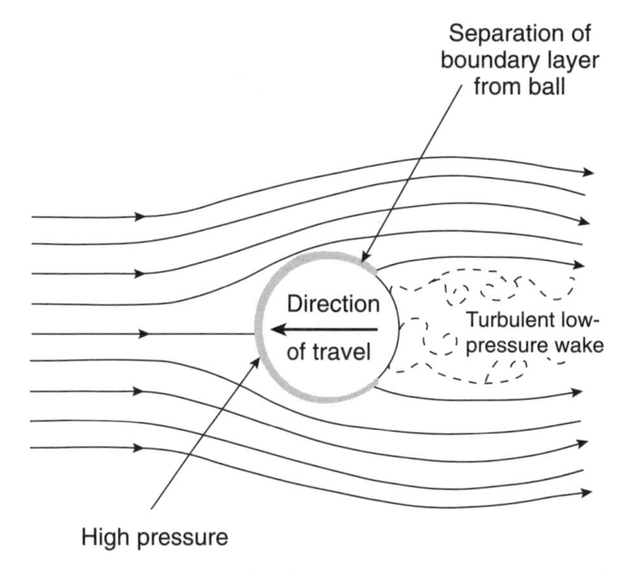

FIGURE 6.10 Turbulent flow pattern. At high velocities, the boundary layer breaks away to the rear of the ball, causing a low-pressure wake.

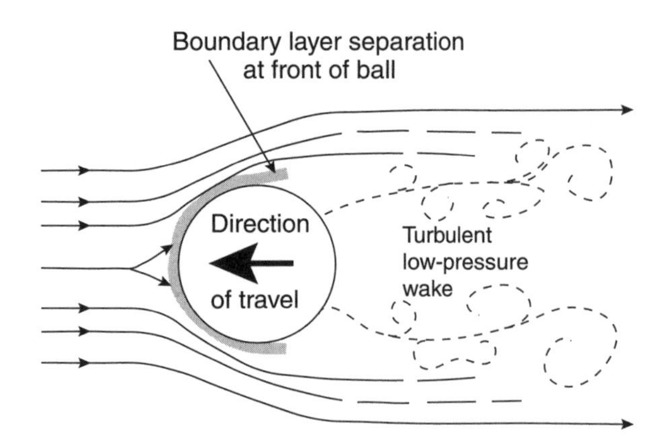

FIGURE 6.11 Turbulent flow pattern. At higher velocities, the boundary layer shifts to the front of the ball.

surface moves from the rear of the ball toward the front. The result is an even bigger wake area at the rear. There's more high pressure at the front of the ball and more low pressure at the rear, so form drag increases even further (see figure 6.11).

Finally, if the ball and air pass each other at extremely high velocity, the boundary layer becomes totally turbulent. Now a surprising change happens. When the boundary layer is totally turbulent, the place where it separates from the ball shifts back toward the rear and in so doing reduces the size of the low-pressure wake. Because the low-pressure wake has been reduced, the ball's form drag is reduced also (see figure 6.12), even though the ball's surface drag will have increased.

These variations in form and surface drag are important in games like baseball and golf because a ball that has seams or dimples causes a turbulent boundary layer to form all around the ball—not only at high velocities but at low velocities as well. So seams and dimples reduce form drag and help the ball fly farther and faster and hold to its flight path better than if the ball were smooth all over. It's true that roughing up the surface of the ball with seams and dimples causes surface drag to increase, but the total drag affecting the ball is less because the bigger drag (i.e., form drag) has been reduced (see figure 6.12). When spin is put on the ball, which is a big part of baseball and golf, a ball with seams and surface roughness curves better in flight and holds to the curve better than if the ball were smooth. Later when we look at how a pitcher throws curveballs you'll see how this occurs.

Wave Drag

The third and final type of drag acts at the interface where water and air meet. It is called **wave drag** and affects swimming (particularly breaststroke and butterfly where there can be a lot of up and down motion) and other water sports like rowing, kayaking, speedboat racing, and yachting.

When a swimmer moves through the water, waves pile up in front of her and create a high-pressure wall of water that resists her forward motion. The faster she travels and the larger the cross-sectional area that she pushes forward against the water, the larger the wall of water that builds up in opposition. So if the swimmer acts like a barge bashing its way through the water, the wave drag will be phenomenal.

At competitive speeds, wave drag is much more detrimental and energy consuming than surface and form drag. Surface and form drag increase according to the square of the velocity. This means that if a swimmer doubles her velocity through the water, both surface drag and form drag are increased fourfold. Wave drag, on the other hand, increases according to the cube of the velocity. If a swimmer doubles her velocity, then whatever wave drag she was generating at the original velocity is increased eightfold (i.e., $2 \times 2 \times 2$). If the swimmer triples her veloc-

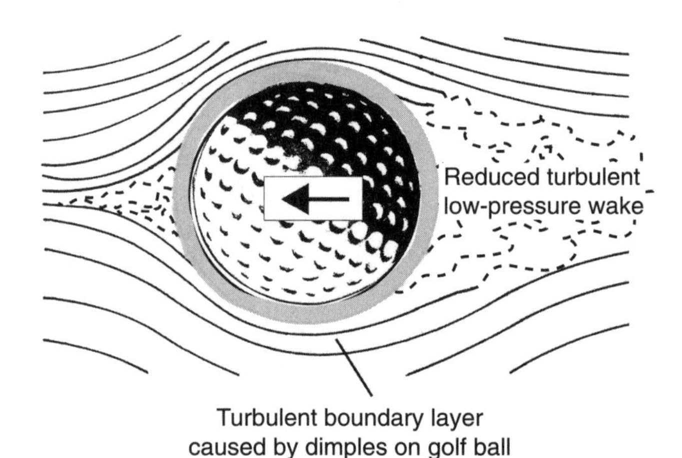

FIGURE 6.12 Turbulent flow pattern. At extremely high velocities, the boundary layer becomes fully turbulent, reducing the size of the low-pressure wake. As a result, form drag is reduced.

ity, the original amount of wave drag is increased 27 times (i.e., $3 \times 3 \times 3$).

Inefficient swimming technique and poor swimming pool designs produce waves. A top-rated swimming pool has specially designed gutters and lane dividers that absorb waves and stop them from bouncing and flowing from one lane to the next.

When an athlete's swimming technique is poor, wave drag is likely to increase along with other forms of drag. In events like the famous Hawaiian Ironman and Iron Woman triathlons, wave drag caused by the ocean and by other competitors can be a significant form of resistance. Expert swimmers hone their swimming technique to cut wave drag to a minimum. They also learn to draft in the wake of a leading swimmer. This simulates drafting used by cyclists, and in this way they can swim with less expenditure of energy. But they risk getting kicked in the face!

There are two ways of beating wave drag. One is for the athlete to swim underwater as long as possible. This technique used to occur in breaststroke and at the start of backstroke races until rule changes restricted how much underwater swimming could occur. The second method of beating wave drag is to get as far out of the water as possible (i.e., to hydroplane). This is a technique used by high-speed boats in which the pressure the water exerts on the planing surfaces of the boat lifts the hull of the boat partially or completely out of the water. This reduces the wave, surface, and form drag caused by the water. Instead, the boat fights against the lesser resistance generated by moving through the air.

Lift

When an athlete throws a discus, the discus applies force to the air, and the air reacts by applying force to the discus. The force that the air exerts can be broken down into two separate forces. One acts in the same direction as the airflow, directly counteracting the implement's forward movement. This is the drag force we discussed earlier. The other force acts at right angles to the drag force and is called **lift.** Drag and lift combine to produce a resultant force that most commonly pushes upward and backward in opposition to the motion of the discus. These forces are shown in figure 6.13.

Experimentation in a wind tunnel with the tilt of the discus relative to the airflow (called the **angle of attack)** indicates that variations in this angle determine how much lift and drag a discus experiences. When the leading edge of the discus is tilted upward as it is in figure 6.13, air is deflected downward and exerts an equal and opposite pressure upward. This equal and opposite pressure produces the upward lift acting on the discus.

New Rules Bring Javelins Back to Earth

Until the rules were changed, technicians realized that if a javelin was made "fatter"— with very little taper from either end to the grip—it had an increased surface area and got more support from the air. If its center of gravity was shifted back close to the javelin's midpoint, the javelin would follow a much flatter and longer trajectory. In the hands of a superb thrower and under perfect wind conditions, a javelin of this type would fly incredible distances. In 1984, Uwe Hohn of East Germany (who held his country's military record for throwing a 21-oz practice grenade over 328 ft) threw such a javelin an incredible 343 ft 10 in (104.8 m)—almost exactly 20 ft (6.1 m) farther than the *current* world record set in 1996 by Jan Zelezny of the Czech Republic. Rule makers flipped! Javelins flying this far would endanger athletes at the other end of the stadium. What did they do? They moved the center of gravity forward toward the tip of the javelin and put specific controls on its shape and taper. Distances temporarily dropped. Now, with talented athletes and improved training techniques, javelin throwers are back up to 300 ft again.

Adapted from D. Wallechinsky, 1991, *The Complete Book of the Olympics: 1992 Edition*, p. 115.

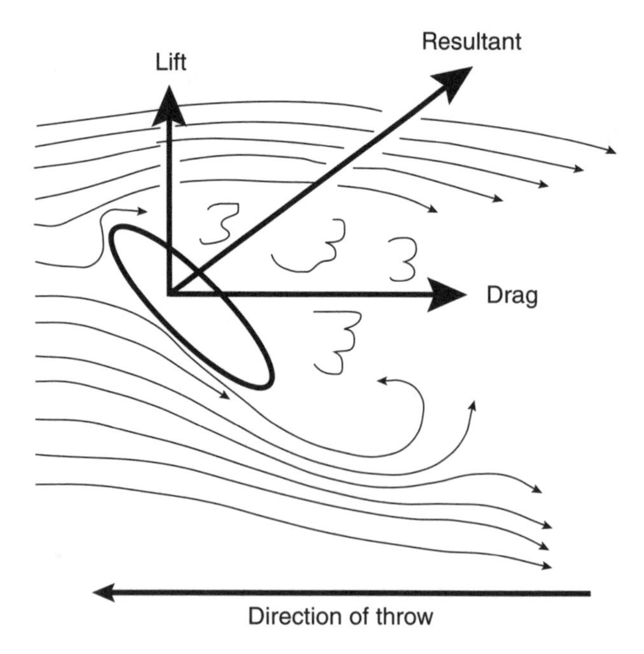

FIGURE 6.13 Lift and drag forces combine to form a resultant force.

There is a limit on how much the leading edge of a discus can be angled upward and still give the discus lift. If this angle is too great, lift disappears and drag increases dramatically, causing the discus to stall. Figure 6.14a shows a worst-case scenario. A discus thrown in this fashion gets no lift at all and quickly drops to the ground. If the leading edge of the discus is tilted downward, the lift force acts downward. As strange as it may seem, this is still called lift. Lift does not have to be upward, although the term itself suggests this. Lift can occur in any direction. In figure 6.14b the resultant of lift and drag forces acts backward and downward, opposing the motion of the discus.

Many factors influence the amount of lift acting not only on a discus but on an athlete as well. Lift is influenced by the athlete's body position because shape influences airflow patterns. The body position assumed by the great Norwegian, Finnish, and Japanese ski jumpers traveling down the inrun is very different from the position they use immediately after takeoff. Accelerating down the inrun, they crouch low with their backs parallel to the inrun. This body position is similar to that used by competitive cyclists in that it minimizes air pressing against their chests and slowing them down. When they take off, their bodies are extended and angled forward and upward, giving an angle of attack that maximizes lift. Holding this position, their time in the air is extended by lift, producing soaring jumps of well over 400 ft (see figure 6.15).

The faster a ski jumper flies through the air, the greater the lift force that pushes him upward. This force depends on the ski jumper holding an

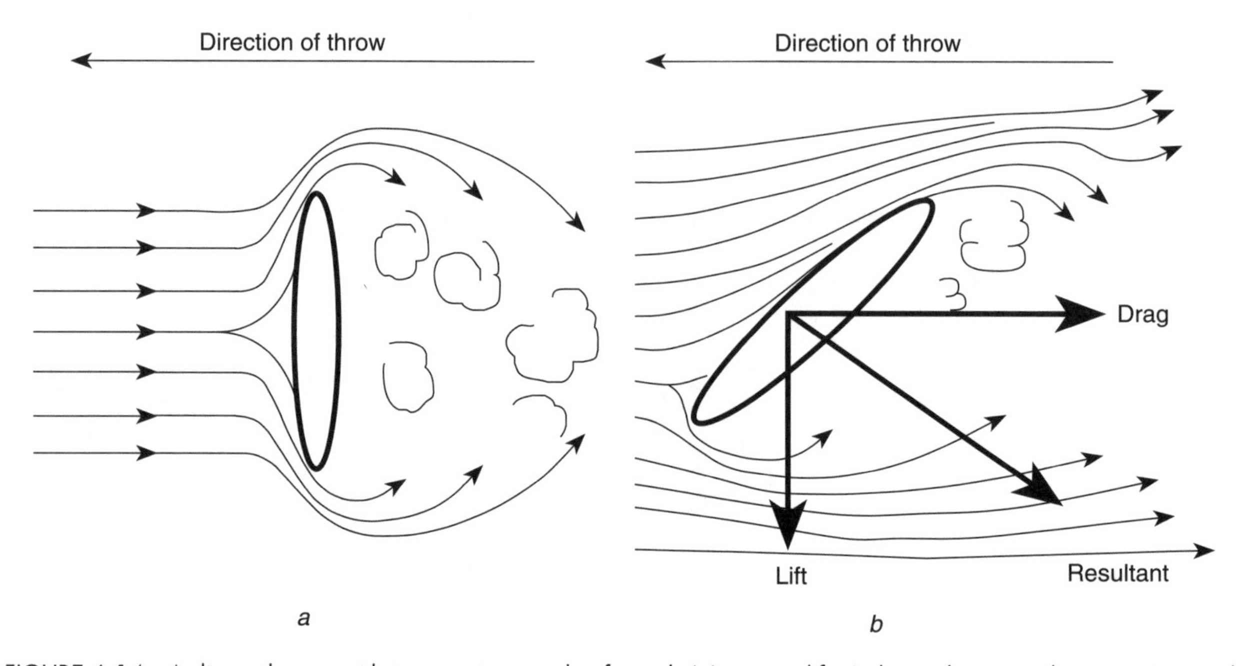

FIGURE 6.14 A discus thrown with too great an angle of attack *(a)* gets no lift. A discus thrown with a negative angle of attack *(b)* generates downward lift.

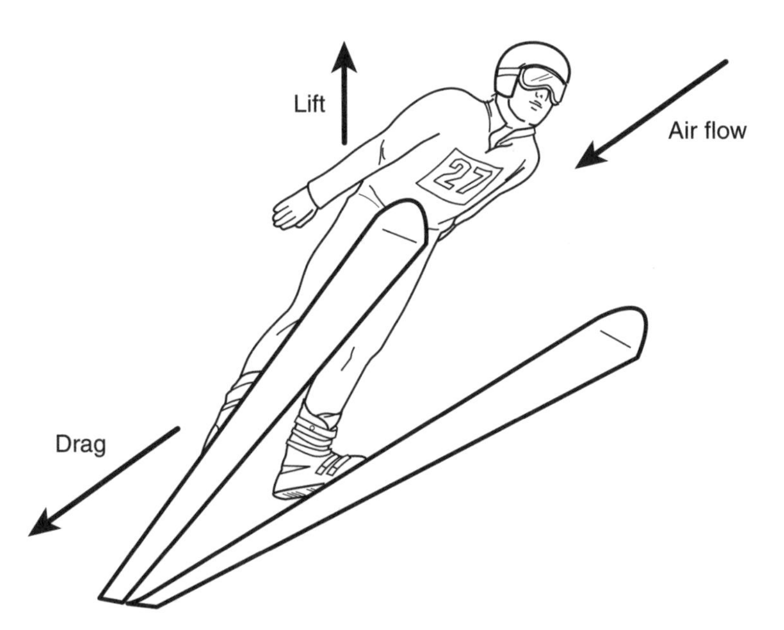

FIGURE 6.15 Lift and drag in a ski jump.

optimal body angle to maximize lift. The jumper must change this angle as he slows in flight. Likewise, the size of the surface area that the jumper angles into the airflow also influences lift. If this area is increased, the lifting force is increased in much the same way that a larger wing surface on a plane increases lift. Ski jumpers now angle their skis in a V shape so that the skis and the jumper's body both contribute to lift. The old style of jumping with the skis parallel beneath the jumper's body is inefficient and you will not see a jumper using this technique anymore.

Finally, the consistency of the fluid and its relative motion dramatically influence lift. When air and water are thicker (denser), lift is increased, and the faster the fluid moves by an object, the greater the lift. This means that a water-skier will experience more lift (but also more drag) skimming across the ocean than across a freshwater lake. Similarly a ski jumper will get more lift at lower altitudes where the air is more dense than in the thinner air of higher altitudes.

The amount of lift an athlete requires varies according to the demands of the sport. Downhill speed skiers compete to see who can reach the greatest speed (often faster than 130 mph) through a measured section of the course. Lift is the last thing these athletes want—it makes them lose contact with the snow. Their skis are purposely long and heavy with tips that have hardly any curl so that they do not lift upward. Speed skiers fight to keep their upper bodies parallel to the ground so that the lift from their upper bodies is minimal. Their body positions and equipment design are intended to keep them locked onto the snow.

The action of the skis in water-skiing differs from speed skiing because lift is a necessity. Unless the athlete gets lift there is no hope of rising onto the surface of the water. The boat must pull the athlete at an adequate velocity because this helps produce lift. The faster the boat the more lift. In addition, water-skiers angle the tips of their wide water skis more at the start than when traveling on the surface. By doing this, the reaction force from the water combines with the pull of the boat to lift the athlete out of the water and into a skiing position (see figure 6.16).

Downward lift (and the resultant forces generated by drag and lift) is commonly made use of in auto racing, where rear spoilers are angled to press the car down onto the road. This improves the friction of the tires with the road surface and gives better traction (see figure 6.17).

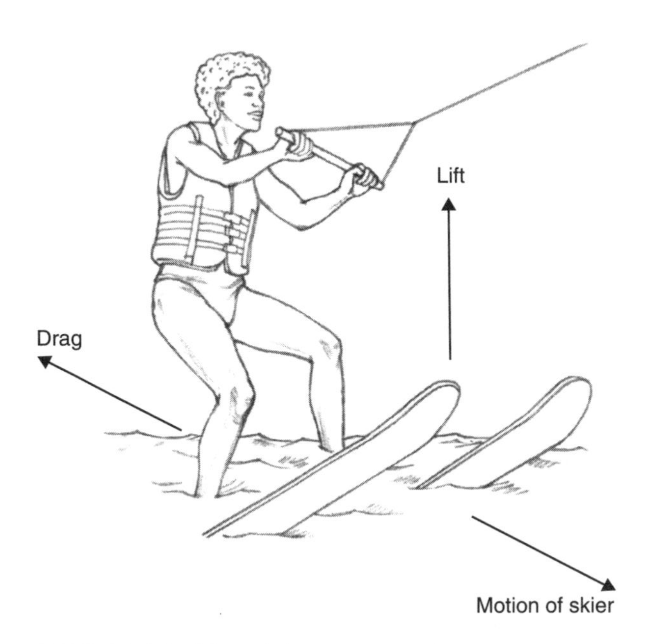

FIGURE 6.16 Lift and drag in water skiing.

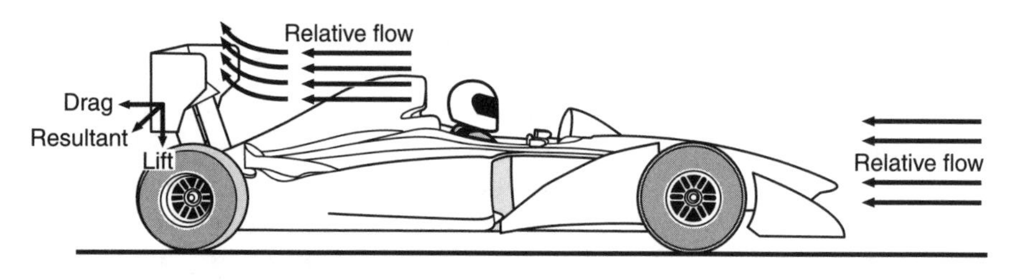

FIGURE 6.17 Lift, drag, and the resultant produced by spoilers press a race car down onto the track and improve its traction.

Earlier in this chapter we looked at how drag forces commonly oppose the motion of an athlete. In this section you'll see that it's possible to get drag and lift to act as propulsive forces. We'll see how this occurs in the sport of swimming.

In the late 1960s it was thought that swimmers should pull and push directly back against the water in order to propel themselves in the opposing direction. If an athlete is swimming to the left, then the thrust from the swimmer's hands and arms to the right against the water generates a drag force (i.e., a reaction force) from the water that propels the swimmer forward. Figure 6.18 shows a lateral view of a swimmer's hand moving to the right and the drag force that propels the swimmer, acting to the left.

In the 1970s and 1980s drag propulsion was challenged by those who thought that hydrodynamic lift provided the dominant propulsive thrust for a swimmer. Rather than push directly backward to swim forward, it was proposed that swimmers concentrate on performing a weaving down-in-up-out S-shaped pull and push motion with the hand and arm to produce lift.

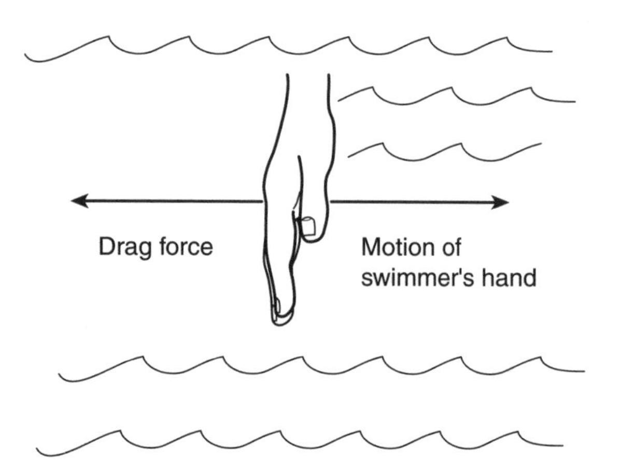

FIGURE 6.18 The motion of the swimmer's hand pushing the water to the right creates a drag force that propels the swimmer to the left.

Adapted from Rushall 1994.

To understand the theory behind the creation of lift as a propulsive force, let's see how it occurs when air flows over an airplane's wing. A cross-section of a plane's wing is called an **airfoil.** Airfoils come in all shapes and sizes. Some are fat, some are thin, some are symmetrical, and many are built with an undersurface that is predominantly straight and an upper surface that is curved. To generate lift, an airfoil is tilted (i.e., angled) relative to the flow of the air passing over it. This angle of tilt is called the angle of attack, with angles of up to 15 degrees considered most advantageous for generating lift. Research has shown that air passing over the upper surface of an airfoil increases in velocity as it flows from the leading edge to the trailing edge of the airfoil. In addition, the airflow is deflected downward by the inclination of the upper surface of the airfoil. On the undersurface, the air reduces its velocity and is deflected downward as it follows the lower surface of the airfoil.

Thanks to the work of Daniel Bernoulli, a Swiss mathematician, we know that a decrease in the velocity of a fluid *increases* the pressure exerted by that fluid and conversely an increase in the velocity of a fluid *decreases* fluid pressure. Applied to the airfoil, this means that air pressure below the airfoil is increased and air pressure above the airfoil is decreased (see figure 6.19). This pressure gradient, acting from below to above, helps produce the force of lift. Lift is further enhanced by a reaction to the motion of the air as it is directed downward from both the upper and lower surfaces of the airfoil. If you direct air molecules downward, there is an equal and opposite reaction in the opposing direction. So **Bernoulli's principle** and Newton's laws work together to apply a lift force to the airfoil (see figure 6.20).

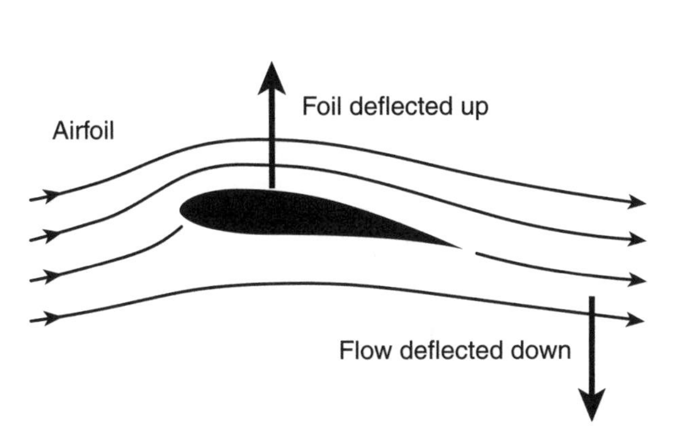

FIGURE 6.19 Lift generated by an airfoil (i.e., a cross-section of an airplane's wing). The pressure above the airfoil is less than that below. Airflow both above and below the airfoil is directed downward, contributing to lift.

FIGURE 6.20 Increasing the angle of attack (i.e., raising the leading edge of the wing relative to the flow of air) up to 15 degrees increases lift.

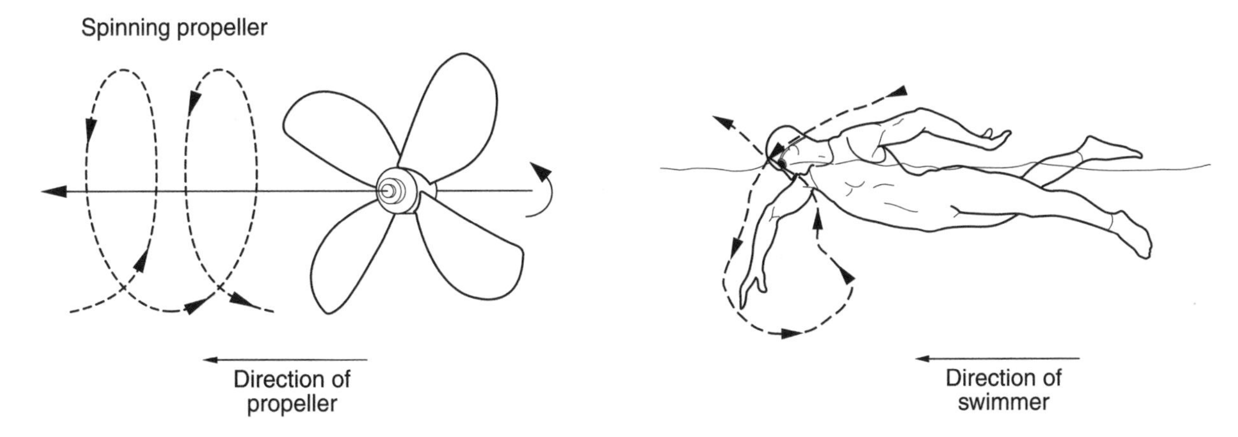

FIGURE 6.21 Mimicking the motion of propeller blades to develop a propulsive lift force.
Adapted from Rushall 1994.

Advocates of lift as a propulsive force in swimming noticed the tendency of swimmers to use a weaving S-shaped pull and push with the hand and arms. This action occurred in all strokes but particularly in freestyle and butterfly. It was thought that during these weaving down-in-up-out motions, the swimmer was using the hands as airfoils and that water flowing over the upper- and undersurfaces of the hands produced a lift force in the same way air does when flowing over and under an airfoil. By varying the angle of the hand during the pull and push of the stroke, the athlete achieved the optimal angle of attack in the same way as an airfoil moving through the air. These hand movements were described as "blading" motions because it was thought that swimmers were using their hands to generate lift like the airfoil-shaped blades of a boat's propeller do (see figure 6.21).

Modern research has shown that although lift acts as a propulsive force during certain phases of a swimmer's hand and arm motions, it does not contribute as much to propulsion as drag propulsion does. Lateral and vertical in-out up-down hand actions still occur as a result of other motions the swimmer makes—such as arm recovery at the end of the stroke and rolling the head and body for breathing. In strokes like freestyle, backstroke, and butterfly, coaches no longer emphasize excessive lateral and vertical hand movements with the idea of generating lift forces. Instead the trend is toward a long straight-line pull-push action. Drag propulsion is again seen as the dominant propulsive force.

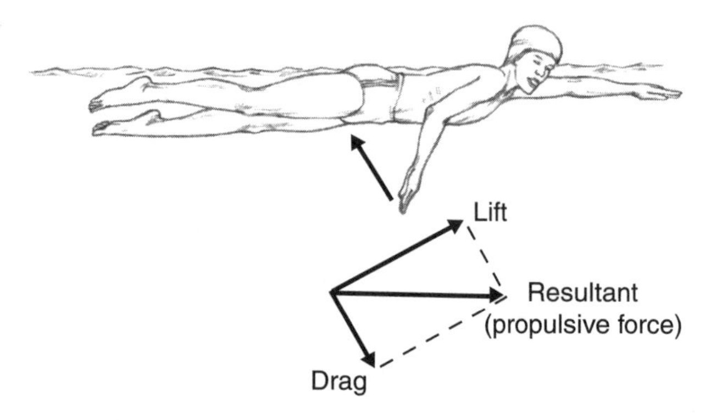

FIGURE 6.22 Drag can combine with lift to produce a resultant force that aids in propulsion during certain phases of hand and arm motion.

Depending on the angle of the swimmer's hand (and other propulsive surfaces like the forearm), drag's powerful propulsive force combines with lift to produce a resultant force that aids in propulsion (see figure 6.22).

Videos of today's elite swimmers show that the majority use a predominantly straight-line pull-push pattern, particularly in freestyle, backstroke, and butterfly. In the breaststroke, drag forces are also considered the dominant propulsive force even though this stroke demonstrates the greatest amount of lateral hand and arm motion. A summary of modern trends in swimming as they refer to freestyle follows:

- Work to improve the efficiency of propulsion and to reduce drag forces to a minimum. Keep your body stretched out and in a horizontal position to minimize form drag, frontal resistance, and wave drag.
- Gain a feel for the water by swimming with long, relaxed, rhythmic actions. Train to achieve long-range strokes with each arm.
- Avoid a high-cadence "water-pounding" arm action. It forces you to shorten your stroke and you will run out of energy. Research indicates that energy consumption in water increases according to the cube of the stroke rate. In other words, doubling the cadence of your arm action through the water demands eight times (2 × 2 × 2) the energy.
- Avoid over-emphasizing lateral (out to the side) and vertical (deep extended motions) with the hand and arm in the propulsive phase of the stroke.
- When the arm is pulling under the water, flex at the elbow. Remember that the forearm is very important as a propulsive surface and that the hand is a mobile extension of the forearm.
- Pull and push against the water for the longest possible time period along a line that is predominantly parallel to the direction of swim. For constant speed, make sure that as one arm pulls, the other is recovering in a fluid, cyclic manner.
- Keep your head down because this will raise your legs into an efficient drag-reducing position. Roll with each stroke so that you continuously present the narrowest profile to the water.
- When your hand enters the water ahead of your body, "catch" the water and try to use your momentum to move forward, past the hand position where you have "caught" the water.
- Eliminate jerky up-down, side-to-side motions that generate the massive resistance of wave drag. Water "rewards" smooth fluid actions and punishes those swimmers who feel that they can smash their way through the water from one end of the pool to the other.
- Consider shaving your skin where it comes in contact with the water and wearing a full-length bodysuit, especially if your leg kick is weak and not doing its job in maintaining a horizontal position in the water. A full bodysuit (with no wrinkles) increases buoyancy and helps hold your body in an efficient drag-reducing position.
- For an illustrated sequence of the freestyle stroke, see figure 9.2 on page 182.

Vortex Propulsion

A **vortex** is a mass of swirling fluids or gases. Vortexes (also called vortices) occur at the wingtips and trailing edges of airplane wings, and they also occur spiraling away from paddles, oar blades, the hulls of boats, and the bodies of swimmers moving through water. In the butterfly stroke a swimmer simulates the undulating

Bodysuits Reduce Drag Forces

In the 2000 Olympic Games in Sydney, athletes in swimming, track and field, and cycling used new bodysuits designed to reduce drag to a minimum. Adidas developed a Teflon-coated full bodysuit and Speedo designed a suit whose surface was designed to mimic the skin of a shark. The objective of the Speedo suit was to allow water to flow more easily over the surface of the material and generate less drag. Speedo claimed that the paneled design of their suit also increased the coordination of the swimmer's muscles and helped promote more efficient movements. The Speedo suit was worn by the "Thorpedo"—super Australian swimmer Ian Thorpe. In track and field, full bodysuits including hoods were supplied by Nike and were designed to reduce air resistance. Adidas has since responded with a hoodless suit for runners and a throwing suit for field events that has no sleeve on the throwing arm. In cycling, the athletes wore slick, form-fitting suits and helmets with a blade shaped tail, all designed to reduce drag.

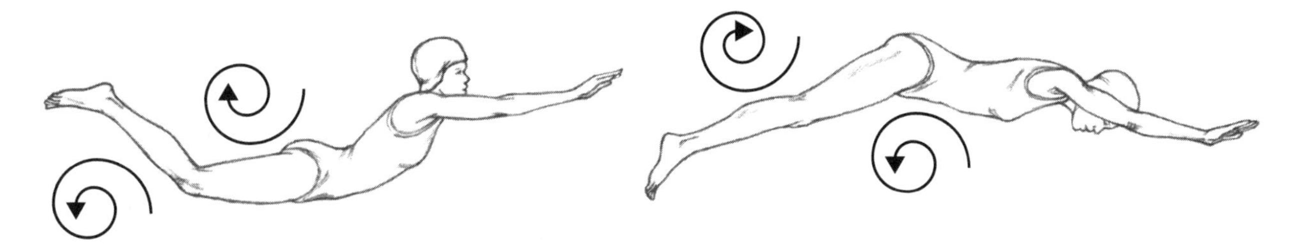

FIGURE 6.23 Vortex propulsion in the butterfly.

tail-lashing motion of a dolphin. Vortexes in the water are produced by this undulating action (see figure 6.23).

Modern research suggests that water that is spun into a vortex by the movement of a swimmer is given rotary inertia and has the tendency to resist lateral movement—instead it rotates in the same spot much like a child's spinning top. The undulating motions of a swimmer, particularly in the butterfly stroke, generate these vortices and a swimmer can push against them in order to move through the water. Each spinning vortex can be viewed as a "post" against which the swimmer is able to apply and exert force. Viewed from the side, the swimmer's actions look like those of a fish as seen from above. The athlete moves through the water, setting up vortices both above and below the body and then thrusts against these vortices. In the 2000 Sydney Olympics, the undulating motions that generated vortices were frequently used at the start of races in strokes other than the butterfly, particularly when the swimmers were temporarily below the surface of the water.

Vortex Generators

In swimming, vortex generators are raised ridges of various shapes and sizes that are positioned across the chest, lower back, and buttocks of full-body and partial-body swimsuits. They reduce both form and surface drag as the swimmer speeds through the water. They are particularly effective when the swimmer is submerged and moving underwater: at the start of a race, after turns at the end of the lap, and during the underwater drive to the wall at the finish. Vortex generators cause spiraling masses of water to flow parallel to the long axis of the swimmer's body. They mix faster moving streams of water that occur outside of the boundary layer with the slower moving boundary layer water that passes in contact with the swimmer. The objective of the mixing action is to delay the breakup of the boundary layer of water flowing over the swimmer's body. In this way vortex generators on swimsuits act like the dimples on a golf ball. In the case of the swimmer, they provide the greatest benefit when the athlete is moving in a streamlined extended body position below the surface of the water.

Magnus Effect

The **Magnus effect** (often called the Magnus force and named after its 1852 discoverer Gustav Magnus) is a lift force of tremendous importance to all athletes who want to bend the flight of a ball. You see the Magnus effect at work in the curved flight path of balls that are thrown, hit, or kicked and at the same time are given a spin. Golfers, baseball pitchers, and soccer, tennis, and table tennis players all employ this effect to curve the flight path of the ball. The game of baseball in particular is made more fascinating by the Magnus effect. The ability of a pitcher to throw curveballs, sliders, screwballs, and knuckleballs that have very little spin—and then have a batter hit these pitches—is the essence of baseball.

The Magnus effect operates in the following manner. As a spinning ball moves through the air, it spins a boundary layer of air that clings to its surface as it travels along. On one side of the ball the boundary layer of air collides with air passing by. The collision causes the air to decelerate, creating a high-pressure area. On the opposing side, the boundary layer is moving in the same direction as the air passing by, so there

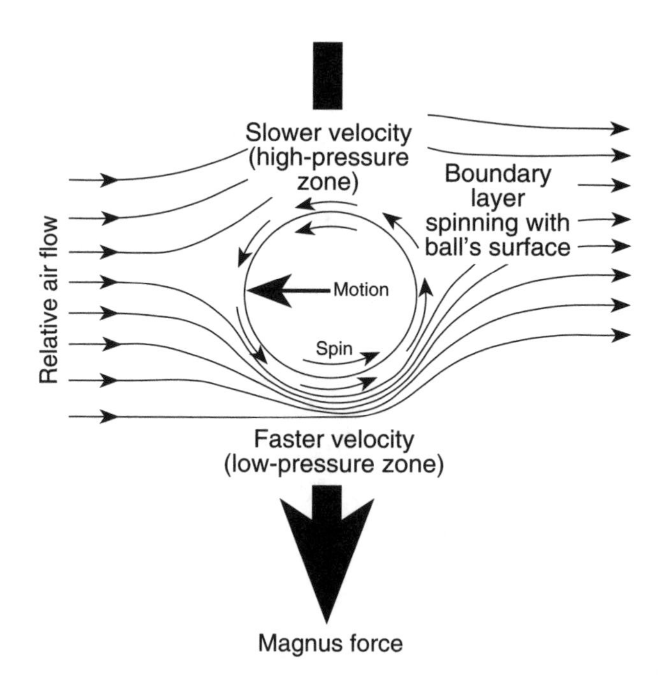

FIGURE 6.24 The Magnus effect.

is no collision and the air collectively moves faster. This sets up a low-pressure area. The pressure differential, high on one side and low on the other, creates a lift force (the **Magnus force**) that causes the ball to move in the direction of the pressure differential (i.e., from high to low) (see figure 6.24).

The Magnus effect can be applied in any direction, and in this way an athlete can create backspin, topspin, and sidespin. Soccer players are well known for the way they use "banana kicks" (i.e., the Magnus effect) to curve free kicks and corner kicks around defenders and

Using the Magnus Effect for Propulsion

In 1922 Anton Flettner, a German engineer searching for a way of making use of the Magnus effect, designed a large two-rotor ship called the *Bruckau*. The ship was originally a three-masted schooner 170 ft long. The masts and sails were removed, and for propulsion it used two vertical rotating cylinders 60 ft tall and 9 ft in diameter. By spinning the huge cylinders with motors in the hull, the ship used the Magnus effect to propel itself forward. It was extremely successful on its trials and subsequently sailed the Atlantic to New York under the name of *Baden-Baden*. As a result of its success, the German navy constructed another rotor ship called the *Barbara*. This ship had three rotors that were spun at 150 rpm by an electric motor. The *Barbara* sailed between Hamburg and Italy for six years.

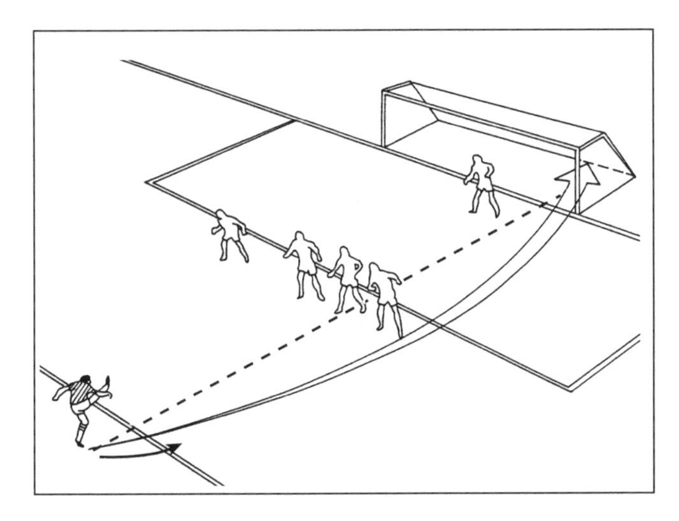

FIGURE 6.25 The Magnus effect in a soccer free kick.
Reprinted, by permission, from, S.J. Hall, 1995, *Basic biomechanics*, 2nd ed. (St. Louis, Mosby-Year Book, Inc.), 487.

into the goal mouth (see figure 6.25). Tennis players and volleyball players use the Magnus effect when they apply topspin to make the ball drop suddenly while in flight. Elite golfers apply Magnus forces to produce draws and fades, and the weekend hacker unwittingly applies spin and uses the Magnus effect to slice the ball off to the left and right.

The Magnus effect can combine with the force of gravity or fight against gravity. A topspin combines with gravity's downward pull, and this is why topspin forehands in tennis (and table tennis) arc viciously over the net and down toward the court. A topspin rotates in the same direction that the ball is traveling, and the spin causes the ball to accelerate once it hits the court surface. A backspin, on the other hand, fights against gravity. The more spin, the more the ball will "hang" in the air. Because the backspin is rotating in the opposing direction that the ball is traveling, the spin causes the ball to slow down and even jump backward once it hits the court surface. Experienced players are able to "read" the spin on the ball from the motion of their opponent's racket.

A golf club like an eight or nine iron is steeply angled to give the ball tremendous backspin. The spin helps the ball fight gravity and gives it terrific lift, plus the possibility of stopping dead or rolling backward after it lands. The raised stitches on a baseball produce the same effect as dimples on a golf ball. The seams grab a thick boundary layer and so a spinning baseball gets plenty of help from the Magnus effect. Pitchers throw curveballs with a powerful snapping action of the wrist, which gives the ball terrific spin. The more spin, the greater the Magnus effect and the greater the curve. Pitchers frequently combine topspin and sidespin so the ball not only drops but also moves laterally across the plate. Spin, gravity, and drag forces all work together to produce this effect (see figure 6.26).

What happens when hardly any spin is put on a baseball? A pitch with a slight spin is called a knuckleball, and in volleyball a serve with hardly any spin is a floater. The characteristics of a knuckleball and a floater can be summed up in one word: unpredictable. In baseball not even the pitcher knows for sure where the ball is going to end up. Both the knuckleball and floater shift and flutter around in flight. It is this erratic movement that is so confusing to the batter in baseball and to the receiver in volleyball. A pitcher gives a knuckleball a lobbing

FIGURE 6.26 A curve ball moves down and laterally across the plate.

or pushing release, so it is traveling slowly and maybe spins one full revolution by the time it reaches the batter. During flight, air flowing past at one instant grabs at the seams and at another instant contacts the smooth surfaces of the ball. The ball may go straight for a while, then suddenly veer to the right or the left and possibly back again. Pitchers have learned that a ball released at a certain speed will first demonstrate a regular flight pattern. Halfway to the plate the ball slows down to a critical level, at which point drag forces build up dramatically and the ball suddenly drops. All this is meant to confuse the batter. But if the pitcher makes the error of throwing the ball with too much spin or too much speed, a knuckleball becomes an easy target for the batter.

It is well known that greasing, cutting, scuffing, or wetting the surface of the ball can produce the strange antics of the knuckleball. A ball that has been treated in this manner will act as a knuckleball but at a faster speed. In a sport already dominated by the pitcher, this strategy gave the pitcher further advantage and as a result has been outlawed.

All the pitches thrown in baseball are affected in one way or another by environmental conditions. As a generalization, dense air helps move the ball around, whereas thin air at high altitudes makes it easy for the ball to go faster. Knuckleballs become less deceptive at high altitudes and as temperature rises. In these conditions, the game belongs to the slugger and the fastball pitcher.

SUMMARY

- An athlete or object moving through a fluid is affected by hydrostatic pressure (exerted by the weight of a fluid), buoyancy (the force opposing gravity that acts on objects partially or totally immersed in a fluid), drag (the force opposing motion through a fluid), and lift (the force acting perpendicularly to motion that deflects an object from its original pathway).
- The pressure that the atmosphere exerts on the earth's surface is 14.7 psi (pounds per square inch). An increase in altitude decreases this pressure.
- Water is virtually incompressible and as a result, the pressure exerted by water increases with depth. In the ocean, pressure is increased approximately one atmosphere (14.7 psi) for every 33 ft of depth.
- The buoyant force acting on an athlete or an object is equal to the weight of the fluid that the athlete or object displaces when immersed in the fluid.
- The center of buoyancy is the place where the buoyant force concentrates its upward thrust on an object immersed in a fluid. The center of buoyancy is usually positioned higher on an athlete's body than is the athlete's center of gravity. The center of buoyancy of an object or athlete is in the same place as the center of gravity of the displaced water.
- Atmospheric pressure, temperature, and the contents of air or water affect how water and air act. The more dense or viscous a fluid, the greater the frictional forces acting on an athlete or an object passing through that fluid.
- The frictional forces occurring when an object or athlete moves through a fluid are the same as when fluids flow at the same velocity past the object or athlete.
- Laminar flow, which is smooth and regular, occurs around an object when fluid flow is slow. Turbulent flow, which is disturbed and rough, occurs at high velocities. Turbulent flow generates more drag than laminar flow.
- The three types of drag are surface drag, form drag, and wave drag.
- Surface drag is also called skin friction or viscous drag. The amount of surface drag is determined by the relative motion of object and fluid, the area of surface exposed to

the flow, the roughness of the object's surface, and the fluid's viscosity. Surface drag increases according to the square of the velocity.

- Form drag is also called shape drag or pressure drag. The amount of form drag is determined by the relative motion of object and fluid, the pressure differential between the leading and trailing edges of the object, and the amount of surface acting at right angles to the flow. Form drag increases according to the square of the velocity.
- Wave drag occurs at the interface between water and air. The amount of wave drag is determined by the relative velocity at which the object and wave meet, the surface area of the object acting at right angles to the wave, and the fluid's viscosity. Wave drag increases according to the cube of the velocity.
- At high velocities turbulent flow produces a low-pressure wake acting to the rear of an object or an athlete. This low-pressure area is used in sport for drafting, or slipstreaming.
- Balls moving through the air are very much affected by form and surface drag. The lower pressure wake to the rear of the ball can be reduced by increasing the surface drag: An increase in surface drag can help decrease the ball's form drag. Dimples on a golf ball increase its surface drag.
- Athletes and objects are affected by lift forces that depend on the relative motion of the object and the fluid, the angle of the object relative to the flow of the fluid, the size of the surface area angled into the fluid flow, and the nature (e.g., density) of the fluid.
- Swimmers angle their hands and feet to create lift, which can act as a propulsive force. Modern research shows that the greatest propulsive force for a swimmer comes from pulling and pushing back against the water as long as possible in a direction parallel to the long axis of the swimmer's body. This propulsive force is called drag propulsion.
- Vortexes (or vortices) are masses of swirling water produced by undulating lashing motions of a fish's body and tail. Athletes mimic this action particularly in the dolphin kick of the butterfly stroke. Vortexes generated in this manner are considered to reduce drag and assist in propulsion.
- Vortex generators are a series of raised ridges placed across the surface of full-body and partial-body swimsuits. They are intended to reduce the form and surface drag of the swimmer particularly when the athlete is moving below the surface of the water.
- A spinning object (e.g., a ball) traveling through the air builds up high pressure on the side spinning into the airflow. Low pressure occurs on the side spinning in the same direction as the airflow. The ball is deflected from high pressure to low pressure. This phenomenon is called the Magnus effect.

KEY TERMS

airfoil
angle of attack
Archimedes' principle
Bernoulli's principle
boundary layer
buoyancy
buoyant force
center of buoyancy
drag
form drag
hydrostatic pressure
laminar flow
lift
Magnus effect
Magnus force
relative motion
surface drag
turbulent flow
viscosity
volume
vortex
wake
wave drag

REVIEW QUESTIONS

1. A scuba diver mistakenly holds his breath on the way to the surface from 66 ft. How many lungfuls of air will a single lungful of compressed air inhaled from his scuba tank at 66 ft become when the diver reaches the surface?
2. A muscular, lean gymnast weighing 100 lb and an obese sumo wrestler weighing 400 lb jump into a swimming pool. Why is the obese sumo wrestler more likely to float?
3. The world record for pedaling a bike behind a pace vehicle is held by Fred Rompelberg of the Netherlands at more than 166 mph. Why was it unnecessary for Fred and his bike to be streamlined?
4. What factors enable recumbent cyclists covered in aerodynamic shells to cycle faster than the world's best sprint cyclists?
5. What reasons were given in the chapter for propulsion in swimming being generated predominantly from drag propulsion rather than lift forces?
6. How does the Magnus effect deflect a spinning ball?

PRACTICAL ACTIVITIES

1. **Atmospheric pressure.** Take a round plastic juice container (1 qt or 1.87 L). Fill it a quarter full with hot water. Swish the water around inside the juice container. Pour out the hot water and immediately screw the cap tightly back onto the container. Stand back and wait. The plastic container will collapse; in fact it will frequently be crushed with a loud bang. Explain what has occurred and include references to atmospheric pressure and its pressure per square inch on the surface of the plastic container.
2. **Lift caused by a rotating cylinder.** Remove all the internal components from a ballpoint pen. Hold the outer cylinder against the edge of a table. Place your thumbs on top of the cylinder, one at one end and one at the other. Snap your thumbs down toward the table so that you give the cylinder a strong rotation. It will fly upward in the air and may even "loop the loop." Explain what occurred by referring to Bernoulli's principle, and draw the cylinder with the airflow passing around it giving it lift.
3. **Buoyancy.** In chest-deep water in the swimming pool and working with a partner, perform the following actions: Inhale as much air as possible, hold your breath, and then reach down with extended arms to simulate a toe-touch position. Do you float at the surface with your back bobbing out of the water or do you sink? Second, if you float, then first exhale and repeat the same body position. Now do you sink? Have your partner observe your actions, then switch roles and observe your partner. Do either of you sink even with full lungs? Finally, do either of you float even after exhaling? In your explanation, show that you understand how the buoyant force operates.
4. **Form drag and propulsion.** Using the freestyle stroke, first swim with your head held out of the water, then swim with your hands as fists. Next, using regular freestyle arm action, swim using no leg action at all. What are the results? Write an explanation. Finally, swim freestyle with arms only using first one pull buoy held between your legs and then two pull buoys. How does the different number of pull buoys affect your leg position in the water and your forward velocity? Explain in terms of form drag.

PART II

Putting Your Knowledge of Sport Mechanics to Work

7

Analyzing Sport Skills

When you finish reading this chapter, you should be able to explain

- how to determine skill objectives,
- how knowing the special characteristics of a skill can help you analyze athletic performance,
- what you gain from an analysis of the performances of elite athletes,
- how to divide a skill into phases and key elements, and
- how to use your knowledge of mechanics in the analysis of a skill and the correction of errors.

Chapters 7 and 8 are closely linked and will show how to put your knowledge of sport mechanics to work. Here you'll find advice on how to break a skill into smaller parts. This process will make it easier when you critically observe your athlete's performance. Chapter 8 gives examples of observation techniques and teaches you how to select errors that need correcting.

One of the greatest challenges you'll face as a coach is watching your athlete perform and deciding which aspect of the skill needs correction. If you don't have a well-planned approach, you're likely to be overwhelmed by the complexity and speed of the skill you are trying to analyze. You won't know what aspect of the skill to look at or what error to correct first. In fact, you may see so many errors at once that you throw your hands up in the air and in desperation give vague coaching tips such as "Hit harder" or "Be more aggressive!" Advice like this is of little assistance to your athlete. What you need to do is gather background information about the skill before you start, and come to each coaching session with a precise plan to guide your observation, your analysis, and your correction of errors. If you understand the mechanics of the skill your athlete is performing and you know how to go after major errors, your athlete benefits immensely and quickly improves in performance.

The following steps provide the information you need before you start correcting errors:

Step 1: Determine the objectives of the skill.

Step 2: Note any special characteristics of the skill.

Step 3: Study top-flight performances of the skill.

Step 4: Divide the skill into phases.

Step 5: Divide each phase into key elements.

Step 6: Understand the mechanical reasons each key element is performed as it is.

If you work your way through each step, you'll learn how to break a skill into important parts (or phases), and you'll know how to use your knowledge of sport mechanics when you analyze each phase. You'll find out how much easier it is to analyze each phase of a skill separately rather than concentrate on the total skill and then try to recollect what happened.

Don't think that you must go through each step *every time* you teach a skill. Once you have read this chapter, you'll understand what information you need, and with a little practice, you'll be able to carry out most of the steps in your head. To begin with, however, write down on a clipboard the information that's required. Then take this material with you and use it as a guide during your coaching sessions.

Step 1: Determine the Objectives of the Skill

The rules of the sport and the conditions that exist when a sport skill is performed determine **skill objectives.** Most skills have more than one objective. It's good to be aware of these objectives because they determine the technique and mechanics that your athlete must use to perform the skill successfully. Let's look at some sport skills to see what we mean by skill objectives.

The dominant objective for an athlete competing in the discus event is to throw the implement as far as possible. The farther the discus travels, the better. However, the discus must land within a sector, so accuracy of flight is an important objective as well. The distance thrown is not counted if the discus lands outside the sector lines. In addition, if the thrower loses balance and falls out of the ring, the throw is declared invalid even if the discus lands within the sector lines.

The objectives of distance and accuracy determine what mechanical principles to keep in mind when you coach your athlete in the discus. The overriding importance of distance tells you that the dominant mechanical objective in the event is maximum velocity at release. This means you should concentrate on teaching your athlete how to make the discus leave the throwing hand as fast as possible.

How the discus leaves an athlete's hand and how it spins determine its flight characteristics and its distance. So you cannot forget that an optimal spin and trajectory are important objectives too. Remember as well that the body positions your athlete uses during the throw influence the distance and flight of the discus and the athlete's stability after the discus is released. It would be heartbreaking if your athlete threw a world record distance only to have it declared a foul because he fell out of the front of the ring or stepped on the rim of the ring during the throw.

In a volleyball spike, your athlete has to jump high enough to strike the ball over, around, or off the blockers. The prime objective of a spike is to make the ball hit the floor in the opponent's court (see figure 7.1). To achieve this objective, jumping ability and timing are tremendously important and so is accuracy in directing the ball.

FIGURE 7.1 The objectives of a volleyball spike are (1) to either land the ball in the opponent's court or deflect it off an opponent's block so it goes out of bounds, (2) to avoid contacting the net, and (3) to avoid landing with one or both feet in the opponent's court.

In addition, your athlete must take care not to contact the net. Keep these objectives in mind when you coach spiking skills. Work on the mechanics of the approach, the jump, and the spiking action and then on control of the body after the ball has left your athlete's hand.

Compare the objectives of the volleyball spike with those required of a high jumper. Height is obviously a prime objective in high jump just as it is in a volleyball spike. However, a high jumper is also required to cross a bar—an objective not required of a volleyball player. So a high jumper needs to jump both vertically and horizontally, then rotate in the air to get into a good bar clearance position. It's no use producing great height if the athlete knocks the bar off on the way up or on the way down.

In Olympic weightlifting, the prime objective of both the clean and jerk and the snatch is to hoist a barbell to arm's length above the head. A secondary objective is to demonstrate control over the barbell once it's in this position. This second criterion is necessary for the judges to pass the lift. Even though the barbell must be held steady for a relatively short time, control and stability are important objectives that must be taught for your athlete to achieve success in this skill.

Whatever sport you coach, whether it is an individual or team sport, be aware of *all* the objectives required of each skill. If you coach to satisfy one objective and forget or deemphasize another, you'll limit the success of your athlete. What use is it if a water polo player learns to fire the ball at phenomenal velocity if no emphasis is placed on controlling and directing the path of the ball? Similarly, what use is it if you teach a diver how to get great height and spin if the entry into the water is a disaster? So be aware of all the objectives required by a skill, and remember that all these objectives play a part in determining the technique that you teach your athlete.

Step 2: Note Any Special Characteristics of the Skill

Sport skills can be divided into different types based on the manner in which your athlete performs the skill and the conditions under which your athlete performs the skill. Both manner and conditions are interrelated, and both dramatically influence the methods you use when you coach. For example, if you consider the manner in which skills are performed, you'll see that some skills are performed once, then a totally different action occurs next. Other skills are different because they repeat cyclically (i.e., over and over). These two types can be called nonrepetitive and repetitive skills.

The conditions under which athletes perform skills also differ considerably. Some conditions are controlled and predictable. You know how the conditions will be before the competition starts. Other conditions vary considerably and are unpredictable, and it's difficult to know how they'll be when the competition begins. Let's first look at nonrepetitive and repetitive skills and then at predictable and unpredictable conditions.

Nonrepetitive Skills

Nonrepetitive skills are often called discrete skills in that they have a definite beginning and an end—even though they can be repeated more than once in a sporting situation. Examples include a tower dive, a shot put, or a baseball bunt. Skills such as these do not repeat in a cyclic pattern. Instead some other action occurs immediately afterward. If your athlete is a diver, she'll land in the pool, climb out, and wait for her turn in the next round of dives. A similar situation occurs for the shot-putter, who after throwing must wait for other competitors to complete their throws before he can perform again. The baseball player follows a bunt with a totally different action. In most cases it's a sprint to first base.

You can easily teach nonrepetitive skills as separate entities. Once the athlete has learned and mastered the skill, add some other skill or action to lead into it or to lead out of it, similar

How Do Elite Athletes Compare?

Although our best athletes produce great performances, they don't compare with animals and insects! In a sprint, a cheetah will finish the 100 meters with an Olympic sprint champion still accelerating at the 30- to 40-m mark. Kangaroos jump 9 ft vertically with ease and impalas have no trouble leaping 40 ft horizontally. Fleas jump over 150 times their own length, vertically or horizontally, and to match this performance, athletes would have to jump close to 1,000 ft! Relative to body weight, the best Olympic weightlifters in the world are easily outmatched by ants, which carry in excess of 50 times their own body weight. In water, the top speed of around 8 mph by our best swimmers is 5 times slower than the bluefin tuna, which has hit speeds of 45 knots (51.75 mph)!

to the baseball player bunting and sprinting to first base or a gymnast performing a handspring followed by a dive roll.

Nonrepetitive Skills in Sequence

Frequently, the momentum generated in one nonrepetitive skill will carry over and assist in beginning another nonrepetitive skill. A young gymnast builds a floor exercise routine in this manner. A front handspring may join a front somersault, and the somersault leads into another skill. Similarly, a triple jumper hops, steps, and finally jumps. The three jumps differ, yet the skill of triple jumping depends on the synchronization of all three skills. For an excellent distance, the hop must contribute to the step, and the step to the jump (see figure 7.2).

When you coach nonrepetitive, or discrete, skills in sequence, it's a good idea to teach each skill separately. Then teach your athlete to adapt to the rhythm pattern and changes that occur when two or three skills are performed in sequence. Be aware that two or three skills in sequence present additional difficulties for your athlete. Novice triple jumpers frequently perform an immense hop only to collapse at the end of it and have nothing left for the step or the jump. There is no balanced effort or flow from the hop to the step and finally to the jump.

In gymnastics, a young athlete can learn to perform a back somersault by itself. Then you can teach her to perform a round-off that leads into the back somersault. If correctly performed, the round-off makes the performance of a back somersault easier. Performed poorly, the round-off positions the gymnast incorrectly for the takeoff into the back somersault. This makes it difficult for the gymnast to get around and safely complete the somersault.

Repetitive Skills

Repetitive skills have a cyclic, continuous nature. For example, the actions that make up the movement pattern of sprinting repeat continuously during the race. This repetitive, continuous feature occurs in many sports such as race walking, cycling, swimming, speedskating, and cross-country skiing.

The most important aspect of repetitive skills is that one complete cycle of the skill immediately leads into the next. This means that a follow-through (which slows down and dissipates energy in a nonrepetitive skill) becomes a recovery in a repetitive skill and is essential for maintaining continuity and rhythm.

In competitive swimming, athletes aim for a fast arm recovery when they perform their strokes. The arms complete their pull in the water, then quickly cycle forward into the next propulsive action. There is no braking action or dissipation of energy as occurs in the follow-through of a discus or javelin throw. Like a cyclist who wishes to keep the pedals spinning at a high rate, a competitive swimmer wishes to do the same thing with the arms after each arm pull (see figure 7.3).

Repetitive skills are frequently taught to young athletes in much the same way as nonrepetitive skills. The freestyle stroke is broken down into leg action, arm action, and breathing. These components of the stroke are taught separately and then molded together to build the complete skill. The number of repeats, or cycles, of the total

FIGURE 7.2 The triple jump is an example of three discrete skills in sequence: the hop, the step, and the jump.

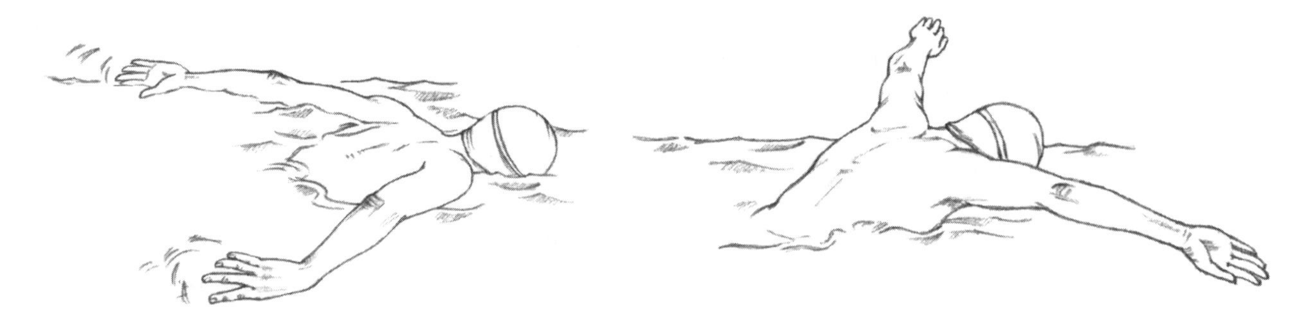

FIGURE 7.3 Swimming strokes are examples of cyclic, repetitive skills in which the recovery of the arms and legs leads into the next propulsive phase.

skill is progressively increased as the athlete's ability improves.

Skills Performed in Predictable Environments

Many skills are performed in a precise and predictable environment. These types of skills are frequently described as **closed skills.** In this situation, your athlete can get on with the job of performing the skill without having to make quick decisions because of a sudden change in conditions. A clean and jerk in weightlifting and the skills in a synchronized swimming routine are examples. The fact that your athlete can concentrate on the lift or on the skills in the routine without worrying about the actions of opposing players or changes in weather conditions makes practice sessions easier for you to plan and training easier for your athlete.

Skills Performed in Unpredictable Environments

Many sport skills are performed in an unpredictable environment. These skills are often described as **open skills.** The most frequent cause of an unpredictable environment is the presence of opposition whose prime purpose is to make your athlete fail in whatever he is trying to do. Consequently, your athlete must respond according to the conditions that occur in any instant during the competition. In baseball, your batter responds (in less than half a second!) to whatever pitch is thrown. Your volleyball player responds according to the serve that comes over the net. Her response is going to be different for a floater serve than for a fast topspin spike serve. In freestyle wrestling and judo, your athlete attacks or defends according to the maneuvers of the opponent. In soccer, a goalkeeper reacts according to the maneuvers and shot fired by an attacking player.

Wind, waves, rain, sun, and varying field and court conditions can also cause uncertainty and unpredictability. A surfer must assess the nature of the wave and perform surfing skills accordingly. Each wave needs to be considered individually when it occurs, and the surfer must develop an ability to cope with these conditions. The variability that exists in all the sports we've mentioned, from baseball to wrestling, forces your athlete to make sudden decisions and to perform skills at varying velocities. The ability to judge the situation and to react quickly is obviously an important element of success.

When you coach open skills, which are performed in unpredictable conditions, begin by making the situation as predictable as possible. For example, wrestlers work repeatedly on the same defensive maneuver against an opponent who is required to repeat the attacking move. In baseball and tennis, players face balls fired repeatedly and predictably from pitching and serving machines, and in rugby, football, and field hockey, athletes practice set plays without opposition. Then other team members work as opposition and the same plays are repeated. In this way the mechanics of a particular skill are practiced in a predictable situation until the quality of the skill performance is good. Then more unpredictability is introduced.

How soon unpredictability is introduced depends on many factors, one of the most im-

Perseverance

Most people remember world and Olympic champion Greg Louganis as an athlete who continuously came up with flawless diving performances. What people tend to forget is that Greg had off days just like any other athlete. As the following list indicates, when divers perform poorly, they get physically punished besides receiving a poor score and being beaten in the competition! A great athlete has the strength of character to work through and learn from these experiences.

1976—Greg Louganis hits the platform and receives two black eyes and a bloody nose

1979—Greg hits the platform and is knocked unconscious

1981—Greg hits the bottom of the pool and breaks a collar bone

1984—Greg lands on his back from a tower dive

1988—Greg hits the board with his head and requires four stitches.

Adapted from Wallechinsky, 1991. *The Complete Book of the Olympics: 1992 Edition.*

portant being how fast the athlete learns the required skill. Many coaches like to move quickly to unpredictable situations. Others mix it up so that in some drills the athlete learns rapidly how to judge what should be done, and in other drills the athlete works on a particular skill repeatedly under predictable conditions.

Step 3: Study Top-Flight Performances of the Skill

Watching elite athletes perform is an activity that you and your young athletes can do at any time. It doesn't necessarily have to be step 3 in the sequence that has been offered in this chapter. But it's certainly worthwhile to watch the best perform the skill or event that you are coaching. For example, when you watch top-class athletes perform a skill, you get a picture of the speed, rhythm, power, body positions, and other characteristics that make up a quality performance. This helps you understand the basic movement patterns in the technique of the skill you intend to coach. Use a video camera to tape these performances from various angles. Then you can watch the skill repeatedly at normal speed and in slow motion. You'll soon notice that, in spite of differences in body type, the techniques top athletes use all show common features. Elite golfers shift their body weight and rotate their hips in much the same way. Great throwers in track and field use similar throwing positions and activate their muscles in a similar sequence. Top-class divers use a similar hurdle step, and they drive up off the springboard with similar arm and leg actions. These identical features exist because top-class athletes use good mechanics. Their coaches taught them to use actions in their performances that produce the optimal force, velocity, spin, and so forth required by the skill.

As you progress through steps 4, 5, and 6 in this chapter, you'll get used to associating mechanical principles with technique. You'll start using your knowledge of mechanics when you look at an elite performance, so that you can say to yourself, "I understand the mechanical reasons these champion athletes shift their weight and rotate their hips when they drive a golf ball, and I understand why their arms are extended when the club head contacts the ball." You'll realize that these technical features are necessary actions that must be taught to all young golfers irrespective of their shape, size, and build. The same principles apply to the skills of any sport. Elite performers use good technique based on sound mechanics and so provide you with a model on which to base your coaching.

Step 4: Divide the Skill Into Phases

Your next task is to divide the skill you're interested in coaching into phases. This process is important because it makes your job much easier when you look for errors in your athlete's performance. Quite simply, it stops you from becoming confused by trying to watch too much of the skill at the same time.

Most skills consist of several phases. A **phase** is a connected group of movements that appear to stand on their own and that your athlete joins together in the performance of the total skill. Many skills, for example, can be broken down into the following four phases:

1. Preparatory movements (setup) and mental set
2. Windup (also called backswing)
3. Force-producing movements
4. Follow-through (or recovery)

If you look at a golf swing, a hockey slapshot, or a baseball pitch, preparatory movements and mental set make up the first phase in the skill, however brief they might be. The second phase consists of the windup (or backswing) and is followed by the third phase, which includes the force-producing movements. The fourth phase, the follow-through, completes the skill (see figure 7.4). Each phase, starting from the preparatory movements and mental set, leads into and influences the next phase in line like a chain reaction. This common characteristic tells you that errors occurring during an early phase of a skill are bound to affect all the phases that follow. So when something goes wrong at the end of a skill, examine not only the last phase but also earlier phases to see if the root of the problem lies there. For example, if a golfer makes an error in setting up and addressing the ball or if he performs the backswing incorrectly, the effect of the error carries into the remaining parts of the drive and, of course, into the flight of the ball. Don't be deceived by thinking that all errors stem from the phase in which they occur. Check out earlier phases—the problem often lies there!

Let's look at each phase individually to see what specific contributions they make toward the performance of the total skill.

Preparatory Movements (Setup) and Mental Set

Preparatory movements and mental set include the motions and mental processes that your athlete goes through when setting up and getting ready to perform. A golfer takes up a stance and addresses the ball. A tennis player gets herself ready to serve and mentally decides where to direct the ball. An offensive lineman will crouch with his muscles in a static-stretch position. When the ball is snapped, his muscles respond with an explosive thrusting motion that immediately leads into the next phase of the skill.

Cyclic, repetitive skills may require preparatory movements at the start of the skill, after

FIGURE 7.4 Phases of a golf drive are the *(a)* preparatory movements and mental set, *(b)* backswing, *(c and d)* force-producing movements, and *(e and f)* follow-through.

which they normally don't occur. For example, a butterfly swimmer doesn't establish a static stance before each propulsive action. She flows immediately from each arm pull and leg beat into the next.

Windup (or Backswing)

Many skills use a **windup**, or backswing, in preparation for the movements to follow. Whatever name is given to this phase, the objective remains the same—to stretch the athlete's muscles and establish a position from which she can apply force over an optimal distance or time frame. Examples include the rotary windup of a discus thrower, the backswing in golf and baseball, and the backward extension of a javelin thrower's arm. In a tennis serve and a volleyball spike, the dropping back of the hitting arm to the rear of an athlete's body fulfills a similar purpose to that of a thrower. In kayaking, the forward reach of the paddler before thrusting the blade in the water acts as a windup.

Force-Producing Movements

Force-producing movements are the specific actions that your athlete uses to generate force. They usually involve the athlete's whole body and may include an approach, but in finer, more discrete actions (such as archery or throwing a dart), they may require use of only the arm and shoulder muscles and minimally involve the muscles of the rest of the body.

Force-producing actions are tremendously important for creating the desired effect of a skill. Your athlete's muscles need to apply force in the correct amount, over the correct range and time period, and in the correct sequence. You'll find that force-producing actions come in many types. They include such sequential actions as the approach, pull, and push of the pole-vaulter; the body extension and arm flexion of the rower; the rotating spins and throwing actions of the hammer thrower and discus thrower; and the approach, takeoff, and arm actions involved in a basketball layup. In contrast, in a power lifter's deadlift, the force-producing actions occur almost simultaneously, with the athlete's leg, back, arm, and shoulder muscles pulling at the same time.

In all skills an important and critical instant in time occurs at the end of the force-producing movements. It happens when a baseball is struck, a takeoff occurs, or an implement is released. At this instant, the athlete has applied the optimal amount of force and set its direction. At this point, there is nothing more that the athlete can do to upgrade the skill.

Follow-Through (or Recovery)

Follow-through and recovery actions occur immediately after the force-producing motions are complete. In throwing skills, the implement has been released, and in hitting skills, the impact has been made. In many skills it is impossible and even dangerous for an athlete to come to a complete stop immediately after completing the force-producing actions. The momentum generated causes your athlete's limbs to continue along their original pathway. The **follow-through** acts to safely dissipate the force of these actions.

In a swimming stroke, a skill in which the movement pattern is repeated in a continuous and cyclic fashion, the recovery of the arms leads quickly to the next repetition of the arm pull. In these repetitious skills, momentum and rhythm are an essential part of the cadence of the complete skill. The recovery actions help maintain balance and continuity of motion. In addition to the swimming stroke, other examples include the leg and arm recovery in sprinting, speedskating, and cross-country skiing.

Step 5: Divide Each Phase Into Key Elements

When you have chosen the most important phases of a skill, direct your attention toward the task of dividing each phase into its key elements. Key elements are distinct actions that join to make up a phase. Try to view a skill as a building that you are erecting. Phases are the walls of your building, and the key elements are the bricks you use to make each wall.

How do you choose key elements? Identify the distinct actions that are essential to the success of each phase in the skill (the same way you identify phases that are essential to the success of the skill as a whole). A windup phase will have its key elements, as will the force-producing phase and the follow-through.

The following examples will give you an idea of what key elements are, although we haven't listed every key element in the phases we've chosen. You'll see these key elements in the techniques used by all top-flight athletes. Why? Because they are essential for good technique and contribute mechanically toward the success of the skill. Without them, your athlete could not produce an optimal performance.

- In the force-producing phase of a golf drive, your athlete shifts his body weight to the rear foot and from the rear to the forward foot. He rotates his hips into the drive and has extended arms when the club contacts the ball. (See figure 7.4.) *Key elements:* weight shift, hip rotation, head position, arm extension.
- In a high jump approach, your athlete leans into the curved path of her approach, which is part of the force-producing phase of a high jump. At the completion of the approach she leans back and lowers her center of gravity when stepping into the takeoff position. Her arms are positioned to the rear of her body in preparation for swinging forward and upward at takeoff. (See figure 9.3 on page 185.) *Key elements:* backward lean, lowering of the center of gravity, arms to the rear of the body.
- In the force-producing phase of a javelin throw, your athlete makes his approach, leans back, and steps forward into a wide throwing position. He then rotates his hips and chest toward the direction of throw. Simultaneously, the athlete shifts his body weight from the rear to the forward leg. (See figure 9.4 on page 187.) *Key elements:* approach, backward lean, wide throwing stance, hip and chest rotation, weight shift.
- In a football punt, after stepping forward with the supporting foot, the athlete swings the kicking leg through a long arc. The kicking leg, which starts partially flexed, is fully extended on contact with the ball. The athlete simultaneously shifts his body weight forward and upward into the punt. His arms, which fed the ball onto the kicking foot, are extended sideways to maintain balance. (See figure 9.6 on page 191.) *Key elements:* extended base, weight shift, long kicking arc, leg extension, arm extension.

Remember that there are more key elements present in each of these skills, and the sequence in which these elements are performed is an important factor in itself. In some phases of a skill, key elements are performed almost simultaneously. In other situations, there is a definite flow from one to the next. With practice and careful observation of elite performances, you will be able to pick out *all* the key elements for each phase of a skill and understand the timing of their performance. Your next job is to understand the mechanical reasons key elements exist and what purpose they serve. This is the final step.

Sport Mechanics Helps Bring About Olympic Gold Medals

Valeri Borzov from the former Soviet Union won the 100 m and 200 m in the 1972 Olympic Games in Munich, Germany. Commenting on Borzov's preparation, his coach Valentin Petrovsky said, "We began with a search for the most up-to-date models of sprinting. We studied slow motion films of leading world-class sprinters both past and present. We figured out the best angle of thrust and body position and went into a whole number of minor details. When the mathematical equivalent of a runner was worked out and given a scientific basis, we began testing our calculations in practice. It was subtle work, which could be compared to the training of a ballerina."

Adapted from D. Wallechinsky, 1991, *The Complete Book of the Olympics: 1992 Edition*, p. 11.

Step 6: Understand the Mechanical Reasons Each Key Element Is Performed As It Is

Understanding the mechanical basis behind each key element is a tremendously important step in your sequence. Chapters 2 through 6 showed you how mechanics form the foundation of all sport techniques. All the fundamental actions an athlete makes in technique are founded on mechanical principles. In other words, technique is based on mechanical laws. So once you've picked out the key elements in the skill you are analyzing, you have to understand the mechanical purposes behind each element. You must be able to answer questions of the following nature with responses like the ones listed here.

Why cock and uncock the wrists during a golf drive?

Cocking and uncocking the wrists during a golf drive causes the golfer's arms and club to simulate the whiplash, or flail-like, action of the high-speed tip segments of a whip. When the wrists are cocked and uncocked, they act as an additional axis around which the club can rotate. The velocity developed from the swing (and length) of the golfer's arms is multiplied along the length of the club shaft. Without the cocking and uncocking action, the arms and club move as a fixed unit. This would not allow the head of the club to reach optimal velocity (see figure 7.4).

Why should a sprinter's legs and arms thrust and swing parallel to the direction of sprint during a 100-m sprint?

If a sprinter's arm swing and leg thrust are in any direction other than parallel to the direction of sprint, the forces that the sprinter applies to the earth in the direction of sprint are reduced. In reaction, the force that the earth applies against the sprinter is lessened as well (see figure 9.1 on page 179). The result is that the sprinter doesn't run as fast as possible.

Why should a freestyle swimmer pull with the hands and forearms along a line parallel to the long axis of the body rather than emphasize an S-shaped "out-in-up-down" pattern of pull?

Emphasizing an S-shaped "out-in-up-down" motion with the hands during the freestyle stroke is now considered to generate less propulsive force than pulling straight back against the water. A modified S-shaped motion still occurs during entry and exit of the swimmer's hand, but these actions occur more from body roll and the anatomy of the swimmer's body than from efforts to generate more propulsion. It is now considered correct technique to pull back against the water as far as possible parallel to the long axis of the body. Under water, the arms flex at the elbows so that the swimmer's hands and forearms provide the major propulsive surfaces (see figure 9.2, page 182).

Why must athletes have their center of gravity positioned behind the jumping foot as they enter a high jump takeoff, or behind both feet as they prepare to jump to block or spike in volleyball?

Positioning the athlete's takeoff foot ahead of her center of gravity gives the athlete more time to apply force with the jumping leg at takeoff. The athlete rocks forward, up, and then over the jumping foot. This large arc of movement gives the athlete time to drive down at the earth (see figure 9.3, page 185). The earth in reaction drives the athlete upward. The same principle applies to a volleyball spike, a volleyball block, a basketball layup, and a basketball block.

Why is it important for athletes to rotate the hips and thrust them ahead of the upper body during a golf drive, shot put, and discus or javelin throw?

Rotating the hips ahead of the upper body and toward the direction of throw serves three purposes:

1. It shifts the athlete's body mass in the proper direction (i.e., toward the direction that the golf club, discus, shot, javelin, or baseball bat will be accelerated). This action extends the distance and time over which the athlete applies force.

2. The rotation of the hips acts as an important link in the sequential acceleration of the athlete's body segments. The movement of the athlete's legs and hips toward the direction of throw (or impact with the ball in golf or baseball) simulates swinging a whip handle ahead of the rest of the whip so the tip of the whip will crack.
3. The rotation of the hips stretches the muscles of the abdomen and chest so that they pull the shoulders and throwing arm in slingshot fashion toward the direction of throw. (Notice the weight shift and hip action in the javelin throw [see figure 9.4 on page 187]) and in the golf drive [see figure 7.4].)

Why should athletes extend their kicking legs when contacting the ball in a football punt?

When the athlete extends his kicking leg, it puts the part of the foot that contacts the ball farther from the kicker's axis of rotation (i.e., the hip joint). Because of this increase in radius, the kicking foot is moving faster than any other part of the leg when it contacts the ball. The flexion of the kicking leg before contact with the ball, together with its extension at impact, simulates a whiplash action (see figure 9.6, page 191).

Why must athletes extend their bodies fully at takeoff in gymnastic and diving skills?

Any time an athlete needs to rotate quickly, she must apply an **eccentric thrust**, or an off-center force, at takeoff to initiate rotation. The athlete must then pull her body inward from a fully extended position. The large reduction in rotary inertia caused by compacting the body mass around the axis of rotation is rewarded by a huge increase in the rate of spin (i.e., angular velocity) (see figure 9.9, page 197).

All phases and all key elements in a skill are performed for specific mechanical purposes. If you know the mechanical reasons they're performed as they are, you can confidently say to yourself, "Okay, I understand what should occur in the technique of this skill, and I understand the mechanical principles behind the movements that the athlete must perform. I'm ready to watch my athlete, and I'm ready to correct any errors that I find."

We have asked you to use elite performances as a model when you coach. Don't make the mistake of trying to mold a young athlete in the exact image of an elite athlete. When you watch a series of elite performances, be sure to study the basic technique that these top athletes use—nothing more. With your knowledge of mechanics, you'll see the purpose behind these actions. As you improve as a coach, you'll learn to disregard some actions that a top-class athlete uses because they are personal idiosyncrasies and of no mechanical value. Accept them as something that makes an individual athlete comfortable, but disregard them as a necessity for good performance.

Remember that the actions an elite athlete performs at high velocity over a great range of movement need to be modified to the maturity, strength, flexibility, and endurance of a young athlete. You cannot and must not expect a young, immature athlete or a novice of any age to assume the body positions or match the explosive actions of an elite athlete. This comes with regular training and good coaching.

SUMMARY

- Six steps are useful in analyzing a sport skill: (1) Determine the objectives of the skill, (2) note any special characteristics of the skill (these two steps highlight the objectives and conditions governing the performance of a skill), (3) study elite performances of the skill (this step recommends careful analysis of elite performances of the skill that you are coaching), (4) divide the skill into phases, (5) divide each phase into key elements (these last two steps show the importance of breaking a skill down into phases and key elements), and (6) understand the mechanical reasons a key element is performed as it is (this step emphasizes the need to understand why the performance of the phases and key elements of a skill should be based on sound mechanical principles).

- The rules of sport and the conditions that exist when sport skills are performed determine skill objectives. Most sport skills have more than one skill objective.
- Sport skills can be divided into different types based on the manner in which the athlete performs the skill and the conditions under which the skill is performed.
- Nonrepetitive skills are also called discrete skills because they have a definite beginning and an end. Nonrepetitive skills are frequently joined in a sequence.
- Repetitive skills have a cyclic, continuous nature, with the movement pattern repeating continuously.
- Sport skills can be performed in predictable and unpredictable environments. Skills performed in a predictable environment are also called closed skills. Skills performed in an unpredictable environment are called open skills.
- A phase in a sport is part of a connected group of movements that an athlete joins together in the performance of the total skill.
- Many skills can be divided into the following four phases: (1) preparatory movements (setup) and mental set, (2) windup (backswing), (3) force-producing movements, and (4) follow-through (recovery).
- Key elements are the finer, distinct actions that together make up a phase. Force-producing movements generally contain the most key elements.
- An understanding of the mechanics of a sport skill's key elements and phases is necessary in order to coach a technically correct performance of a sport skill.

KEY TERMS

closed skills
eccentric thrust
follow-through
force-producing movements
nonrepetitive skills
open skills
phase
repetitive skills
skill objectives
windup

REVIEW QUESTIONS

1. The first steps in analyzing a sport skill are to determine the objectives and note any special characteristics of the skill. Why is it important to carry out these steps?
2. What is meant by a predictable environment as it relates to the performance of a sport skill? Give two examples of skills normally performed in predictable environments.
3. What characteristics must you always consider when you use an elite performance as a model for your coaching?
4. What is the value of dividing a skill that you are analyzing into phases?
5. What is the purpose of dividing each phase of a skill into its key elements?
6. Why is it important to understand the mechanical reasons for performing each key element in a particular manner?

PRACTICAL ACTIVITIES

1. **Identifying objectives for sport skills.** Make up a list of 10 different sport skills, and write down the objectives for performing each of these skills. Pick as wide a variety of sport skills as you can and refer to your text for assistance. Remember that many sport skills will have more than one objective.
2. **Identifying fundamental patterns and individual characteristics in sport skills.** By watching videos or attending sporting events, watch the performances of five elite or high-level athletes all performing the same sport skill. Write down the fundamental movement patterns that these athletes perform in a similar manner. Then write a list of individual movement characteristics that distinguish these athletes one from the other.
3. **Comparing elite athletes to novices.** Compare and contrast elite athletes and novices performing the same sport skill. Choose five different sport skills. List the characteristics of technique that make the elite athlete so much better than the novice performer.
4. **Analyzing high-velocity sport skills.** By watching videos or attending sporting events, watch the performance by elite athletes of the following high-velocity sport skills: (a) discus throw, (b) baseball batting, (c) javelin throw, (d) golf drive, (e) tennis serve, and (f) badminton smash. Compare and contrast the sequence of limb movements that the athletes use, how they shift their body weight, and other factors that are characteristic in producing high-velocity movements. In addition, note any outstanding similarities and differences that occur.
5. **Equipment development.** List equipment developments and changes that have occurred over the last 10 to 15 years in five sport skills of your choice. When selecting these sport skills, be sure to consider both winter and summer sports. Some excellent examples for consideration are swimming, cycling, speedskating, ski jumping, and golf.

8

Identifying and Correcting Errors in Sport Skills

When you finish reading this chapter, you should be able to explain

- how to critically observe the performance of a skill,
- how to analyze each phase of a skill and the key elements within each phase,
- how to make use of your knowledge of sport mechanics in your analysis,
- how to determine the order in which to correct errors, and
- how to select the appropriate coaching methods for correcting errors.

In chapter 7 you went through the first stage in getting ready to correct errors in an athlete's performance. You saw how to break a skill into phases and key elements, and you understood how sound mechanical principles form the foundation of good technique. This chapter will give you advice about observing an athlete's performance and using your knowledge of mechanics to pick out errors that need correcting. We have laid out what you must do as a series of five steps:

Step 1: Observe the complete skill.

Step 2: Analyze each phase and its key elements.

Step 3: Use your knowledge of sport mechanics in your analysis.

Step 4: Select errors to be corrected.

Step 5: Decide on appropriate methods for the correction of errors.

Step 1: Observe the Complete Skill

It's a good idea to plan how you intend to observe a skill. By deciding what to look at and where to stand, you can watch the skill or the elements in a phase of the skill as accurately as possible.

Begin by observing (and videotaping) your athlete's performance from several different positions. Watch from the left and right, from the front and rear. In this way you can cross-reference and double check the information you gather. Characteristics of the performance that are hidden from one point of view will be revealed from another. A good approach is to watch the whole skill in this manner several times and then home in on the skill's phases and key elements.

Observing Skills That Involve Distance and Height

Observing skills that involve height and flight (e.g., gymnastics vaulting, ski jumping, and pole vaulting) can be more demanding than observing skills that contain much less movement (e.g., archery or power lifting). A gymnastics vault contains a long and fast approach, a takeoff, flight onto and off the horse, and finally a landing. These phases of the skill occur at high speed and cover considerable distance and height. To critically observe all aspects of the action, observe from positions that are at right angles and about 15 ft from the flow of the skill. Stand at right angles to the board, then the vaulting horse, and finally position yourself to the rear of the approach. You can also stand beyond the landing pads so that the athlete runs toward you. In this way you'll get several viewpoints of the takeoff, flight, and hand positions on the horse (see figure 8.1). This observational technique also works well for track events. Figure 8.2 shows a track coach observing from different directions while the athlete practices hurdle clearances. Get closer when skills cover less distance and height and when you are focusing on particular phases and elements of the skill.

Modern television gives us excellent slow-motion coverage of athletes viewed from above. You're probably familiar with the dramatic replays of hand changes on the high bar, swings to a handstand on the rings, and the incredible rotary skills of gymnasts on the pommel horse. In swimming, cameras on tracks on the sides

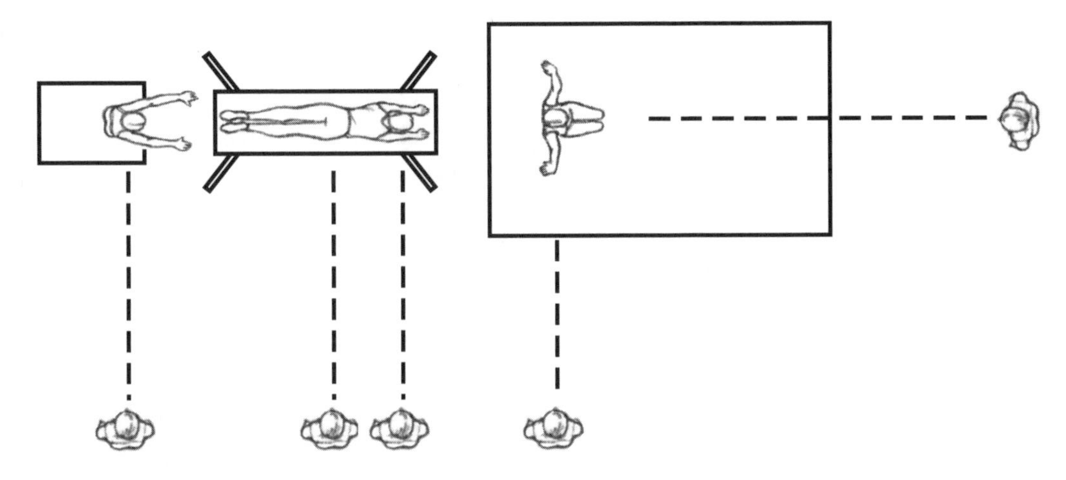

FIGURE 8.1 Various viewpoints for assessing a vault in gymnastics.

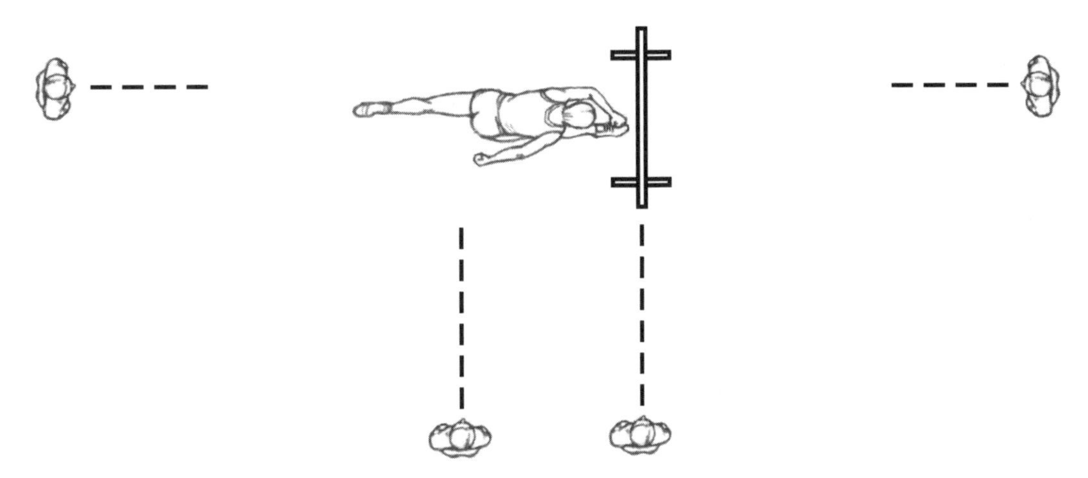

FIGURE 8.2 Various viewpoints for assessing a hurdle clearance.

and bottom of the pool give superb coverage of swimming strokes. If this additional visual information is available, it will help you tremendously with the assessment of your athlete's performance.

Ensuring Safety When Observing

With many skills, you'll pick up much worthwhile information when you observe from in front of your athlete. But be particularly careful in this situation. You may center your concentration on your athlete's movements and not on what happens after. Viewing from the front is not recommended for the throwing events in track and field or for sports such as golf, where the velocity of the ball is exceptional. Unless you have a specially designed protective screen as is used in baseball, be satisfied with viewpoints that are from the side and the rear.

With rotational skills such as discus, shot put, and hammer throw, it is best that you are well back behind a safety cage officially approved for the event. This is highly recommended for the discus and hammer throw. In the hammer throw, the 16-lb ball travels at phenomenal velocity; in the hands of a novice there is no certainty it will fly in the required direction. If you have no safety cage available, stand well back to the left rear (as you view the athlete from the rear) for hammer and discus throwers who rotate counterclockwise across the ring. In the shot put event (which normally does not use a cage), stand well back to the left rear of right-handed throwers

throwers and vice versa for left-handed throwers. Be sure to stand well to the rear of the throwing ring if the athlete is learning the rotary shot putt technique.

If you are marking distances while you assess your athlete's performance, be aware that an implement in flight is extremely deceptive. Javelins viewed head-on have a habit of momentarily seeming to disappear from sight, and wind can dramatically alter flight paths. You must also allow for the distances that implements skid and bounce. A discus skidding across wet grass is extremely dangerous!

Finding Settings for Observation

Try to avoid conditions that distract you and your athlete. Physical education classes and recreational settings disturb your concentration and that of your athlete because there are too many other activities going on. Any movement in the background can disrupt your attention from the details you want to analyze. If you are instructing a group, you cannot, and should not, pay attention to one athlete for very long. Other athletes need your supervision and encouragement. The best setting is one in which there are no distractions at all. Your athlete can concentrate on performing and you on observing and analyzing.

Getting Started With Your Observation

Have your athlete warm up and perform the skill several times so you get a good overall impression. Don't concentrate on specific phases, even though a poor windup or a poor force-producing phase will obviously catch your eye. Try to get a feel for your athlete's rhythm, flow, and general body positions from the start of the skill to the finish. Your main objective at this stage is to get an overall impression of the athlete's performance.

Making Comments During Observation

When you observe, don't distract your athlete by continually offering instruction. Watch without making any comments other than an occasional encouraging remark after the skill is completed. Try to keep him relaxed and enjoying having you as an enthusiastic and knowledgeable spectator. Your athlete should not struggle to impress you, be discouraged, or become so casual that he loses concentration. You want an accurate impression of his abilities, not a performance altered by tension or insufficient concentration. Above all, don't start listing aloud all the actions your athlete is doing wrong! This serves no purpose and destroys morale. You don't want your athlete to become tense or stressed in any way. Your job is to get a true impression of how he performs.

While you observe, make a mental note if you think your athlete is lacking in strength, flexibility, or endurance. But remember that your athlete cannot change these characteristics during one training session, just as he cannot gain or lose weight on command. Take these factors into account by modifying your demands when

Character of a Champion

2003 was the 100th anniversary of the Tour de France, and Lance Armstrong—with support from the great U.S. Postal Service team—made it five wins in a row. That Lance Armstrong has won any of the Tours at all is amazing. In 1996 he was diagnosed with testicular cancer with tumors in his lungs, lymph nodes, and abdomen. After surgery, which also required removing tumors from his brain, followed by what seemed like endless treatments of chemotherapy, Lance Armstrong claimed that his competitive cycling career was far from over. Few people believed him. But in 1999 he won his first Tour de France, and followed this win with four more. Lance Armstrong is the epitome of determination and shows the value of having a positive outlook on life. A better role model for young athletes in any sport could not be found.

you start correcting errors. In other training sessions you can get him to work on improving these areas.

Observing Skills Performed at Normal Speed

When you observe a complete skill for the first time, it is best if your athlete performs the skill at normal speed. The reason for this recommendation is that skills performed at unnaturally slow speeds are dramatically different from those that occur at normal speed. Timing, coordination, and the feel for the skill are different. Slow-speed performances serve little purpose when you are looking for errors to correct. They give you a false picture of what is occurring. However, reducing speed is helpful when you teach new movement patterns. When the fundamentals are learned, the speed of movement can be increased.

Determining the Number of Times You Observe a Skill

How often you watch your athlete perform depends on the physical demands of the skill. Skills that take considerable time, concentration, and effort for each repetition, such as diving and ski jumping, are by necessity viewed fewer times than a place kick, a volleyball serve, a pass in soccer, or the repetitive paddling actions of a kayaker. It is important, however, that you view the performance enough times so your athlete's pattern of movements becomes apparent. Skills such as diving and ski jumping may require more than one training session to develop an accurate impression of the athlete's abilities.

Noting Differences Between Beginners and Experienced Athletes

As you observe, expect that a beginner's performance will change dramatically from one repetition of the skill to the next and that a beginner will tire more quickly than an experienced athlete. Novices make gross errors in which they miss several key elements in a phase, or even a whole phase of a skill. During your general observation you'll notice that their foot positions are incorrect at one moment and correct the next. You'll notice that the beginner may not use the large muscles of the body or shift the body weight in the correct direction. You may even feel, after completing the observation, that the best course of action is to totally rebuild the skill. With novices you must accept this situation. It's part of coaching beginners!

In comparison to beginners, elite athletes make fewer apparent errors. You'll see errors when you analyze a slow-motion video of the skill, or you'll catch them when you concentrate on specific elements in the skill. Perhaps you'll discover that your athlete's line of vision is incorrect or that his head is in the wrong position, upsetting his balance. You may discover that your athlete's overall performance is good except that the wrist action at the end of a pitch, throw, or hit is not as it should be. Unfortunately, you may also have to struggle through many training sessions to get an elite athlete to eliminate these seemingly minor errors. The reason for this difficulty is that an elite athlete has probably been performing the skill in the same way for years, and the incorrect action has become ingrained. How different it is when you coach a young novice. Every coaching session can be a giant leap forward! To your delight you'll find that the technique of most novice athletes is like clay that you can mold. Each coaching session can make a massive change in the quality of their performance. This is why many coaches find great pleasure in coaching novices.

Observing and Spotting

If you are a novice coach, you may find it difficult to critically observe a performance when you are also involved in spotting. Your attention tends to focus on where you should give support (and perhaps on protecting yourself from the flailing arms and legs!) rather than assessing whether certain movements are performed correctly. In gymnastics, dividing your attention can be a risky practice. Experienced coaches can carry out both jobs at once, but even these coaches must concentrate on their spotting when more complex skills are attempted. If you are starting as a coach in a sport that has a high level of risk, play it safe and use competent spotters if you want to be free to observe. If spotters are not available, have an onlooker videotape the performer while you give the necessary assistance.

Afterward, analyze and discuss the performance with the athlete.

Looking for Other Clues

Part of your observation technique will be to look beyond your athlete for clues about the performance. The flight path, rebound, and roll of balls result from the movements and actions your athlete uses in the skill. Skate marks on ice, ski patterns on snow, and footprints on approaches, takeoffs, and landings are all clues to what's going on in the skill.

Don't forget to use your ears as well as your eyes when you look for clues! The rhythm of footfalls during an approach or during the repetitive bounding of a triple jump is an indication of stride length and stride cadence. The overemphasized thud of one of your athlete's feet during throwing events is a sure sign of poor balance and weight distribution. (It's also a sure sign that the hop or the step in the triple jump is too large.) The noise of bat and club on ball helps determine a direct or sliced impact. In volleyball, a slapping noise is a giveaway to the coach of a carried ball or some other incorrect contact. Almost every sport will give you visual and auditory signals that you'll be able to associate with good or bad performance. Use every source of information. Don't limit yourself in any way.

Step 2: Analyze Each Phase and Its Key Elements

After you have watched the complete skill several times, you are ready to concentrate on individual phases and their key elements. There are two ways to approach this task.

Start With the Result

One method coaches commonly use with an athlete who is competent is to start with the end product and work back from there. Here's an example: A player in rugby is trying to spiral the ball for distance. There's enough force put into the kick but there's no spiral. So you concentrate on the action of the foot as it contacts the ball. Is the ball fed onto the player's foot correctly? Is the foot drawn across the long axis of the ball to produce the torque necessary to generate a spiral? On the other hand, if the ball's spiral is satisfactory but distance is lacking, then shift your attention to other phases and key elements in the skill. Ask yourself, "Is there an adequate shift of the player's body into the kick?" "Are the lower leg and kicking foot allowed to swing freely, or is the kicker tightening the leg muscles and eliminating any chance of a whiplash action occurring?" "Is flexibility a problem?" Inflexibility will restrict the range that the kicking leg swings through and reduce the force applied to the ball.

In throwing, kicking, and striking skills, checking the result gives you a wealth of information. You might have a big, powerful athlete who ought to throw the shot a long distance, but there's no force behind the shot and you know the athlete should be throwing 5 ft farther. So you concentrate on the throwing stance that the athlete assumes after the glide across the ring is complete. When you examine the throwing stance, you ask yourself, "Is the athlete's body angled correctly?" "Are his shoulders still facing the rear of the ring when the glide is complete?" "Is the foot placement correct?" "Is the athlete rotating his hips toward the direction of throw, and are the massive muscles of the legs, seat, and back used before the chest, arm, and fingers?" (Figure 8.3 shows the actions that you should be looking for in shot put.) After critically examining the throwing stance, you may decide that the problem lies in a poor glide across the ring. Well-performed standing throws confirm your suspicion. The athlete's glide across the ring is ruining the remaining part of the throw! With the problem diagnosed, you and your athlete can now work on correcting the errors in this phase of the throw.

Observe Each Phase of the Skill in Sequence

Another method of observation that coaches commonly use is to start by critically observing the first phase in the skill, then progress to the second, third, and so on. The first phase contains preliminary movements and your athlete's establishment of a mental set. In the first phase, look at such elements as your athlete's stance

FIGURE 8.3 Excellent technique using the glide method in the shot put.

and weight distribution. Take note of her head position, her line of vision, and the way she concentrates for the actions that will follow.

In the second phase, when your athlete winds up or performs a backswing, examine her weight transference from one foot to the other. (Figure 7.4, a-b, on page 152 illustrates proper body position, weight shift, and backswing in a golf drive.) Check the position of the implement and your athlete's body at the end of the windup. Make a mental note if she appears stiff and needs to improve flexibility.

When you examine the force-producing phase, remember that in many skills this phase is made up of several distinct sections, such as an approach and a takeoff in jumps and vaults; an approach, glide, spin, and throw in throwing events; or an approach, hurdle step, board flexion, and takeoff in springboard dives. In freestyle, it can be the "catch" of the water at hand entry followed by a long pull and push from the hand and forearm. Break these complicated force-producing phases into key elements, and concentrate on each key element in sequence.

In most skills, the follow-through is the least important of all phases. Your athlete has applied force, and the follow-through safely dissipates her momentum and kinetic energy. But be sure to observe what happens during the follow-through and, of course, what happens to the implement and your athlete immediately afterward. Her actions and the implement's flight are clues to what has happened earlier. Check your athlete's arm and hand actions during the fol-

Franz Klammer Combines Innovative Technique With Superior Coaching and Equipment

At the 1976 Winter Games in Innsbruck, downhill skier Franz Klammer thrilled the world with his hair-raising gold medal performance. Franz's rocketing do-or-die performance is still considered one of the most exciting and brilliant downhill runs in the history of modern ski racing. In remembering that run many years ago, Franz said he was helped immensely by superior coaching and equipment. He also said that a large part of his success came from the innovative way that he carved his turns. At the time, ski racers used to skid through their turns on the flats of their skis. Franz started his turn with the ski on its edge. With less skidding at the beginning of the turn, Franz came out of his turns with more speed than his rivals. After Innsbruck, in a World Cup career that spanned nearly 15 years, Franz amassed 25 downhill victories, the most of any downhill ski racer. In 1975 and 1976 he won 13 World Cup downhills in a row, a record that remains unbroken.

low-through on a jump shot in basketball or a volleyball spike. In these skills, a follow-through can indicate the amount and direction of force and spin that's applied to the ball.

In some sport skills, insufficient control during the force-producing phase produces a follow-through that violates the rules of the sport. A field hockey player swings the stick too high, or a volleyball player hits the net after spiking or blocking. So don't disregard the follow-through; consider it an important phase that gives you clues about what happened during the windup and force-producing phases that occurred earlier.

In cyclic, repetitive skills such as swimming, remember that the follow-through is a recovery action that sets your athlete up for another force-producing phase. Check that these recovery actions are mechanically efficient and not wasting your athlete's energy. A misdirected arm recovery in freestyle, in which the swimmer's arm swings across the midline of the body, produces poor body alignment, generates excessive form drag, and affects the efficiency of the force-producing phase that follows (see figure 8.4). Cyclists talk of the need to spin the pedals, meaning that proper pedaling technique is a rotary motion, not just a push downward with a rest on the way up. Check that your athlete is pulling up and around with one leg while pushing down and around with the other. Proper pedaling technique is the ultimate in cyclic, repetitive action.

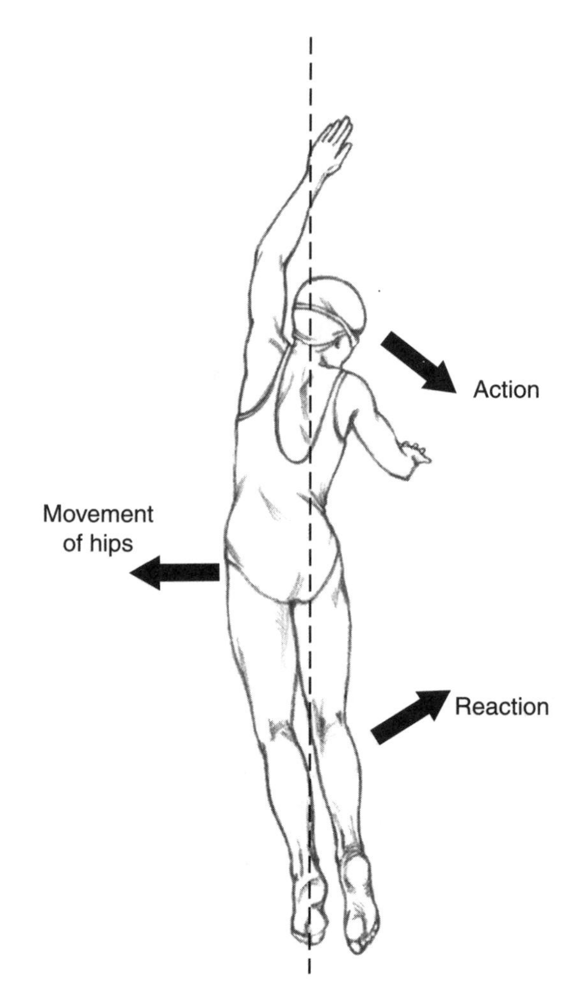

FIGURE 8.4 Overreaching and crossing the midline at arm entry causes the hips to react and move in the opposing direction.

Adapted, by permission, from E. Maglischo, 1983, *Swimming faster* (Palo Alto, CA: Mayfield Publishing Company), 55. Reproduced with permission of The McGraw-Hill Companies.

Step 3: Use Your Knowledge of Sport Mechanics in Your Analysis

As you observe each successive phase and its key elements (from preliminary movements and mental set to follow-through, or recovery), you must put your knowledge of sport mechanics to work. In particular, concentrate on how your athlete applies muscular force to produce a desired action in a skill. You must carefully assess the mechanical efficiency of your athlete's actions and the way he competes against gravity, friction, drag, air resistance, and the forces generated by opponents, whatever the opposition might be. In this way you can pick out technical (i.e., mechanical) errors that your athlete commits. What should you be looking for as you examine the elements in each phase? Here is a series of important questions you can ask yourself as a guide:

Does your athlete have optimal stability when applying or receiving force?

A wide base and correctly positioned center of gravity are essential for applying and receiving force. Check the position of your athlete's center of gravity and the way she sets up a base of support. Ask yourself the following questions:

"Is her base of support extended in the direction it should be?"

"Is the base too narrow or too wide?"

"Is she standing too erect instead of squatting down?"

"Is her center of gravity too close to the edge of the base when it should be centralized?"

If your athlete stumbles or gets thrown one way as the implement goes the other, or if she is too easily knocked off balance by an opponent, you'll know there's an error in this area. So check through the mechanical principles associated with balance and stability. Remember that stability is one turning effect (i.e., torque) battling another. To remain stable, your athlete may have to reposition her feet and center of gravity to apply more leverage and more torque.

In many skills, your athlete must be able to move quickly and react in an instant. When she is receiving a serve, playing in goal, or reacting to the moves of an opponent, the objective is not maximum stability but rather a level of stability that allows her to move in a flash in any direction. We discussed these principles in detail in chapter 5.

In particular, be sure to carefully check the size and alignment of your athlete's base of support and the position of her center of gravity during a skill's force-producing phase. An inadequate base not only makes your athlete unstable but, equally important, also reduces the distance and time over which she can apply force.

Is your athlete using all the muscles that can make a contribution to the skill?

Athletes produce inferior performances when they do not apply force with all the muscles they can and should use in a skill. This may seem a strange state of affairs! After all, why not use the leg muscles, or any other muscle group, if they can make a contribution to the performance? If the performance of a skill requires the muscles of the legs, trunk, chest, and arms, and your athlete uses only the muscles of the chest and arms, then the total force put into the skill will be below optimal level. How can you tell if your athlete is using all the necessary muscles? Usually this is easy because in dynamic skills muscle contractions produce actions. If a limb segment or some other part of the body moves, you know muscles are contracting. Here's an example of what we mean. When a child throws a ball for distance for the first time, the youngster frequently stands still with his feet close together, then throws with his arm alone. There's no wide throwing stance. The throwing arm is not taken back, and the shoulders aren't rotated away from the direction of throw. In the force-producing phase of the throw, the muscles in the legs, trunk, and chest make no contribution. This error doesn't occur only among children. You'll see it happening among adults as well.

Top-flight athletes always aim to have all the required muscle groups contributing to the skill. Elite rowers make sure that the muscles of the legs, back, shoulders, and arms play their part in the stroke. Elite speed skaters make sure that their leg muscles contribute optimally in powering them along the ice. The muscles working their arms and shoulders make their own contribution in counterbalancing the actions of the legs. Imagine how poor the performance would be if a speed skater failed to use her quadriceps muscles adequately to extend her legs, or if she skated with her arms hanging straight down instead of forcefully swinging them back and forth! The same principle applies to all skills. Make sure that all members of your athlete's "muscular team" are making a contribution by moving the body segments that they are responsible for moving. Think of your athlete's muscles as a tug-of-war team. If a team is made of eight members, why have one or more of the team resting on the rope without pulling?

Is your athlete applying force with the muscles in the correct sequence?

If a world champion weightlifter performs a clean and jerk, and you critically examine the key elements of the clean (in which the athlete pulls the bar up to the chest), you'll see that the muscles of the legs, back, shoulders, and arms are contracted at about the same time. The extension of the legs is closely linked to an extension of the back and a strong upward pull with the arms. On the other hand, if you examine the key elements in the force-producing phase of a pitcher's fastball, you'll see a well-defined sequence of actions, starting from the big muscles that accelerate the athlete's body and the large, more massive body segments and finishing with the high-speed movement of smaller, less massive body segments (i.e., the throwing arm and hand). All great pitchers step forward as the throwing arm is drawn back and the shoulders are rotated away from the batter. When they have stepped out into the pitching stance, their bodies rotate toward the hitter in a whiplike sequence that starts from the legs, shifts to the hips, then to the chest, and ends with a tremendous acceleration of the throwing arm. The pitcher's body acts like the handle of a whip that is being cracked. The hand gripping the baseball is the tip of the whip. A volleyball spike or a tennis serve uses a similar whiplike sequence of actions.

This comparison between the actions used in a clean and jerk and a baseball pitch demonstrates opposite extremes in the sequence that an athlete's muscle contractions occur. When you examine each phase in your athlete's performance, check that the movement of his limb segments occurs in the correct sequence. If it does, then you know that muscle contractions are occurring in the correct sequence as well. A common fault for many athletes in throwing and hitting skills is to use the small muscles of the shoulders and throwing arm long before the big muscles of the legs, back, and trunk have done their job. The result is that the big muscles never get the heavier parts of the body moving ahead of those that are lighter. It's impossible to crack a whip if you don't accelerate the whip handle first.

Is your athlete applying the right amount of muscular force over the appropriate time frame and distance?

You'll recognize this statement as our old friend impulse, which we discussed in chapter 3. Remember that impulse refers not only to the amount of force that your athlete uses but also to the time period that your athlete applies force.

If your athlete uses the right amount of force for the right amount of time, his limbs move at the required speed through the required range of movement. When this occurs and all muscle contractions are sequenced correctly, you'll see movements that are fluid, smooth, rhythmic, graceful, and well coordinated. When an athlete applies force indiscriminately and haphazardly (and this is what novices do!), you'll see actions that are jerky and awkward. What you are seeing is the difference between a polished and well-practiced technique and one that is not.

Practice helps your athlete establish how much force each muscle involved in a skill must exert. When they are learning, many athletes apply too much or too little force at the wrong time, and as a result, their technique will look jerky and awkward. You can help your athlete correct this situation by giving rhythmic cues that provide an idea of the speed at which he should perform the actions. You can also provide coaching tips such as "Step out long and low, and as soon as your foot hits the deck, thrust your hips toward the direction you are throwing" or "Stretch up at takeoff and swing your arms upward as fast as you can."

Keep in mind that what is correct in a mechanical sense is not always possible in an anatomical sense. In other words, mechanical principles must fit with the design of your athlete's body. For example, athletes *do not* apply maximum force over the *longest* possible time frame even in hitting skills that require maximum velocity at impact or in jumping skills that require maximum velocity at takeoff. In these skills, good technique is characterized by limbs that are slightly flexed at the start of the skill and fully extended from maximal muscular contraction when impact and takeoff occur. Look for this action when you examine the final elements of the force-producing phases in golf driving, baseball batting, tennis serving, and track and field throwing.

In skills that require accuracy, such as a volley in tennis or a drop shot in squash, look for controlled force applied over a specific range. Too much force or too great a range of movement defeats the purpose of the skill. When too much force or too great a range of movement occurs, the volley puts the ball out of the court. Likewise a drop shot in squash rebounds too high from the front wall, making the "get" easy for the opponent.

A word of advice in relation to force and the time frame that force is applied. No one expects young novices to be able to produce the same force as adults, and likewise, you cannot expect novices to assume the same body positions and apply force over the time frame and distances elite athletes use. Make allowances when you watch your young athlete perform. Less force applied over a limited range of movement is not necessarily an error but a stage in the developmental process. More force and a greater range of movement will come with increased strength, flexibility, endurance, and coordination—all of which are carefully molded by your good coaching.

Is your athlete applying force in the correct direction?

This principle may seem hardly worth mentioning, yet you should look for this factor, particularly in the force-producing phase of the skill. Elite sprinters drive down and back with each leg thrust so that they travel forward at the greatest possible velocity toward the finish line. The direction of each leg thrust gives these superb athletes the exact amount of vertical and horizontal thrust required by each sprinting stride. The result is optimal forward propulsion. You'd have no trouble deciding that there was something terribly wrong with a sprinter's technique if her leg thrust was directed out to the side, or if her arms swung sideways across her body rather than forward and backward! Errors like these in the force-producing phase of sprinting indicate that the athlete is wasting force and not applying it in the correct direction.

In your analysis of the performance of sport skills, you'll notice that inexperienced athletes apply force in many different directions. They waste much of their muscular effort, and so it makes no contribution to their performance. Look also for inexperienced athletes to thrust and push in the correct direction with one part of their bodies and in an incorrect direction with another part. You see this often with novice downhill skiers. Young shot-putters often complain that putting the shot bends their fingers backward. A careful examination of the arm action in their throws indicates that they are not pushing directly behind the center of gravity of the shot. The thrust is in some other direction, and the result is that the shot bends their fingers back. Likewise, a hammer thrower or discus thrower who spins on the spot instead of traveling across the ring, or who falls sideways out of the ring when he releases the implement, is obviously misdirecting his force! Poorly directed force produces inadequate rotation at takeoff for gymnasts, divers, and figure skaters; mishits in sports that use clubs, bats, and rackets; miskicks in sports that use kicking skills; and poor propulsion in swimming skills.

Is your athlete correctly applying torque and momentum transfer?

Many sport skills require your athlete to generate and control rotation. Rotation is applied to your athlete's body, an opponent, a ball, or an implement such as a discus. To initiate rotation your athlete must apply the turning effect of torque. The more spin required, the more torque he has to apply. In your analysis, check how much force he is generating and the distance this force is applied relative to the axis of rotation. In judo, look at the position of the axis of rotation that your athlete sets up for a hip throw and where he applies force to the opponent. Is your athlete strong enough to apply tremendous force? If not, is there any way of increasing the force arm (i.e., the distance from the axis to where force is applied)? The larger this distance, the less effort he must apply.

In gymnastics, diving, and figure skating the number of rotations your athlete performs depends on how much torque he generates and how much momentum transfer he uses at takeoff. Momentum comes largely from arm and leg actions that are transferred at takeoff to his body as a whole. A skater who performs a double spin when intending to perform a triple may say, "I don't think I got enough spin." However, after you've analyzed the skater's takeoff you may disagree. Check the actions of the arms and the free leg to see whether they make an adequate contribution to rotation. You might decide that sufficient torque was applied at takeoff, and in fact it was overemphasized. Your analysis may indicate instead that there was not enough upward thrust at takeoff and that the arms and free leg were not swung vigorously enough to provide any transfer of momentum in an upward direction to the skater's body.

Is your athlete decreasing rotary resistance to spin faster and increasing rotary resistance to spin slower?

If a skill requires your athlete to spin faster, turn quicker, or swing the limbs at high speed, she must decrease her rotary resistance (i.e., rotary inertia) by pulling her body in toward her axis of rotation. The requirements of the skill determine the tightness of this position. Arms flexed at the elbows help produce a fast and efficient arm swing for sprinters and speedskaters. A tight tuck for gymnasts and a compressed body position for figure skaters when they spin around their long axis helps produce the required number of rotations. Extended body positions oppose a fast spin and slow down rotation. Is your athlete not tucking tight enough because of insufficient flexibility or lack of muscular strength, or is the problem inadequate knowledge of the correct timing in the skill? Your careful analysis of each phase and its key elements can pinpoint the source of the problem.

Keep in mind that when your athlete rotates and extends the arms, her body slows but her arms and hands travel faster. Put a racket or a bat in your athlete's hand and the head of the implement travels fastest of all. Many skills require a combination of hip and shoulder rotation coupled with full extension of the arm (particularly at impact or release). For example, a tennis serve and a golf drive require full arm extension when the racket or club hits the ball. A discus thrower must release the discus with the implement as far from the body as possible. In your analysis, look for extension at release and impact and for some flexion and tighter body positions earlier in the skill.

Step 4: Select Errors to Be Corrected

After you've analyzed each phase and its key elements, you have the task of deciding what sequence to follow in correcting errors. Like any enthusiastic coach, you'd like to correct every error in the first training session. The difficulty you'll face is that inexperienced athletes commit numerous errors, some of which are major and some minor. What is a major error and what is a minor error?

A major error will be the absence or poor performance of any item that we've discussed under step 3 in this chapter. So, errors that destroy your athlete's stability or the optimal use of his muscular force are major errors. A minor error is an action that only partially detracts from the performance of the skill. Examples of minor errors include a backswing in golf that needs to travel a few degrees farther back, a throwing stance that needs to extend a little farther, and an arm swing at takeoff in a jumping event that needs to be more vigorous. To an elite athlete,

A Coach's Responsibility

For many female gymnasts, the toughest piece of equipment they face is not the beam or the uneven bars, but the scale where they must constantly monitor their body weight. Some female gymnasts have been told that they are overweight even at 100 lb and because of the pressure have suffered from bulimia and anorexia. Competing at a high level, young female gymnasts are well aware of the importance of a superior strength-to-weight ratio, and equally important, that gymnastics, like synchronized swimming, figure skating, and diving, is an aesthetic sport where looks count! So they have to keep their weight down and look great! Male gymnasts suffer less from bulimia and anorexia, but the problem exists in sports like wrestling, boxing, and judo where making the weight is essential. As a coach, it is important to understand how talent, training, technique, and determination combine to produce a superior performance. At the same time don't forget that you have a responsibility for your athlete's well-being. Winning is great, but a well-adjusted and happy athlete after a sport career is over is more important!

minor errors of this nature make the difference between a good performance and a world record. To a coach working with a beginner, these errors can be placed on the back burner while other more important errors are corrected.

The simple method to follow when selecting errors is to forget those that are minor and pick out those that are major. When you've picked out the major errors, select the one that has the most adverse effect on the skill and work on this error first. If you still have trouble deciding which error to choose, home in on major errors in your athlete's stance and body position—particularly in the preliminary stance and in the force-producing phase. Get the preliminary stance straightened out and then shift to the force-producing phase. Why? Because an athlete cannot apply force correctly unless stance and body position are correct. In throwing, hitting, and striking skills and in contact sports, poor position and lack of balance destroy everything! An incorrect stance will ruin a golf stroke, and an incorrect body position simply sets a wrestler up for a countermove. In swimming, a body position where the athlete is not horizontal in the water creates tremendous form drag. Keeping the head down and improving the leg kick can correct this. It's amazing how much faster a swimmer will travel when drooping legs are raised into a horizontal position.

Step 5: Decide on Appropriate Methods for the Correction of Errors

The final step leads you from sport mechanics into methods of teaching and coaching sport skills. Errors in skill performances vary in complexity. At their most complex, errors can be mistimed sequences of high-speed arm actions occurring in flight as a diver combines a somersault with a twist. At their simplest, a young novice might put the wrong leg forward when throwing a softball. If you are coaching the diver, you have the options of discussion, video analysis, and demonstration of the correct arm actions from the side of the pool. If you are lucky enough to coach at a pool with high-tech equipment, with a press of a button you can foam up the water with air bubbles so your diver can work on the arm actions knowing that she's not going to get hurt if she fails. You might also decide that the correct arm actions need to be reinforced over and over on the trampoline with your diver in a spotting belt. But once your diver is in flight there's no possibility of hands-on help from you, and your athlete has no chance of slowing the skill in any way. It's the same when you coach a basketball layup or a start in swimming. It's obviously easier when you teach a youngster to throw a softball. You can demonstrate and say, "Put your left leg forward, and turn your right shoulder to the rear as you take your arm back." You can even move the youngster's limbs into the correct position. It's impossible to do this with a diver in flight or a swimmer leaving the blocks.

Because sport skills vary so much and errors are so diverse, it's impossible to offer a single method that works for the correction of all errors. But we can provide a step-by-step sequence that will help in most situations you'll encounter.

Steps in Error Correction

- If possible, separate the phase that contains the error from the rest of the skill. Treat this phase and its key elements as a skill in itself.
- Break the phase and its elements into smaller parts. For example, if the error is poor synchronization of footwork and arm actions, consider teaching the footwork first. Then teach the arm actions. Later add the two together. Use verbal counts and rhythmic cues to assist your athlete.
- Design a practice or a specific activity that is useful for teaching the correct movements. This practice should be easy to perform and, if possible, novel and interesting to the performer. Above all, be creative and flexible in your approach. If the practice activity you have designed doesn't help correct the error, then change it. What works well for one of your athletes may not work so well for another.
- Whenever possible have your athlete perform new movements slowly.

- Walk your athlete through the required body positions, pausing wherever appropriate. Again, use a verbal count for rhythm.
- Increase the speed of performance slowly. Always be prepared to repeat a step in this progression if speed reintroduces errors.
- When you have decided that the actions you wanted to correct are learned well enough, put them back into the phase they came from, and check your athlete's performance. If you are satisfied, attach additional phases from the skill at either end of the one containing the correction. Check how your athlete integrates the new movements. If problems persist, reinforce earlier steps in this sequence.
- Attempt the complete skill at reduced speed and effort with the corrected movements in place.
- Progressively increase speed and effort.

Safety in Coaching High-Risk Skills

As we mentioned earlier, in many skills it's impossible for the athlete to pause halfway through to rethink movements. These skills usually involve flight and often have a high element of risk. A back somersault in gymnastics floor exercises is such a skill. With skills of this nature we recommend the following sequence:

- Maximize safety with spotters, overhead spotting rigs, safety belts, crash pads, pits filled with foam rubber, or any other specialized equipment that fully protects the performer. In this way your athlete can perform the required actions with confidence and without danger.
- With highly complex skills, go back to a known skill that contains elements of the movement patterns you wish to correct. Use this skill to reinforce the correct actions.
- Progressively remove spotters and other specialized equipment as the skill is learned.
- When your athlete attempts the skill alone, provide experienced spotters who can assist if necessary.

Your Attitude During the Correction Process

During whatever process of correcting errors you use, remain positive and praise good effort and correct performance. Your objective is to help your athlete persevere through those difficult periods when he feels that the correct action will never be mastered. The progress he makes will depend on the amount of practice, the complexity of the required actions, and how long it takes you and your athlete to mold the corrected action into the total skill.

Giving Advice to Your Athlete

How you and your athlete communicate determines how much success you gain when you attempt to correct errors. Don't befuddle a young athlete with needless technical jargon. Translate your mechanical know-how into instructions that fit the age, intelligence, and physical ability of the athlete. Some athletes will be genuinely interested in the mechanical principles behind a movement. Statements such as "Push forward when you contact the ball—it'll help you apply more force" are excellent because they indicate in easy-to-understand language the mechanical reasons behind a body position. But for most athletes, comments about angular momentum, momentum transfer, kinetic energy, and rotary inertia are meaningless. For most athletes, the less cluttered their minds, the freer they are to perform. This recommendation also relates to the amount of information you give at any one time in your instructions. Give simple, short, easy-to-understand instructions that are to the point. No athlete wants to stand around while the coach talks endlessly about what could or should be done.

Improving Feedback

Few sports use mirrors the way that bodybuilding does. Besides providing bodybuilders with continuous aesthetic assessment, mirrors provide instant visual feedback on how they perform an exercise. In the highly dynamic actions of most sports, the nearest you can get to a mirrored image is an immediate playback on a video machine. The machine tells your athlete, "This is what you looked like." However, neither the machine nor a mirror can say, "This is what the movements should feel like." Your athlete is the only one who actually feels the movement. As a coach observing and assessing the skill, you cannot feel what your athlete feels although you can describe what it should feel like to your athlete. To be able to provide this kind of information requires considerable experience from you, first as an athlete and then as a coach.

Sometimes you'll be surprised at the response when you ask, "What did you feel at that moment in the skill?" Many novices have no idea what happened, and they have no sense of where their limbs are as they attempt a skill. Elite athletes differ considerably. Most of them have a well-developed kinesthetic sense and are aware of what their bodies are doing during a performance. Teach young athletes to develop this sensory awareness. It becomes an invaluable resource in helping them master sport skills.

Consider the Time Available for Correcting Errors

One factor that considerably influences error correction is the time you have available to work with your athlete. Can you plan a training program that stretches over several months or a year, or are you working with a three- to six-week block before the competitive season starts? When you are restricted by time, it affects your choice of errors and the methods you use to correct them. You may feel that all you can do is correct some minor errors because major changes initially make your athlete perform poorly, and this downtime can carry into the competitive season. Whenever an athlete has to think about what he is doing in one or more phases of a skill, the effect is a drop in performance. The correct action must become second nature—an unconscious action. Remember that correcting a minor error can affect performance considerably. Be satisfied with that! Save large changes in technique for the off-season.

Getting Additional Help

You must gather information from many different sources if you want to become a good coach in your chosen sport. In the area of analysis and skill correction, you should be ready to do the following:

- Read texts on your sport that offer successful teaching methods and techniques for correcting errors. The best texts should offer excellent illustrations of lead-ups and teaching progressions. A quality text contains lists of common errors that occur in the skills of your sport, with explanations on how to correct them. Look also for safety recommendations, not only for individuals but also for group activities. This is particularly important when you are teaching skills that have a high element of risk.
- Plan to attend coaching seminars and workshops. Here you can listen to presentations on coaching from experienced coaches and experts who do research in your area. You can discuss problems in your sport with coaches who are aware of the latest coaching techniques. Join your local and national coaching association so you can regularly receive the latest newsletters.
- Modern video and computer technology now gives you the opportunity to analyze performances in superslow motion and to place elite performers on the same screen as novices. This is of tremendous assistance when you want to look at differences in technique. Computer advances in virtual reality now offer you and your athlete a "virtually real" method of experiencing the movement patterns of a skill while remaining static. Previously, downhill skiers and luge competitors would close their eyes and imagine steering through the curves and straightaways. With virtual reality your athlete can put on a specially designed helmet that feeds a 3-D image containing an accurate representation of the bends, twists, and straightaways that exist on the course. When applicable, a computer will have worked out the best course to follow relative to temperatures and other weather conditions.
- Read beyond this book and improve your knowledge of mechanics as it applies to your sport. Take an interest in other sports as well. It will make you more of an expert in your own!

SUMMARY

- There are five steps to the effective observation, analysis, and correction of errors in sport skills: (1) Observe the complete skill, (2) analyze each phase and its key elements, (3) use your knowledge of sport mechanics in making an analysis, (4) select errors to be corrected, and (5) decide on appropriate coaching techniques for the correction of errors.
- Observe skills from several different positions; avoid settings in which you and your athlete are distracted. Use video recordings to assist in your analysis.
- Do not concentrate on skill analysis while you are involved in spotting the performer.
- Once you have a good overall impression of an athletic performance, analyze each successive phase of the skill together with its key elements.
- Use your knowledge of sport mechanics for analyzing performance. Ask yourself mechanical questions (e.g., Does the athlete have maximum stability when applying or receiving force?).
- Divide the performance errors you see into major and minor categories. Major errors seriously detract from the optimal performance of the skill, whereas minor errors have minimal effect on performance. Follow the sequence outlined in this chapter for correcting errors.
- Use the very best safety techniques when correcting errors in skills that contain a high level of risk.
- Maintain a positive attitude during the correction process. Avoid using excessive technical jargon during coaching sessions.
- Teach athletes to develop sensory awareness to assist them in error correction.
- Be aware of the time you have available for correcting errors. Do not attempt massive changes in technique if time is limited.
- Attend coaching seminars and read texts on your sport that offer top-quality teaching methods and techniques for the correction of errors. Be alert to any advances in computer and video technology that can assist you in coaching. Expand your knowledge of sport mechanics, not only in your sport but in other sports as well.

REVIEW QUESTIONS

1. What is a good sequence to follow when analyzing each phase of a sport skill?
2. What is the mechanical difference between the recovery in a swimming stroke and the follow-through in a throwing skill?
3. What is the common appearance of a skill when an athlete tries to use more than the optimal number of muscles in a performance? Give a common example of an athlete applying muscular force in the wrong sequence to perform a skill.
4. Why is it important for an athlete to apply the right amount of force over the appropriate time frame with each of the muscles involved in a sport skill?
5. What does this text refer to as major errors in the performance of a skill?
6. What resources can you use to help you become a better coach in your sport?

PRACTICAL ACTIVITIES

Using video clips, analyze the following sports (and sport skills) using the questions listed underneath each sport as a guide to your analysis.

1. **Freestyle wrestling competition.** Watch one competitor for the duration of the clip and then repeat the clip watching the other competitor. The following questions will guide your analysis.
   a. How did the attacker increase or decrease his stability during an attack?
   b. How did the attacker attempt to increase the torque and lever action applied against the opponent?
   c. What did the defender do to increase his stability and counter the attack?
   d. When there was a successful attack that overcame the defense, what mechanical factors made the attack successful?
   e. When there was a successful defense that overcame an attack, what mechanical factors made the defense successful?
2. **A mount on the beam in women's gymnastics.** Watch five elite athletes run and mount onto the beam in a gymnastics competition. The following questions should guide your analysis.
   a. How big was the base of support of each athlete at the instant of landing on the beam?
   b. Did the skills used for the mount provide any specific methods for enlarging the athlete's base of support? If so, describe what these actions were.
   c. Did the athletes regulate their height and line of gravity to improve their stability? If so, what actions did they make?
   d. What actions did the athletes make with their arms, head, or legs to maintain stability?
   e. Did the athletes move quickly into the next skill in their routine? If so, did this help or detract from the performance of the mount?
3. **A fastball pitch in baseball.** Watch an elite athlete pitch several fastballs. Then use the following questions to guide your analysis.
   a. How did the athlete increase the size of the pathway over which the baseball was accelerated?
   b. What was gained from stepping forward toward the hitter? How does this compare with the actions used by a javelin thrower?
   c. Outline the sequence of limb and body movements used by the athlete to achieve the great release velocity of the baseball.
   d. What relationship is there between the sequence of limb movements that the athlete uses and the mass, inertia, and length of these limbs?
   e. Explain how linear and rotary motion are combined in the pitch.
4. **A volleyball spike.** Watch elite volleyball players spiking the ball. Then use the following questions to guide your analysis.
   a. What is gained from the steps used in the approach? Why perform the approach to the takeoff in the volleyball spike?
   b. What changes occur in body position and also the position of the athlete's center of gravity in the penultimate (next to last) and last steps (or strides) of the approach?
   c. How is momentum transfer employed in the takeoff?

d. What particular differences exist between the volleyball spike and the high jump? (Compare the approach and the takeoff).
e. Is all of the athlete's force used to direct the athlete in a vertical direction? If not, what occurs?
f. What happens to other parts of the athlete's body when the striking arm is positioned ready to spike, and immediately after contact has occurred?
g. How does the athlete generate the tremendous velocity of the striking hand?
h. If the athlete spun the ball, explain how it was achieved.

5. **Long jump.** Watch elite athletes performing the long jump takeoff. Then use the following questions to guide your analysis.
   a. What is gained from the approach in the long jump?
   b. What changes occur in body position in the penultimate and last steps (or strides) of the approach?
   c. How does the body position of the athlete performing the long jump takeoff differ from that of athletes performing the high jump and the volleyball spike?
   d. What are the mechanical reasons for the changes in body position of the long jumper in the penultimate and final strides prior to takeoff?
   e. How is momentum transfer employed in the takeoff of the long jump? Compare with the high jump and the volleyball spike.
   f. What is the approximate angle of takeoff of the long jumper? Compare with the angle of takeoff used by the high jumper and the volleyball spiker.
   g. Is all of the athlete's force used to direct the athlete through the air in a horizontal direction? If not, what occurs?
6. **Sprint starts in a 100-meter race.** Watch elite athletes performing a sprint start in the 100 meters. Then use the following questions to guide your analysis.
   a. Is the athlete in a position of maximum stability or minimal stability in the set position?
   b. Comment on the line of gravity, the position of the center of gravity, and the area of the supporting base in the set position.
   c. What is the position of the leading leg and the arms immediately after the gun is fired? Why is there an acute angle at the knee and at the elbows?
   d. What is the position of the leg driving back at the blocks immediately after the gun is fired?
   e. Where is the center of gravity of the athlete just prior to the recovery of the leg that is driving back at the blocks? What is the reason for its particular position?
   f. Are the accelerative strides of the athlete long or short as she comes out of the blocks and sprints the first 5 meters?
   g. What is the cadence of the accelerative strides of the athlete coming out of the blocks? Is it fast or slow?
   h. What differences exist in the length and cadence of the strides between those used over the first 5–10 meters and those used in the middle portion of the race?
   i. Comment on the rise and fall of the athlete's center of gravity once the athlete is running at full speed.

9

Mechanics of Selected Sport Skills

In this chapter we analyze a number of sport skills to show you how technique and mechanics are inseparable. You'll find that the following pages are a good review of the mechanical principles you read about earlier in this book.

The technique and mechanics of 11 sport skills are described on the following pages. Under the heading "Technique," you'll find the skill's most important technical characteristics. Technique tells you what movement patterns should occur when athletes sprint, swim, throw, hit, jump, lift, and so on. Under the heading "Mechanics," you'll find a description of mechanical principles at work during the technique. This is where you'll find the mechanical reasons a skill is performed the way it is. In this way you can immediately read the mechanical reasons that each phase of a skill is performed as it is.

This analysis does not cover every sport skill that exists, nor will it satisfy the needs of every coach. But we've tried to cover a wide range of sports. If this review highlights for you how technique and mechanics are intrinsically tied together, and how you cannot teach technique without knowing mechanics, then it's done its job!

Table 9.1 lists the sport skills you'll find in this chapter and our rationale for including them.

TABLE 9.1

Sport Skills

General skill	Specific skill	Rationale
Running	Sprinting	Athletes use sprinting as a form of locomotion. It is the most dynamic and vigorous of all running techniques. Walking and middle- and long-distance running obey the same mechanical laws as sprinting but use less vigorous actions.
Swimming	Freestyle	Freestyle is the most popular and fastest of all major swimming strokes. The mechanical principles that govern the technique used in freestyle apply to other swimming strokes as well as to the sculling actions used in water polo and synchronized swimming.
Jumping	High jump	The mechanical principles that control how an athlete gets up in the air in the high jump also apply to other jumping skills (e.g., a volleyball spike, a basketball layup, and a receiver's leaping catch in football). Once in flight, the same mechanical laws apply to the long jumper (or any athlete in flight) as they do to the high jumper.
Throwing Striking Kicking	Javelin throw Baseball batting Football punting	All three skills require the athlete to generate high velocity. The javelin thrower wants the throwing hand moving as fast as possible, the baseball hitter wants the bat moving quickly, and the punter wants his foot moving at high velocity when it contacts the ball. In all three skills, athletes try to simulate a whiplash action with their limbs.
Pulling, pushing, lifting, and carrying	Clean and jerk	The clean is a lifting-pulling action and the jerk is a push. Carrying and supporting actions occur when the athlete pauses with the bar at the chest and again when the bar is at arm's length above the head. The mechanical principles involved in the clean and jerk apply to all lifting, carrying, and spotting techniques. Laws controlling stability also play an important role in the clean and jerk.
Swinging Rotating	Back giant on the high bar Front somersault	The back giant is a rotational skill controlled by many of the same mechanical principles that govern the front somersault. However, there are some differences. The back giant is a rotary swinging motion performed around a high bar, which acts as an axis external to the athlete's body. In contrast, the axis for a front somersault passes through the athlete's body from one hip to the other. There are similarities and differences in the mechanics of these two skills.
Balance and stability	Judo hip throw	A judo hip throw represents the battle that occurs when athletes try to maintain their own stability and at the same time destroy the stability of their opponents. Judoka (i.e., Judo practitioners) use pulling, pushing, and rotary actions to get their opponents off balance. The mechanical principles governing the front giant, the front somersault, and the clean and jerk appear again in the judo hip throw.

General skill	Specific skill	Rationale
Arresting motion	Judo breakfall	A judo breakfall demonstrates an athlete arresting the motion of his body. The principles that judoka apply in dissipating the forces that occur when they slam into the mat also apply when athletes catch balls, take hits in contact sports, and land after jumping. The breakfall is a technique used to increase the area and lengthen the time frame that impact forces are applied to the athletes' bodies when they drop on the mat.

BEFORE YOU START

As you read the following pages, look for mechanical principles that constantly reappear from one skill to the next. Watch for the turning effect of torque, the battle against inertia as athletes accelerate, and the athletes' use of impulse. It doesn't matter what skill you consider—these mechanical principles (and many others) are always present. Get to know them, to recognize them, and to understand their effect. It will help you immensely in your coaching.

Running Skills

Skill Highlights

1. The time an athlete takes to run a set distance depends on the athlete's stride length and stride frequency. The length of the athlete's legs and the forward thrust that occurs with each stride determine stride length. Forward thrust is produced by the earth's reaction force responding to the athlete's backward thrust against the earth's surface. Stride frequency is the cadence that the athlete uses (i.e., the number of strides that occurs each second).
2. A runner's technique changes the faster the athlete runs. Sprinters spend more time in the air than distance runners. In addition, they flex and swing their arms more vigorously. Sprinters also have a higher knee lift, a greater leg thrust, and a higher flexed leg. Distance runners use less arm action but tend to swing their shoulders more than sprinters. The longer the distance, the greater the reliance on cardiovascular endurance and pacing. All elite runners hold their torsos close to perpendicular.
3. Tension is detrimental to all runners because tension saps energy and restricts muscle action and limb movement. Sprinters try to run explosively while still relaxing their faces, necks, shoulders, and hands. Distance runners use the same relaxation techniques.

Spotlight on . . . Sprinting

FIGURE 9.1 Sprinting technique.

Technique	Mechanics
1. Good sprinting technique demands an optimal blend of stride length and stride frequency. A predominance of fast-twitch muscle fibers is essential for top-level sprinting.	1. A combination of optimal leg power, stride length, and stride frequency produces the best sprinting times. Power, good reactions, and excellent flexibility are all essential. Stride length depends on hip flexibility, leg length, muscle power, and range of movement. Training optimizes the forward thrust that occurs at each stride. Training also brings more muscle fibers into action and teaches the athlete to relax opposing muscle groups. Leg drive is improved by related power training. Overemphasis on stride frequency (i.e., stride rate or cadence) or on stride length produces inefficient sprinting.
2. Sprinting requires excellent leg, hip, and shoulder flexibility. Flexibility in the hip and pelvic area is particularly important.	2. An ability to rotate the hips around the long axis of the body helps produce an optimal stride frequency and stride length. Flexibility in the shoulder girdle promotes good arm swing.
3. A sprinter's arms are flexed at 90 degrees and swing powerfully forward and backward. The hands are relaxed and swing hip high to the rear and shoulder high in front (see figure 9.1, a-e).	3. Forward and backward arm swing counterbalances the twisting motion produced by each leg thrust on either side of the sprinter's long axis. Flexing the arms at the elbows reduces their rotary inertia and makes their pendular movement easier for the muscles involved. A vigorous forward swing of each arm transfers momentum to the athlete's body as a whole. This adds to the athlete's leg thrust and helps drive the athlete forward. Forward and backward arm swing parallel to the direction of sprint (rather than across the body) helps hold the torso and the shoulder girdle steady. This aids balance and relaxation and assures that the athlete runs a straight line toward the finish.
4. The driving leg extends to near full extension (see figure 9.1b). When the driving foot leaves the ground, the leg flexes and the heel kicks up to buttock level (see figure 9.1e).	4. A powerful leg extension via the hip, knee, and ankle joints provides the athlete with optimal thrust in the direction of the sprint. Thrust backward and downward at 50 to 55 degrees produces an equal and opposite reaction from the earth, which drives the athlete in a predominantly horizontal direction along the track. Flexion of the legs (like the arms) reduces their rotary inertia and makes their recovery and forward movement easier for the muscles involved.
5. After thrusting backward and downward, the driving leg flexes at the knee and is brought directly forward and upward so that the thigh swings to just below horizontal (see figure 9.1c).	5. The swing and upward thrust of the driving leg as it is brought forward is counterbalanced by the action of the opposing arm. Forward thrust of both arm and leg generates momentum transfer. This helps produce a greater thrust back at the earth with the driving leg, and in response, from the earth propelling the sprinter's body along the track.
6. When the sprinter's driving leg is recovered and becomes the supporting leg, it flexes slightly on landing. The supporting foot lands below the sprinter's center of gravity. The first contact of the foot with the ground is on the outside edge of the foot. The heel is lowered but does not contact the track (see figure 9.1a).	6. A slight flexion of the supporting leg extends the time of impact that force is applied to the sprinter's body and so cushions the landing. Flexion stretches the leg muscles, ready to extend the driving leg backward and downward against the earth. When the supporting foot lands vertically below the sprinter's center of gravity, it eliminates deceleration that would occur if the foot was placed ahead of the sprinter's center of gravity.
7. A sprinter's forward body lean is extreme during a sprint start. At top speed the torso is perpendicular and the shoulder girdle held square to the direction of run (see figure 9.1, a-e).	7. During a sprint start, forward body lean and shorter high-frequency strides overcome the inertia of the sprinter's body mass and help the sprinter gain momentum. At full speed, a perpendicular torso coupled with vigorous forward and backward arm action counterbalances the movement of the legs.

Technique	Mechanics
8. A sprinter's body rises and falls very little when running at full speed (see figure 9.1, a-e).	8. An elite sprinter's center of gravity follows a low wavelike pattern as it travels forward. Slightly more time is spent in the air than in a support position. Too much time in the air is time wasted and indicates that too much thrust is directed in a vertical direction.
9. The sprinter's head is held in a natural alignment with the torso. Vision is horizontal and directly ahead (see figure 9.1, a-e).	9. Proper position of the head and vision assists in maintaining the stability of the sprinter's torso. Tilting the head back increases tension and restricts stride frequency and stride length.
10. Good sprinting combines power with relaxation. Face, neck, shoulders, and hands are relaxed.	10. Tension in the body reduces the velocity of muscle contraction and reduces sprinting velocity. Good sprinting requires a rapid change from muscle contraction to relaxation. A technically superior sprinter is mechanically efficient. Unnecessary tension is avoided, and in this way the athlete uses energy efficiently.
11. An athlete's sprinting speed is influenced by environmental conditions.	11. The greatest expenditure of energy in sprinting occurs when the athlete is thrusting back at the earth. Energy is also required for the knee lift and support phase of sprinting. The faster the athlete runs, the more energy the athlete must spend fighting air resistance. Headwinds add to this resistance. The condition of the track influences sprinting speed. Lightweight, high-quality spikes produce good traction, and driving back on a firm rubberized track helps thrust the athlete forward better than running on soft ground. Friction-reducing running tights and slick body suits help minimize drag from the air.

Swimming Skills

Skill Highlights

1. The velocity at which a swimmer moves through the water depends on stroke length, stroke rate, and the reduction of fluid frictional forces to minimal levels. Stroke length refers to the distance over which the swimmer pulls and pushes with each stroke. Stroke rate refers to the number of strokes per minute.
2. In the freestyle, back crawl, and butterfly, the swimmer's hands and forearms produce the greatest propulsive forces. In these strokes the legs also make a major contribution by aiding in propulsion and by holding the swimmer's body horizontal, which minimizes form drag. In the breaststroke, both arms and legs provide powerful propulsive forces.
3. Swimmers fight against several frictional forces: (a) form drag, caused by the swimmer's shape and body position in the water; (b) surface drag (skin friction), caused by the water rubbing against the swimmer's body surfaces; and (c) wave drag, caused by the resistance of the water piling up in front of the swimmer. A horizontal body position in the water combined with smooth surfaces (i.e., resulting from shaving and the use of slick swimsuit fabrics) help to reduce form and surface drag. Swimmers reduce wave drag and attempt to be energy efficient by controlling up and down body motions and avoiding thrashing, splashing arm and leg actions.
4. Swimmers emphasize stretching and reaching forward to increase the length and range of the stroke as well as maintaining a high stroke rate. Swimmers aim for rhythmic, relaxed swimming and try to develop a feel for the water.

Spotlight on . . . Freestyle

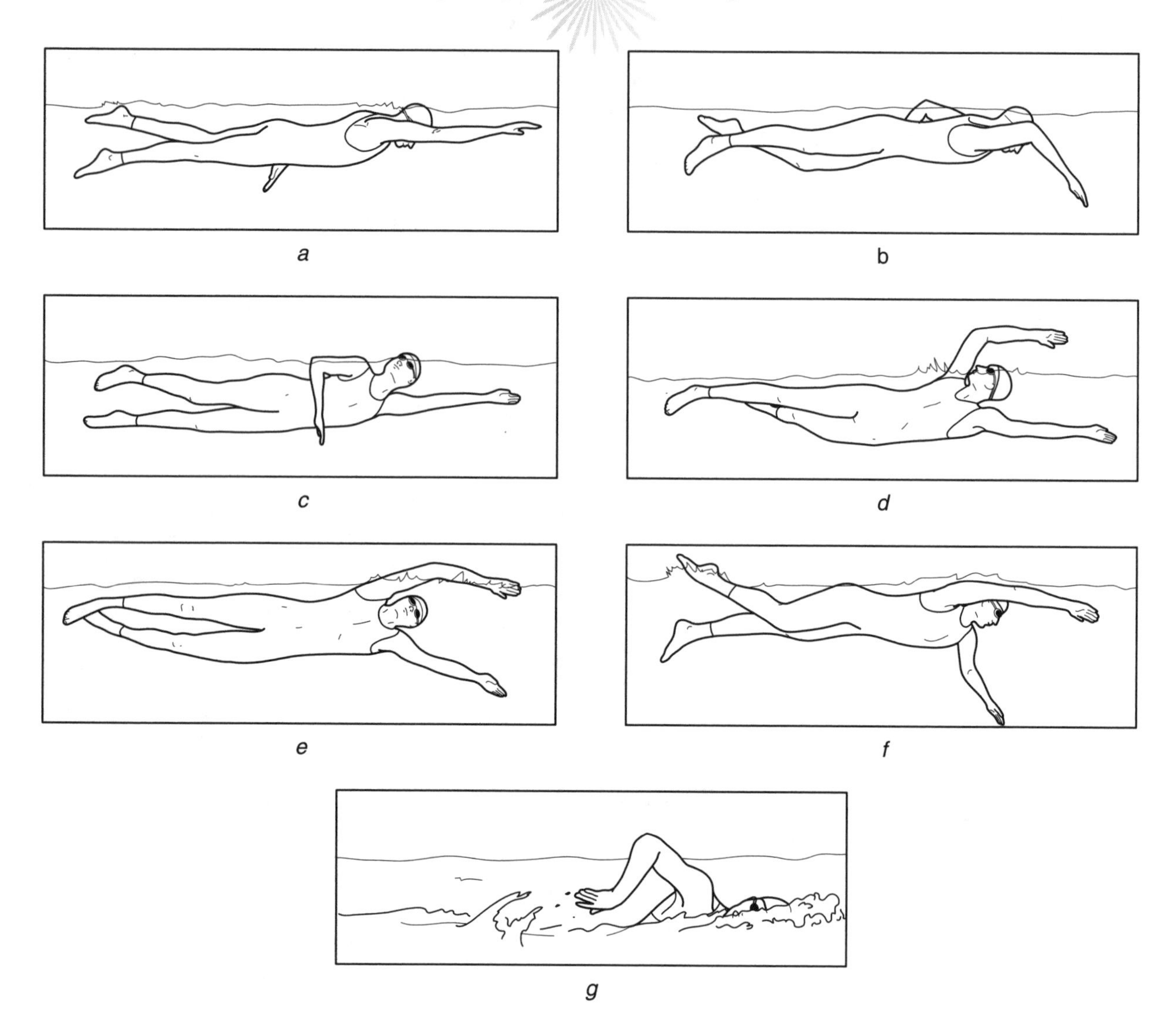

FIGURE 9.2 Freestyle.

Technique	Mechanics
1. The freestyle is the fastest of the four swimming strokes. Arm action, leg kick, body position, and breathing are synchronized and contribute to the velocity of the swimmer through the water.	1. In the freestyle, one arm pulls and pushes while the other recovers and reaches forward for the next propulsive action (see figure 9.1c). Both freestyle and back crawl use this cyclic rotary action, whereas in the breaststroke and butterfly both arms pull, push, and recover simultaneously. In the freestyle the swimmer's hands and forearms produce most of the propulsive force. Excellent joint mobility in the shoulders and ankles help to maximize propulsion from the arms and legs. A good leg kick coupled with the use of full or partial body swimsuits help maintain a low drag horizontal position in the water. Speed through the water is based on increasing the impulse that is applied to the water (i.e., increasing the applied force for the duration of the stroke); increasing the rate of force application (i.e., stroke rate); and eliminating as much as possible all body positions and actions that increase form, surface, and wave drag.

Technique	Mechanics
2. The swimmer's body lies horizontal in the water. The chest is pressed down and the face is immersed (see figure 9.2a and b). Swimmers will often look at the bottom of the pool in order to raise the legs. The head and shoulders roll together as a unit for breathing. (see figure 9.2 c) The leg kick occurs just below the surface (see figure 9.2 a and b). During the arm recovery, the arm is flexed at the elbow (see figure 9.2 g). Stretching forward with the leading arm and rolling along the long axis of the body is coupled with breathing and the recovery of the propulsive arm (see figure 9.2 c and d).	2. Keeping the head down maintains good body alignment and reduces form drag in the water. It also enhances the ability to glide with each stroke. (Raising the head causes the legs to drop, which increases wave, form, and surface drag.) If the arm recovery is swung across the swimmer's long axis, the swimmer's body will "snake" through the water, increasing both form and surface drag. A flexed arm during recovery reduces its rotary resistance and makes it easier for the muscles involved to bring the arm forward. Slamming the arm and hand into the water at entry causes the body to react in the opposing direction and bounce up and down. This increases wave, surface, and form drag. Rolling along the long axis presents a narrow profile to the water, which minimizes drag.
3. Entry of the hand is made in front of the head at a position ahead of the shoulder joint. At entry, the arm flexes slightly, with the elbow positioned above the hand. The fingers slice into the water before any other part of the hand and arm, and the swimmer attempts to "catch" or "fix" the pulling surfaces of the hand and forearm into the still, nonturbulent water ahead and draw the body past this "catch" or fixed point (see figure 9.2 e and f).	3. At the hand and forearm's entry into the water, lift forces generated by the flow of water from the fingertips up the hand and arm aid partially in propulsion. The swimmer uses momentum to move past the "catch" position. Still, nonturbulent water provides an area of water in which the swimmer can pull and push more effectively. The swimmer uses the reaction of the water for propulsion as it responds equally and in the opposite direction to the pull and push exerted by the hand and forearm. This form of drag propulsion is greatest when the hand and forearm pull and push backward parallel to the direction of swim.
4. As the body moves forward past the "catch," the pulling arm flexes at approximately 90 degrees with the elbow high and the hand positioned lower in the water (see figure 9.2 b and c). The degree of flexion at the elbow varies throughout the arm's pull-push phase. The propulsive pull-push of the propulsive arm is timed to occur as the opposing arm is recovered and thrust forward in the water (see figure 9.2 d). The swimmer attempts to produce a predominantly straight-line horizontal pull-push with the immersed hand and forearm pressing back against the water. The propulsive arm exits the water at hip level (see figure 9.2 b and c).	4. Partial flexion of the propulsive arm at the elbow reduces its lever length relative to the shoulder joint, but does not reduce the active surface area of the swimmer's arm as it pulls through the water. Flexion at the elbow also makes it easier for the swimmer's muscles to pull the hand and forearm along their propulsive pathway. The pull-push with the hand and forearm is completed at hip level and optimizes the impulse (i.e., force × time) over which the propulsive force is applied. The swimmer's forward movement results mainly from drag forces generated from the backward thrust of the hand and forearm. The resultant of lift and drag acts also as a propulsive force predominantly at the beginning and end of the pull-push action.
5. Some lateral and vertical motion of the propulsive hand and arm occurs during the pull and push. The swimmer rolls along the long axis of the body as the leading arm is stretched forward. Rolling assists with breathing and with recovery of the propulsive arm after it has completed its pull-push action (see figure 9.2, c-e).	5. Anatomical limitations in the shoulder and elbow joint coupled with reactions to head movements and arm recovery cause some lateral and vertical movement of the propulsive hand and arm as they move backward through the water. The angle of attack of the hand varies continuously throughout the stroke to maximize lift and drag propulsion.
6. The flutter kick has an arc of movement of about 1 to 2 ft in size and occurs just below the surface of the water (see figure 9.2, a-f). It has a downbeat-upbeat action—plus some rolling lateral motion—which is caused by the rolling of the body itself. The kick starts at the hip and works down to the ankle in a whiplash motion. The flutter kick is synchronized with breathing and the arm action.	6. The flutter kick cannot compare with the hands and arms in terms of propulsion but it plays an important role in holding the swimmer's body in a streamlined, horizontal, drag-reducing position. Much depends on the swimmer having great mobility and flexibility in the ankles. Kicking too deep increases drag without contributing much to propulsion. The depth of the kick depends on the swimmer's build, strength, and stroke rate. A stiff-legged flutter kick consumes too much energy and is incorrect.

Technique	Mechanics
7. Breathing depends on the stroke pattern developed by the swimmer. Many swimmers breathe every stroke cycle, and breathing every two strokes is a common pattern in races that are longer than a single length (50 m).	7. Breathing should not in any way hinder efficient propulsive technique. Lifting the head to breathe causes the legs to drop, increasing form drag.
8. The athlete tries to swim with a smooth, rhythmic, nonjerky, cyclic action. Up-down, side-to-side motions are eliminated with the torso held fairly rigid. The athlete's arm action is long rangy, and precise, and gives an impression of relaxation even at high speed.	8. Jerky, up-down, and side-to-side motions increase the resistance of the water, in particular wave drag (which increases according to the cube of the velocity). An increase in wave drag demands an immense increase in energy expenditure from the athlete and generates a phenomenal resistance the faster the athlete swims. The athlete should aim for fluid, relaxed motions. From a biomechanical and physiological standpoint, these actions are efficient and superior in terms of propulsion.
9. Crawl swimmers frequently shave the skin exposed to water flow and use bodysuits that are patterned with vortex generators. Dolphin kicks are used for several strokes immediately after entering the water from the start and also when coming off the wall after completing a turn. The laws of competitive swimming limit the distance that can be swum underwater using dolphin kicks.	9. Shaving reduces surface drag. Vortex generators are lines of small bumps or specifically located seams on bodysuits. The objective of vortex generators is to reduce both form and surface drag caused by the swimmer both on and below the surface of the water. Because body suits increase buoyancy, the swimmer is more easily able to maintain a horizontal position in the water, which dramatically reduces form drag. In the freestyle, dolphin kicks below the surface at the start and after turns are a highly efficient form of propulsion that avoids the adverse effects of wave drag occurring at the surface.

Jumping Skills

Skill Highlights

1. To get up in the air, jumpers exert a force against the earth's surface well in excess of their own body weight. The earth's reaction force then drives the athletes upward. The more forceful the athlete's thrust against the earth, the greater the earth's response.
2. Immediately before takeoff, a jumper's center of gravity is lowered, the body tilted backward, and the athlete's arms and free leg positioned to the rear of the body. Lowering the body prestretches the big muscles of the jumping leg, preparing them for the leg's explosive thrust downward at the earth. Leaning back combines with lowering the body so the athlete can spend more time over the jumping foot applying force to the earth. Swinging the arms forward and upward adds to the downward thrust of the athlete's jumping leg against the earth.
3. The path that a jumper's center of gravity follows during flight is determined by the velocity at which the athlete is propelled upward at takeoff and the takeoff angle used.
4. When in flight, movement of one part of a jumper's body causes other parts to move in the opposing direction. In high the jump, this characteristic helps in bar clearance. In the long jump, rotary actions of the arms and legs are used in flight to counteract the unwanted forward rotation that inevitably occurs when the athlete takes off. In a volleyball spike, drawing the arm back and arching the body in a counterclockwise direction will cause the legs to move in a clockwise direction.

Spotlight on . . . High Jump

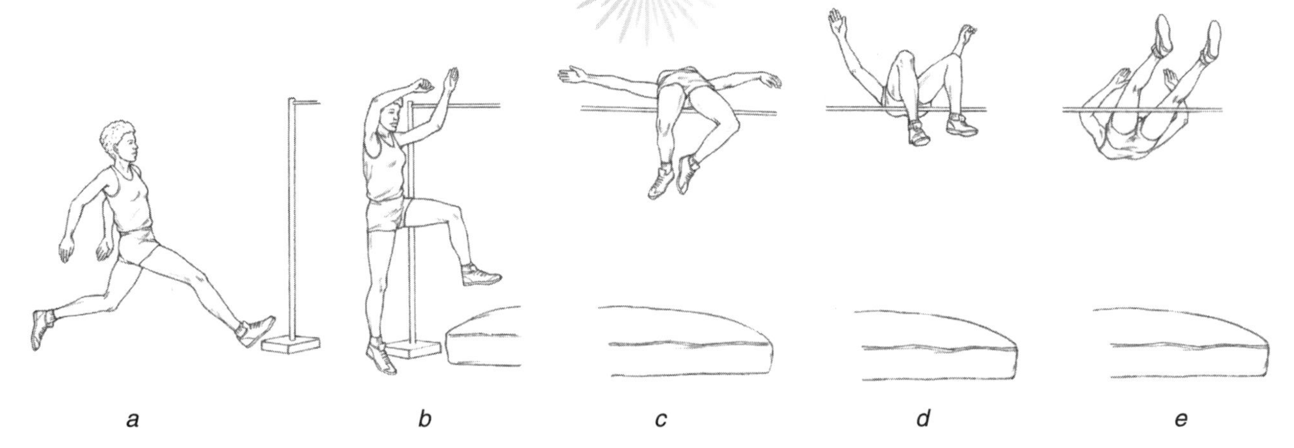

FIGURE 9.3 High jump technique.

Technique	Mechanics
1. The approaches used by elite athletes in the high jump range from 10 to 13 strides and commonly extend over a curved approach of 60 to 110 ft. Jumpers approach the bar at high velocity and accelerate to even greater velocity during the last 3 strides before takeoff.	1. A high jump approach should be long enough so that sufficient velocity is reached to carry the athlete through the actions performed at takeoff.
2. The greater the velocity of the approach, the greater its potential for helping the athlete jump high. Too little velocity in the approach can be detrimental to the takeoff. However, too much velocity can also be detrimental.	2. The speed of the approach must be fast enough to carry the jumper through all the body positions required in the takeoff. There also must be sufficient time to generate the optimal vertical, horizontal, and rotary thrust to get up and over the bar. If the approach speed is too great, there will not be enough time to optimize all of these actions. On the other hand, if the approach speed is slow, the athlete will move far too slowly through the takeoff and will have a difficult time getting from a back leaning position at the start of the takeoff to an upward rotating thrust that occurs just before leaving the ground.
3. Most high jumpers use a curved approach that has a large radius curve (or is almost straight) during the first 6 or 7 strides, followed by a small radius curve during the final 3 to 5 strides. Jumpers lean into the curve of the approach. They also lean backward as they plant the takeoff foot (see figure 9.3a).	3. The athlete's inward lean during the approach produces a centripetal force. A tighter curve and greater approach velocity requires more inward lean. Inward and backward lean as the jumping foot is planted lengthens the time that the athlete spends thrusting down at the earth with the jumping leg. Inward and backward lean before takeoff counterbalances the outward pull of inertia. A body position with lean away from the bar prevents the athlete from committing the error of leaning into (i.e., toward) the bar.
4. During the last 2 or 3 strides of the approach the athlete's center of gravity is lowered and the arms and free leg positioned to the rear of the body (see figure 9.3a). The penultimate (i.e., next to last) stride and the final stride are longer than previous ones. The athlete steps forward onto the jumping foot with the hips positioned well to the rear of the jumping foot.	4. Lowering the center of gravity and stepping well forward onto the jumping foot allow the athlete to apply force against the earth over a large arc (i.e., a long time frame) and in reaction have the earth spend more time driving the athlete upward. Lowering the center of gravity prestretches the jumping muscles in preparation for their powerful extension of the jumping leg.

Technique	Mechanics
5. The arms and free leg are positioned to the rear of the body as the athlete steps forward onto the jumping leg (see figure 9.3a). The forward and upward swing of the arms and free leg is then coordinated with the extension of the jumping leg (see figure 9.3b).	5. The upward swing of the arms and the free leg toward the bar is a form of momentum transfer. The momentum of their upward swing combines with the downward thrust of the takeoff leg. All unify to produce a greater reactive response from the earth, which drives the athlete upward.
6. The free leg and arms flex and accelerate during their upward swing. These actions occur while the takeoff leg is extending vigorously in contact with the earth (see figure 9.3b).	6. Flexing the arms and legs brings their mass closer to their respective axes and reduces their rotary inertia. This action makes it easier for the athlete's muscles to move them upward at high speed. For greatest effect the free leg and arms move at maximum velocity at that instant when the athlete is last in contact with the earth. The muscles in the jumping leg must have enough power to extend the jumping leg explosively.
7. The takeoff for elite high jumpers is from a point 3 to 4 ft directly out from the near high jump standard. The takeoff foot is placed at an angle of 15 to 20 degrees to the crossbar. The free leg is first thrust upward and then rotated away from the bar and back toward the direction of approach. The shoulders rotate parallel to the bar. Vision is usually along the bar in the direction of the far standard.	7. Rotation of the jumper's body in preparation for a back-lying position over the bar is initiated while the takeoff foot is still in contact with the ground. The athlete is then able to rotate around the body's long axis by pushing against the earth's surface. In the air, the same movement causes a twistlike equal and opposite reaction to occur. Elite jumpers try not to compromise vertical thrust by overemphasizing rotation during the takeoff.
8. At takeoff, the high velocity of the approach coupled with the plant of the takeoff foot causes the athlete to move forward over the takeoff foot.	8. Moving up and over the takeoff foot initiates rotation, which continues in flight. Once in flight, the athlete can rotate more quickly (i.e., increase angular velocity) by pulling the body inward, or slow down rotation by stretching out and extending the body
9. After takeoff the athlete's upper body is flexed backward over the bar. The free (i.e., leading) leg, which was swung up at the bar, is lowered from its elevated position (see figure 9.3c).	9. Lowering the leading leg, combined with the act of flexing the upper body backward, produces an equal and opposite reaction that elevates the athlete's hips. This clears the hips over the bar. The athlete's vision, directed along the length of the bar, helps time this critical maneuver.
10. When the athlete's seat has crossed the bar, the head and shoulders are lifted upward. The athlete's torso and the legs (which are now flexed at the knee) are pulled toward each other by contraction of the abdominal and quadriceps muscles (see figure 9.3d).	10. The elevation of the head and shoulders causes two equal and opposite reactions to occur: (1) The flexed legs move upward toward the torso, and (2) the athlete's seat (which has crossed the bar) drops downward. Flexion of the legs reduces their rotary inertia, making their movement easier so they are quickly drawn over the bar.
11. Once the athlete's thighs have cleared the bar, the legs are extended at the knees so that the heels avoid clipping the bar (see figure 9.3e).	11. Even though the legs extend, the fact that the athlete's upper and lower body have moved toward each other causes the athlete's body to increase its rate of spin (angular velocity) around the athlete's transverse (hip to hip) axis. This action, combined with some rotation around the long axis, continues as the jumper drops toward the pit.
12. The athlete relaxes and drops onto the shoulders on the high jump landing pad.	12. The continued rotation of the athlete's body in flight causes the athlete to drop onto the shoulders on the landing pad. The focus of the eyes on the bar during the clearance of the feet pulls the chin into the chest and prevents a headfirst landing. The height of the high jump landing pad relative to the bar is designed to stop over-rotation and prevent the athlete landing on the back of the neck. The foam rubber of the landing pad extends the time frame and area of impact and gradually reduces the forces applied to the athlete during the landing.

Throwing, Striking, and Kicking Skills

Skill Highlights

1. High-velocity throwing, striking, and kicking skills require the athlete to simulate the cracking of a whip. The high-speed movement of a racket in a badminton smash, a 100-mph tennis serve, a golf drive, a fastball in baseball, or a javelin throw all require the athlete to generate phenomenal velocity. This flail-like action is called a **kinetic link** or **kinetic chain** action because the slower motions produced in the athlete's longer, larger, and heavier limbs are made faster and quicker as they pass their motion on to lighter, less massive body parts.
2. In these high-velocity throwing and striking skills, there is a rapid acceleration of the athlete's body segments, beginning with those in contact with the earth. A whiplash, or flail-like, sequence progresses upward from the legs to the hips, from hips to chest, and culminates in the tremendous velocity of the striking or throwing arm. To achieve the greatest possible velocity, it is important that antagonist muscle groups are completely relaxed and that agonist muscles contract in sequence, helping to make each body segment move faster than the previous one.
3. Kicking skills differ from throwing and striking skills in that the last and fastest moving segments, or links, in the whiplike sequence are the athlete's lower leg and kicking foot. The progressive acceleration of body segments in kicking skills is similar to that of throwing and striking skills.
4. Each movement pattern in throwing, striking, and kicking skills contains some preparatory actions commonly called an approach, setup, windup, or backswing. These actions help position the body and implement (such as a ball in baseball pitching) in the optimal position for the application of force. A backswing in a golf drive provides additional distance over which the club is accelerated. The backswing prestretches the golfer's muscles in preparation for their explosive recoil. Relaxation and flexibility help produce an optimal windup.

Spotlight on . . . Javelin Throw

FIGURE 9.4 Javelin throw.

Technique	Mechanics
1. An approach toward the direction of throw leads into a wide, powerful throwing stance. The approach is relaxed and increases in velocity. The athlete accelerates during the first two thirds of the approach. This leads into the final one third of the approach, which includes drawing back the javelin and stepping into the final throwing stance.	1. An approach builds momentum and generates enough velocity to carry the athlete through the throwing stance and into the follow-through. Too much velocity in the early part of the approach may cause the athlete to slow down during the throwing actions or may not give the athlete enough time in the throwing stance to apply an optimal amount of force to the javelin. Efficient use of an approach can increase the distance thrown by 90 to 100 ft compared with a standing throw.
2. Before entering the throwing stance, the athlete's shoulder girdle is rotated away from the direction of throw, and the throwing arm is taken back to arm's length (see figure 9.4a). By use of one or more crossover steps, the lower body moves forward under the torso so the athlete's body is angled backward away from the direction of throw (see figure 9.4, a-b).	2. Rotating the shoulder girdle and extending the throwing arm prepare the athlete for the application of force to the javelin over the largest possible distance and time frame. The backward body lean makes this distance and time frame even greater.
3. The athlete steps into the throwing stance with the leg on the opposite side of the body to the throwing arm. This step is usually larger than those prior (see figure 9.4c).	3. Stepping forward with the opposing foot (e.g., the left foot if the javelin is held in the right arm) sets up a large base of support for the application of force to the javelin. This allows the athlete's hips and shoulders to be rotated back toward the approach and away from the direction of throw.
4. The athlete's body is tilted backward and his center of gravity lowered over a partially flexed rear leg. The rear leg is flexed at the knee and angled outward 45 degrees from the direction of throw.	4. Flexing the rear leg stretches the leg muscles in preparation for their explosive rotary thrust toward the direction of throw. This rotary motion is the first stage in the athlete's whiplash action that starts from the athlete's base (i.e., where the athlete's feet are in contact with the ground) and progresses up to the throwing arm.
5. The rear leg is vigorously rotated toward the direction of throw. This action thrusts the hips in the same direction (see figure 9.4, c-d). The muscles joining the hips to the torso stretch and contract explosively.	5. More massive, slower moving parts of the body shift forward into the throw while lighter body segments (e.g., the throwing arm) complete their backward extension. This motion stretches the muscles in the abdomen, chest, and shoulders, getting them ready for their explosive contraction during later phases of the throw.
6. The athlete's torso rotates and pulls the shoulders and the throwing arm toward the direction of throw. Opposing muscle groups are relaxed. As the shoulders are pulled forward, the muscles of the shoulders stretch and then contract vigorously. A relaxed throwing arm follows the shoulder with a flail-like action. The athlete's body tilts sideways, away from the throwing arm. The free arm rotates backward to help pull the chest and throwing arm around and into the throw (see figure 9.4, c-d).	6. Each of the athlete's body segments, from the legs through to the shoulders and throwing arm, sequentially accelerates. This sequence sets up a flail-like whip-cracking action that progressively builds and ends in the tremendous velocity of the throwing arm. A sideways inclination of the athlete's torso allows for greater height of release. The free (i.e., non-throwing) arm is pulled backward to help rotate the athlete's torso around the long axis of the body. Rotation of the torso makes a contribution in pulling the throwing arm at high velocity into the throw.
7. As the athlete's throwing arm is pulled forward, the upper arm and elbow lead, with the throwing hand and javelin trailing well behind. Flexion occurs at the elbow of the throwing arm (see figure 9.4d).	7. Flexing the throwing arm at the elbow serves two purposes: (1) It helps the athlete become even more whiplike, and (2) the elbow acts like the axle of a wheel with the throwing hand rotating around at its rim. This wheel-axle arrangement increases the velocity of the throwing hand and the javelin.
8. The athlete thrusts forward toward the direction of throw. The torso moves forward beyond the supporting leg, which has been straightened (see figure 9.4e).	8. Forcefully driving the body as far as possible toward the direction of throw extends the application of force to the javelin over the longest possible distance and time period.

Technique	Mechanics
9. The athlete uses a follow-through to complete the throw (see figure 9.4e).	9. The follow-through applies force to the javelin for as long as possible. After the javelin has left the athlete's hand, the follow-through allows for safe dissipation of momentum from the athlete's body.
10. The angle of release of the javelin varies according to the throwing ability of the athlete, the type of javelin used, and environmental conditions at the time of throwing.	10. A javelin is dramatically affected by lift and drag forces. The trajectory angle and the angle of attack at release relative to environmental conditions, such as head- and tailwinds, all determine the path and the distance of the javelin's flight. A moderate headwind of 5-10 mph provides excellent lift to the javelin because it pushes up on the "undersurface" of the javelin. A tailwind is detrimental because it pushes down on the upper surface of the javelin.

Spotlight on . . . Baseball Batting

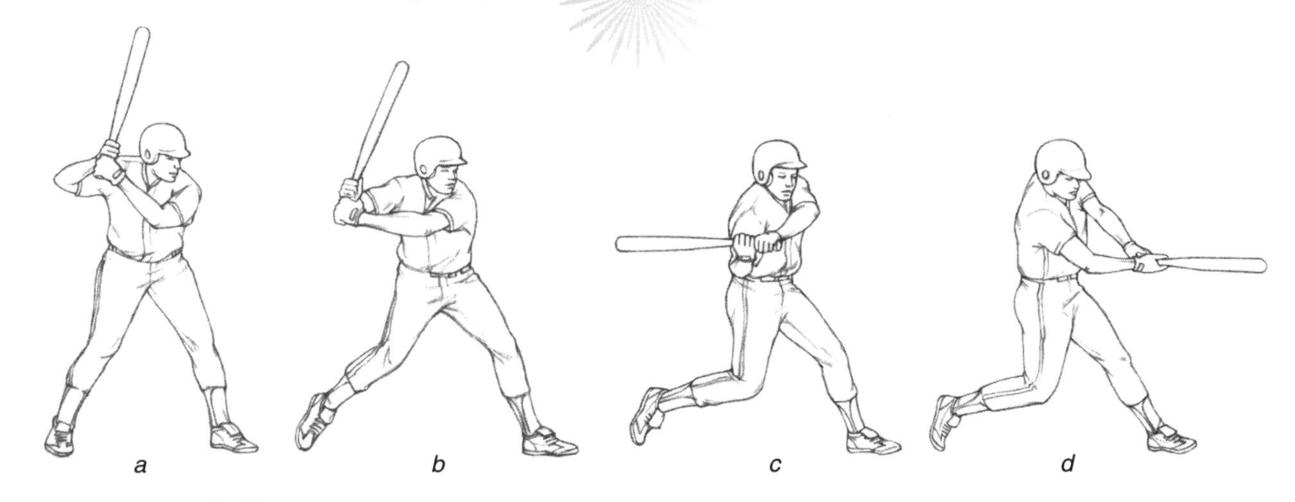

FIGURE 9.5 Baseball batting.

Technique	Mechanics
1. In baseball, a batter faces an approaching ball that is usually spinning and varies in velocity and direction. A baseball bat is cylindrical and has a curved striking surface. Hitting the ball effectively is extremely difficult. Fastballs can have velocities of close to, and even over, 100 mph. Batters have less than a second to react to the pitch.	1. When a batter wants to strike a ball and send it in the opposing direction, the momentum of the bat must be greater than that of the ball. Arm extension, bat length, and the rate of swing determine the velocity of the barrel end of the bat. Timing and coordination are essential for hitting the ball through its center of gravity rather than hitting it off center (i.e., out of line with the ball's center of gravity) and producing a pop-up.
2. Batters take a stance that is slightly wider than shoulder width. Their body weight is close to, or over, the rear foot, which is at right angles to the approaching pitch. Right-handed batters stand with the left side and left hip toward the pitcher. The head is turned so that the batter can concentrate on the pitcher's actions. The bat is frequently held in a ready position, with the barrel end pointing skyward (see figure 9.5a).	2. The initial shoulder-width stance with the body weight above the rear foot allows the batter to shift the center of gravity forward an optimal distance into the hitting stance. With the left side facing the pitcher, right-handed batters can rotate their hips and torsos more than 90 degrees as part of the batting action. Moving the body toward the pitch (plus rotating the hips and torso) increases the distance over which force is applied to the bat and subsequently to the ball.

Technique	Mechanics
3. As the ball approaches the plate, right-handed batters step approximately an additional half-shoulder width toward the pitcher. This puts them into a wide, powerful hitting stance. The batter's arms extend partially away from the body. There is commonly some backward rotation of the shoulders and the bat as the batter shifts forward toward the ball (see figure 9.5b).	3. Movement of the batter's body mass toward the pitch stretches the muscles that subsequently pull the bat around toward the ball. The trailing action of the shoulders, arms, and bat in relation to the movement of the batter's hips will set up a whiplash action characteristic of high-velocity throwing, kicking, and striking skills. Body rotation, coupled with arm extension, increases the angular velocity of the bat.
4. As the batter (right-handed) shifts forward from the right rear foot toward the left, the hips rotate around an axis formed by the left side of the body and the left leg. This action begins with turning in the rear knee and rotating on both feet toward the pitcher. The exact pathway followed by the bat is determined by the batter when reacting to the flight path of the pitch (see figure 9.5, b-c).	4. The batter initiates the sequential rotation of the legs, hips, and torso from the ground up. More massive body segments (i.e., legs, hips, torso) are vigorously rotated then suddenly decelerated. Their angular velocity is multiplied up through the batter's body, along the length of the batter's arms, and out to the barrel section of the bat. The bat rotates in flail-like fashion around an axis formed by the left shoulder and the batter's wrists.
5. When the bat is swung around at high velocity, the batter is forced to lean backward away from the bat.	5. The tremendous angular velocity of the bat requires the batter to generate considerable centripetal force. The batter leans away from the bat and in doing so combines lean with gravity's downward pull on his body mass to counterbalance the inertial and centrifugal pull of the bat.
6. The batter attempts to strike the ball at the bat's "sweet spot." To drive the ball a great distance, the batter tries to avoid a slicing impact, aiming instead to hit the ball through its center.	6. It is best if the batter strikes the ball at the bat's center of percussion (i.e., the bat's sweet spot). If the force applied by the bat fails to pass through the ball's center of gravity, the ball can be undercut and given backspin. A Magnus effect is applied to the ball, counteracting gravity and giving it lift. The ball hangs up in the air, making it an easy target for a fielder. An overcut will give the ball topspin. In this case the Magnus effect joins gravity's downward pull and the ball arcs quickly down toward the earth.
7. The batter's arms swing forward and away from the chest. The left arm will extend as the bat swings around the body into the follow-through. The batter rotates around the left leg and the left side of the body (see figure 9.5d). Vision is on the flight of the ball.	7. The batter's body rotates around the axis of the left side of the body. This extends the radius of rotation from the left shoulder out to the barrel of the bat. A follow-through completes the striking action and dissipates the momentum built up in the rotary action of the batter's body and the bat.
8. The flight of the ball depends on what occurs when the ball is hit plus the effect of environmental conditions.	8. The velocity, direction, and distance that the ball travels after it is hit depend on a large number of factors, including the following: • The momentum of the ball at the instant of impact • The momentum of the bat at the instant of impact • The elasticity (recoil) of the ball • The direction that the bat and ball are moving at the instant of impact • The point of impact between the bat and the ball • The spin put on the ball after impact • Environmental conditions such as altitude, temperature, humidity, and airflow

Spotlight on . . . Football Punting

FIGURE 9.6 Punting a football.

Technique	Mechanics
1. Kicking is a striking action used to apply force with the foot. Punting a football uses the kicking leg in much the same way a javelin thrower uses his throwing arm or a batter uses both arms and a bat to strike a baseball.	1. Punting relates mechanically to striking actions made with a club or bat. It also relates mechanically to throwing actions in which the object thrown is lightweight and moves at high velocity. Kicking also requires the athlete to simulate the actions of a whip. Long legs, great muscular strength, and a huge range of movement help generate a tremendous impact force when the kicking foot contacts the ball.
2. An athlete punts a football by stepping forward onto the supporting foot. This foot is positioned in front of the athlete's body, which is inclined backward (see figure 9.6a). The ball is positioned so that it contacts an extended kicking leg. Vision is on the ball, and the arms are extended from the body to counterbalance the powerful swing of the kicking leg (i.e., left leg).	2. In punting, an athlete overcomes the inertia of his body (and that of the ball) by moving forward toward the direction of kick. This gives momentum both to the athlete and the ball. The supporting foot is placed well ahead of the athlete's center of gravity so that the athlete can move forward, up, over, and beyond this foot, which acts as an axis. The athlete's forward movement extends the time frame that force is applied to the ball. The outstretched arm assists the athlete in maintaining stability throughout the kicking action.
3. As the athlete steps into the kicking stance, the kicking leg (which is flexed at the knee) trails well to the rear of the athlete's body (see figure 9.6a).	3. The trailing action of the kicking leg and the backward tilt of the athlete's body extend the time and distance that force is applied to the ball. The kicking foot travels over a huge arc from its backswing position to the point where it contacts the ball.
4. As the thigh of the kicking leg is swung forward, the lower leg trails to the rear and temporarily rotates in the opposing direction. A 90-degree angle occurs at the knee (see figure 9.6b).	4. The forward swing of the thigh and the backward motion of the lower leg simulate the initial action of a whip, where the more massive lower part of the whip moves forward while lighter sections momentarily move in the opposing direction.
5. After accelerating forward and upward, the thigh of the kicking leg slows down. The lower leg and kicking foot now accelerate. The athlete will have positioned the ball ahead of the body so the kicking foot meets it while moving at high velocity.	5. The high angular velocity of the thigh is arrested. This causes its angular velocity to be passed on to the lower leg, which is less massive and has less rotary resistance. The angular velocity of the lower leg is dramatically increased. The ball is positioned so the impact occurs when the kicking foot has maximum momentum.

Technique	Mechanics
6. The kicking leg is extended when it contacts the ball (see figure 9.6c).	6. There is a powerful extension of the kicking leg as the lower leg rotates around the axis of the knee joint. The kicking foot, simulating the tip of the whip, now moves at maximum velocity.
7. The athlete rises up on an extended supporting leg. The athlete's body tilts backward away from the direction of the punt (see figure 9.6, d-e).	7. Rising onto the toes of an extended supporting leg, coupled with a backward body tilt, enlarges the arc through which the kicking leg swings and through which force is applied to the ball.
8. The athlete's body moves forward past the supporting foot and in the direction of the kick. After the foot contacts the ball, the kicking leg continues with a follow-through. Flexion reoccurs at the knee of the kicking leg (see figure 9.6e).	8. Forward movement of the athlete's body continues the application of force to the ball. Maximum angular velocity of the lower part of the kicking leg occurs just before contact with the ball. Partial flexion of the kicking leg allows the foot to swing upward through a large arc to a position above the athlete's head. The arms are often extended sideways to maintain stability.
9. The distance the ball travels depends on how much force is applied to the ball, the trajectory angle of the ball, environmental conditions, and whether the ball spirals around its long axis or tumbles end over end.	9. The velocity of the kicking foot at the instant it contacts the ball, coupled with the angle of release, determines the distance the ball travels. Feeding the ball onto the kicking foot so it has a high trajectory angle when contact is made is appropriate when kicking with a following wind. A low trajectory angle is used when kicking against the wind.
10. For a spiraling kick, the athlete positions the ball so its long axis is slightly offset from the intended direction of the kick. The kicking foot is swung directly ahead and is drawn across the long axis of the ball.	10. Contacting the ball so the kicking foot moves across the ball's long axis applies torque to the ball, causing it to spin. A spiraling ball will have greater stability and travel a greater distance than a ball that tumbles end over end.
11. The distance the ball travels in flight will be affected by environmental conditions.	11. Wind direction and altitude (air resistance) have a considerable effect on the distance the ball travels.

Pulling, Pushing, Lifting, and Carrying Skills

Skill Highlights

1. Sports that use pull-push actions (e.g., rowing, kayaking, archery, and weightlifting) require the athlete to apply force continuously throughout the desired range of movement. If an athlete wants to apply maximum force to a heavy resistance, the athlete simultaneously uses the largest number of body segments that can be applied to the task (e.g., the legs, back, chest, shoulders, and arms). This simultaneous action differs from high-velocity throwing, kicking, and striking skills in which a sequential movement of body segments occurs.
2. Spotting in gymnastics is a lifting-carrying motion. It requires the spotter(s) to be as close as possible to the athlete without hindering the performance of the skill. Oppositional, or resistive, torque produced by the performer increases in proportion to the performer's body mass, speed of movement, and distance from the spotter(s). Spotters must counter this torque by moving as close to the performer as possible.

Spotlight on . . . Clean and Jerk

FIGURE 9.7 Clean and jerk.

Technique	Mechanics
1. The two Olympic weightlifting events are the clean and jerk and the snatch. The clean and jerk is used to hoist the heaviest weight and is a two-phase lift. The bar is first pulled to the chest, where a pause occurs. The athlete then jerks (pushes) the bar to arm's length above the head. The snatch differs from the clean and jerk in that it is performed with a continuous pulling action with no pause at the chest.	1. The clean and jerk and the snatch are both power events in which strength and speed are combined. In these events the athlete applies great force in a limited time frame to accelerate the barbell upward. At the height of the barbell's upward movement, the athlete must move at high speed into positions where the barbell is stabilized at the chest (for the clean) and above the head (for the jerk and the snatch).
2. The clean and jerk uses two types of leg action. With a heavy barbell, an athlete cleans the bar by first pulling on it and lifting it as high as possible. Without pause, the athlete either squats under the bar or splits the legs forward and backward. The athlete then rises up out of this position, with the bar held at the chest ready for the jerk. To complete the jerk, an upward jerking action is combined with a lunging motion in which the legs are split forward and backward.	2. The quicker the bar moves upward, and the higher it travels, the more time available for the athlete to perform the squat or leg split. The more powerful the athlete or the lighter the resistance, the faster and higher the bar will move upward. A bar pulled upward to a high position does not require a high-speed performance of a deep squat or a leg split as the athlete moves under the bar.
3. The starting position in the clean sees the athlete in a shoulder-width stance with the bar in front of the shins. The athlete grips the bar with the hands slightly wider than shoulder width and equidistant from the plates. The arms and back are straight and the legs flexed no more than necessary. The athlete's back is angled at 45 degrees to the horizontal (see figure 9.7a).	3. The starting position for the clean is such that the bar will be pulled predominantly in a vertical manner. The athlete's partially flexed legs (i.e., approximately a half squat) are in a mechanically efficient position for a powerful extension. The arms are extended and immediately transmit force from the athlete's legs and back to the bar.

Technique	Mechanics
4. The clean begins with a powerful extension of the athlete's legs and back to accelerate the bar upward.	4. The resting inertia of the bar is overcome by a powerful extension of the athlete's legs and back. Both legs and back extend simultaneously, transferring the pull to the bar via the arms. The bar rises in a vertical direction.
5. When the legs have completed their extension, the athlete continues to pull upward by flexing the arms. The athlete rises up onto the toes and hyperextends the back. The pull of the arms completes the vertical movement of the bar, which travels upward close to the athlete's body. The head is thrown back (see figure 9.7b).	5. The large muscles of the legs and back apply the greatest force to the bar. Driving up onto the toes and pulling upward with the arms allow the athlete to continue applying force to the bar over a long time frame. The athlete pulls the bar upward as close to his line of gravity as possible. In this way the center of gravity of the bar stays above (and within) the athlete's supporting base, and the athlete and bar are optimally stabilized.
6. When the barbell has risen to just below the pectoral muscles, the athlete squats down under the bar, simultaneously rotating the arms forward so the bar is pulled in toward the upper chest and held in place with the arms (see figure 9.7c).	6. The greater the mass (weight) of the barbell, the more critical becomes the upward pull on the bar and the velocity with which the athlete rotates the arms forward and squats under the bar. A sufficiently high pull, fast arm rotation, and equally fast squatting action are absolutely essential.
7. The athlete uses leg strength to lift up from a front squat position (see figure 9.7c-d).	7. With the bar at the chest and in a squat position, the athlete must use great leg strength to battle the deadweight (i.e., inertia) of the barbell (plus the inertia of his own body mass) so he can rise to a standing position. With insufficient strength or too massive (i.e., too heavy) a barbell, the athlete will either fail to get up out of the squat or be physically drained from the effort of standing up and fail to complete the jerk.
8. In a standing position and with the bar held at the chest, the athlete dips and flexes the legs slightly. The weight discs at either end of the bar cause the bar to flex downward then upward.	8. A slight flexion prestretches the leg muscles before the extension of the legs. The upward recoil of the bar is timed to coincide with the leg extension to assist in thrusting the barbell upward.
9. The instant the athlete's legs have completed their thrust and the bar is rising upward, the legs split forward and backward. One leg is extended directly backward at an angle of about 45 degrees. The opposite foot steps forward 10 to 11 in. This allows the athlete to lunge forward under the bar (see figure 9.7e).	9. The athlete's leg thrust and arm extension (with no leg split) is adequate to elevate a lightweight barbell. With a heavier resistance (which cannot be elevated to arm's length above the head with leg thrust and arm extension alone), the athlete is forced to split the legs at high speed and drop very low under the bar.
10. With the bar held above the head at arm's length, and with the legs split forward and backward, the athlete carefully brings both legs back toward their original position. The athlete finishes the lift standing with the feet shoulder-width apart and with the barbell held at arm's length above the head. For approval from the judges, the athlete must demonstrate control over the barbell for 3 sec.	10. With the barbell above the head, the combined center of gravity of athlete and barbell is raised upward, and with increased height, both athlete and barbell become progressively unstable. The athlete must struggle to keep the line of gravity of the barbell centralized above his small, narrow base in order not to lose control and fail in the lift.

Swinging and Rotating Skills

Skill Highlights

1. The same mechanical principles that control angular motion govern swinging. In swinging skills, an athlete stretches out on the downswing. This action moves the athlete's center of gravity as far from the axis of rotation as possible. By carrying out this action, the athlete allows gravity to exert maximal accelerative torque to his body on the downswing.

2. To rise high on the upswing, an athlete counteracts the decelerating effects of gravity by flexing at the hips and shoulders. This action pulls the athlete's center of gravity closer to the axis of rotation, which reduces gravity's decelerative torque.
3. When performing somersaults and other rotary skills in the air, an eccentric force (i.e., a force that does *not* pass through the athlete's axis of rotation) is applied to the athlete's body at the start of the skill. An eccentric force applies the turning effect of torque to the athlete's body. A slight body lean in the direction of rotation at the instant of takeoff is a common method for applying this eccentric force.
4. Reducing the athlete's rotary resistance during flight increases angular velocity (i.e., the rate of spin). Rotary inertia is reduced by using muscular force to pull the athlete's body mass in toward the axis of rotation. Angular velocity is decreased by extending the athlete's body mass outward from the axis of rotation. This action increases the athlete's rotary inertia and reduces the athlete's rate of spin.

Spotlight on . . . A Back Giant on the High Bar

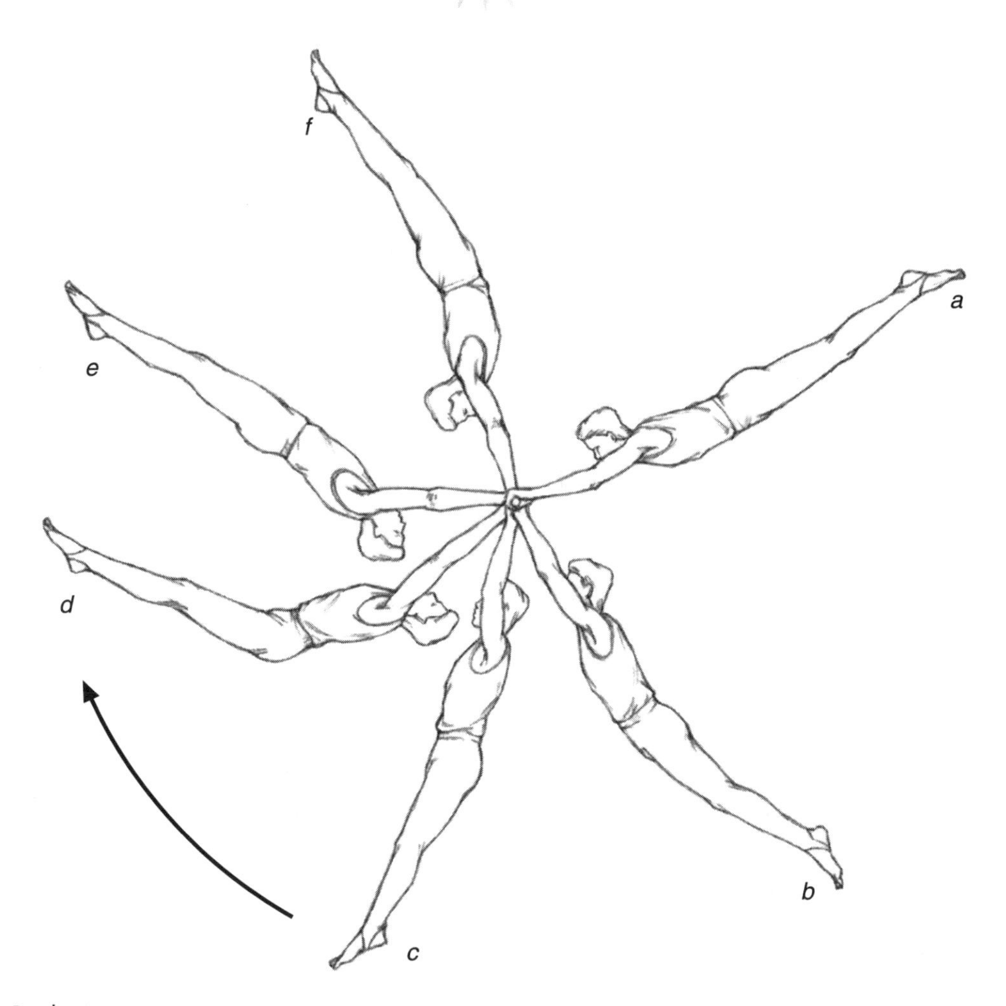

FIGURE 9.8 Back giant.

From M.J. Adrian and J.M. Cooper, *Biomechanics of human movement*, 2nd ed. Copyright © 1989. (Dubuque, IA: Times Mirror Higher Education Group Benchmark Press), 509. Reproduced with permission of The McGraw-Hill Companies.

Technique	Mechanics
1. A gymnast frequently begins in a front support position and "casts" (i.e., thrusts against the bar) to elevate his body to a position where his center of gravity is well above the bar.	1. By elevating his center of gravity high above the bar, a gymnast increases his potential energy. Gravity then has the opportunity to accelerate him from greater height on the downswing. The higher a gymnast raises his center of gravity, the greater the time frame during which gravity can accelerate him downward. This increases the athlete's angular velocity and angular momentum.
2. The handgrip for a back (i.e., backward) giant is an overgrip. For a front (i.e., forward) giant, the handgrip is an undergrip (a reverse grip).	2. Standing facing the bar, an overgrip has the knuckles above the bar and thumbs below. In a back giant, the thumb "leads" the athlete through the skill and the bar "rotates" into the hand for a strong, safe grip. An undergrip, or reverse grip, is used for a front giant. Standing facing the bar, the hands are turned out so that the thumbs are above the bar and the knuckles below. Again, the thumb "leads" the athlete through the skill and the bar "rotates" into the hand for a strong, safe grip. Failure to use the correct grip will cause the athlete to come off the bar during the performance of either of these skills. For greater safety with the more advanced high bar skills, gymnasts use handgrips with dowels inserted in them, which help lock the gymnast's hand on the bar.
3. Once well above the bar and beginning the descent, the gymnast stretches to full extension (see figure 9.8a).	3. By stretching to full extension, a gymnast pushes his center of gravity as far as possible from the axis of rotation (i.e., the bar). This allows gravity to apply the greatest amount of accelerative torque to the gymnast's body on the downswing. Just before reaching a position directly below the bar, the gymnast flexes the muscles of the spine. This action prestretches the abdominal muscles, which are ready to flex the hips when the athlete rises on the upswing (see figure 9.8b).
4. Gravity accelerates the gymnast's body downward.	4. Gravity applies an accelerative torque to the gymnast's body. This torque increases until it is maximal when the gymnast is stretched horizontally from the bar. From that point, although the gymnast's angular velocity and momentum continue to increase, the amount of torque applied to the gymnast's body gets progressively less until it is zero when the gymnast passes directly below the bar.
5. After passing below the bar, the gymnast flexes at the hips and at the shoulders (see figure 9.8, c-d). Beginners may also flex at the knees.	5. Flexion at the hips and shoulders brings the gymnast's center of gravity closer to the axis of rotation. Gravity's decelerative torque is reduced. By flexing, the gymnast also pulls his mass closer to the bar. This reduces the gymnast's rotary resistance (i.e., rotary inertia).
6. In a flexed body position, the gymnast rises upward. The flexion in the body is progressively eliminated as the athlete moves to a position vertically above the bar (see figure 9.8, e-f). Some pulling action is applied to the bar to get the gymnast's body over the top of the bar.	6. With a reduction in the gymnast's rotary inertia and a reduction in gravity's decelerative torque, the gymnast is able to rise to a position vertically above his axis of rotation (i.e., the bar). Immediately before reaching the vertical position, the athlete can extend his body to slow down (i.e., control) the rate of spin around the bar.
7. Once above the bar and beginning the descent, the gymnast extends the body again to repeat the process described in steps 1 to 5.	7. Above the bar, the gymnast straightens out to shift his center of gravity as far from the bar as possible. This allows gravity to reapply maximum torque to the gymnast's body on the downswing. In this way, continuous giants are performed.

Spotlight on . . . Aerial Front Somersault

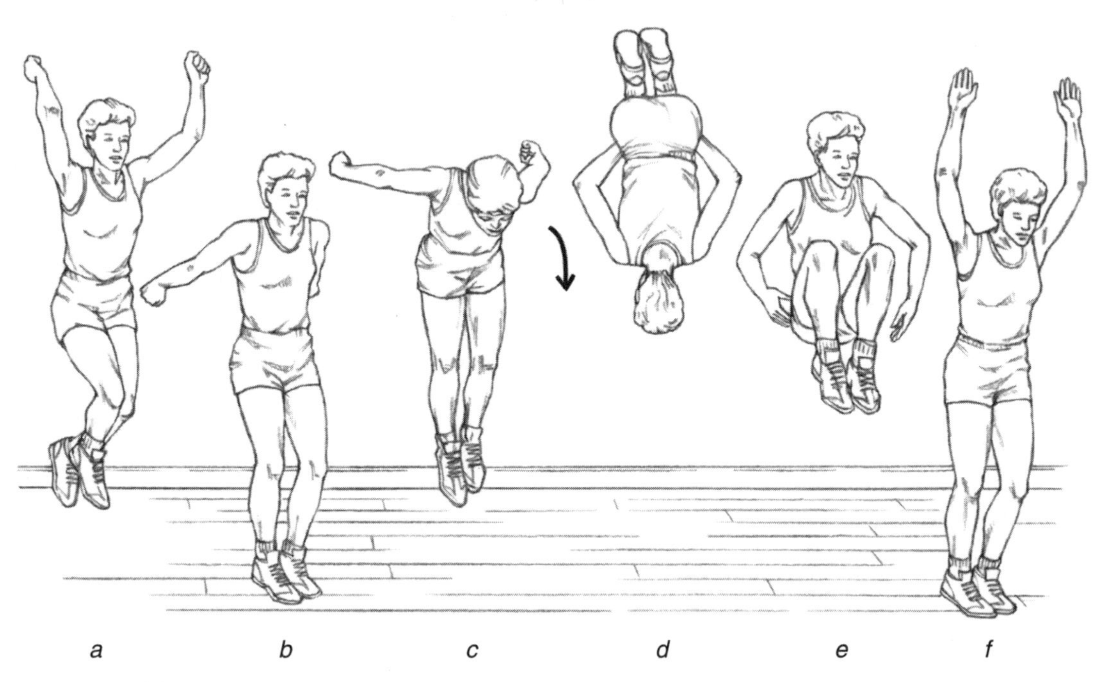

FIGURE 9.9 Aerial front somersault.

Technique	Mechanics
1. To perform aerial rotations, the athlete can begin either with an approach or from standing. An approach allows the athlete to use the same technique as a high jumper for gaining height. The feet are placed ahead of the center of gravity, the quadriceps muscles are prestretched, and the arms are swung upward to assist at takeoff (see figure 9.9, b-c).	1. An approach before takeoff allows the athlete to position the center of gravity to the rear of the takeoff foot. This allows the athlete to rock forward over the takeoff foot and apply more impulse (force × time) against the takeoff surface. A hurdle step coupled with a drop onto a sprung surface (e.g., a tumbling mat, beatboard, or trampette) prestretches the quadriceps muscles and stores strain energy in the apparatus. This is added to the athlete's leg thrust to propel the athlete upward.
2. During the takeoff the arms are swung upward at the same time that the legs thrust down at the supporting surface (see figure 9.9b). Arm swing can be forward and upward or, as in this sequence, backward and upward.	2. Arm swing during the takeoff provides momentum transfer to the athlete's body as a whole. The action of the arms adds to the downward thrust of the legs and increases the reactional force of the takeoff surface (e.g., floor exercise mat, beatboard, or springboard) thrusting upward against the athlete.
3. The thrust of the legs can be made against a sprung surface (e.g., springboard, beatboard, trampette, or gymnastic floor exercise mat).	3. Downward thrust by the athlete produces flexion in a sprung surface. A sprung surface stores strain energy. The recoil of the flexed surface helps drive the athlete up in the air.
4. An angle of takeoff close to the vertical, coupled with a powerful upward thrust, guarantees the most airtime for somersaulting (and twisting) skills (see figure 9.9b). In diving, horizontal movement is necessary to clear the end of the board. This is not the case in floor exercises or during trampoline skills.	4. Vertical thrust is always partially compromised when horizontal movement and rotation are required. To perform the most somersaults in the air, the athlete must thrust downward as close to vertical as possible (to produce the greatest airtime) while still initiating rotation. In the sport of diving, the athlete must also apply some horizontal thrust to move a safe distance from the board.

Technique	**Mechanics**
5. To initiate rotation, the thrust that drives the athlete upward must not pass through the athlete's center of gravity. Swinging the arms and shifting the head and trunk can assist in promoting rotation (see figure 9.9, b-c).	5. An eccentric force causing rotation occurs when the upward thrust from the athlete's legs and the supporting surface is directed at a distance from—rather than through—the athlete's center of gravity. This produces the turning effect of torque (force × distance from axis of rotation), which causes the athlete to rotate. Rotation can be enhanced when the arms, head, and trunk are also thrown in the direction of rotation. This form of momentum transfer must occur while the athlete is still in contact with the supporting surface.
6. The athlete's flight path is set at takeoff and cannot be changed in flight.	6. Any amount of body movement in the air will not change the flight path of the athlete's center of gravity once it is set at takeoff. Likewise, an athlete's angular momentum is also set at takeoff. Angular momentum at takeoff depends on the mass of the athlete, the distribution of the athlete's mass at takeoff relative to the athlete's axis of rotation, and the angular velocity (i.e., rate of spin) initiated at takeoff.
7. To perform somersaults in the air the athlete must have the following: (a) Vertical thrust to provide optimal airtime (b) a takeoff with the body in an extended position, (c) as much rotation at takeoff as possible without compromising airtime or an extended body position, and (d) sufficient muscular strength and flexibility to pull the body into the tightest tuck possible while in the air	7. A rotating body in an extended position produces considerable angular momentum. Since angular momentum is conserved in flight, pulling the athlete's body mass in toward the axis of rotation produces an increase in angular velocity (i.e., rate of spin). A shift from a fully extended body position to the tightest tuck possible results in the greatest increase in the athlete's rate of spin (i.e., angular velocity). In reverse, a shift from a tight tuck to an extended body position reduces the athlete's rate of spin.
8. During flight, axes for somersaults and twists always pass through the athlete's center of gravity.	8. Irrespective of changes in the athlete's body position (e.g., shifting from an extended body position to a tuck, or whatever combination of twist and somersault the athlete performs), axes of rotation always pass through the athlete's center of gravity. Even when apparently out of control, the athlete's body mass is balanced and in a state of equilibrium around her center of gravity.
9. All actions performed in the air cause equal and opposite reactions to occur. A counterclockwise movement of one part of the athlete's body causes a clockwise movement of some other part. These occur simultaneously. Pulling the head and shoulders forward (clockwise) into a tuck causes the flexed legs to move in the opposing (counterclockwise) direction, which is toward the head and shoulders (see figure 9.9, c-d).	9. The arc of movement of the athlete's body segments producing the action and the arc of movement producing the reaction depend on their respective rotary resistance. More massive body segments that are extended farther from the axis of rotation shift a smaller distance (or arc) than those that are less massive and closer to the axis of rotation.
10. Variation of body position in flight allows an athlete to control the number of somersaults (and twists) that are performed. By extending the body, the rate of rotation can be reduced for a headfirst entry (in diving) or a feetfirst landing (in floor exercises and many skills on the trampoline) (see figure 9.9, d-f).	10. Compressing the body mass in toward the axis of rotation reduces its rotary resistance. Because angular momentum is conserved in flight, a reduction in rotary resistance causes an increase in angular velocity. Extending the body increases rotary resistance and decreases the athlete's angular velocity.

Balance and Stability

Skill Highlights

1. Athletes in the sport of judo and other combative sports use combinations of rotation, pulling, pushing, and lifting to lessen an opponents' stability and set them up for a throw. Opponents counter an attack by leaning toward a push and away from a pull. To increase stability and make themselves less vulnerable, athletes widen their base and lower and centralize their center of gravity. Weight divisions in combatives are intended to negate advantages from body mass.
2. Maintaining stability and destroying the stability of an opponent in judo and other combative sports are predominantly a battle of one torque versus another. Pushing, pulling, and lifting are used to spin an opponent around axes formed by the athlete's feet, hips, back, and shoulders. Leg sweeps in judo are a common method of eliminating an opponent's base of support.

Spotlight on . . . Judo Hip Throw

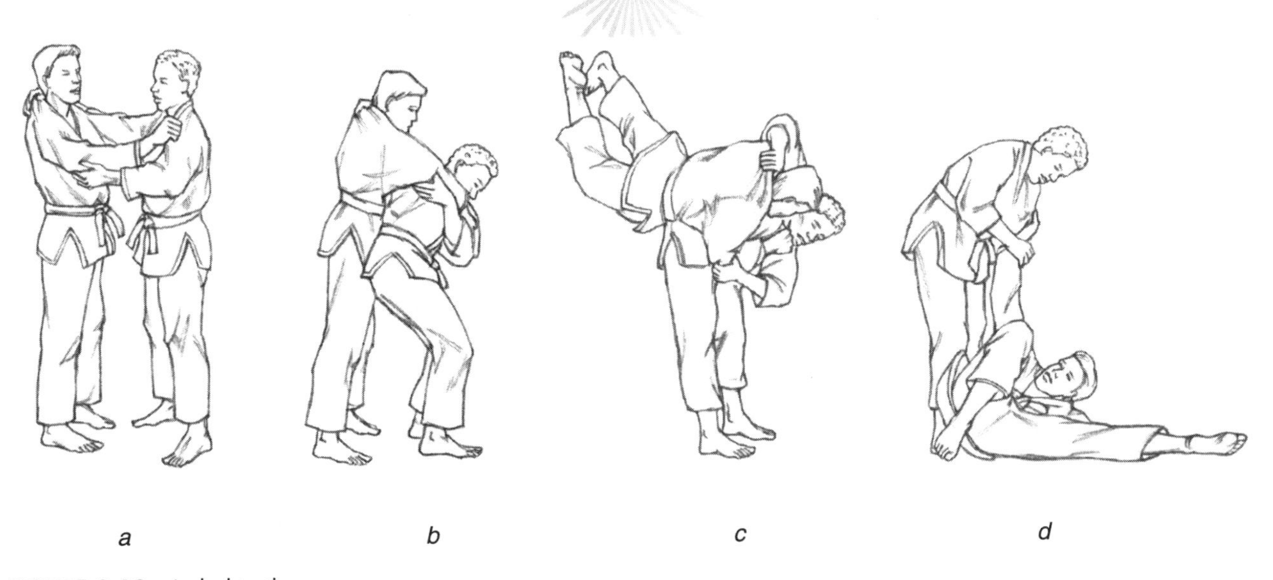

FIGURE 9.10 Judo hip throw.

Technique	Mechanics
1. Judo involves pushing, pulling, lifting, and rotating, all of which are designed to maintain the attacker's stability while disrupting that of the opponent. Precise timing, coordination, and superfast reactions are essential.	1. The intent of the attacker is to shift the opponent's center of gravity outside of the opponent's supporting base and so destroy his stability. Superfast combinations of pushing, pulling, lifting, and rotating are used to achieve this end.
2. Judoka (i.e., judo practitioners) face each other in a standing position with their bodies slightly lowered and with their legs partially flexed. Their feet are at right angles to each other and shoulder-width apart. Fast, shuffling, flat-footed steps are taken, with the athlete's body weight frequently positioned closer to the front foot.	2. Lowering the center of gravity increases the stability of the athlete. Positioning the feet at right angles gives good stability side to side and front to back. Fast, shuffling steps increase stability because they limit time spent on one foot. Having the line of the center of gravity closer to the front foot readies the rear leg for leg sweeps and other destabilizing actions used against the opponent.
3. Judoka begin their attacking and defensive maneuvers by grasping each other's tunics on the collar at shoulder level with one hand and at the sleeve with the other (see figure 9.10a). Grips are suddenly altered according to the throw being attempted.	3. Grips on the tunic are designed to facilitate pull-push and rotational movements. Any combination can be used. Applying force at the sleeve rotates the opponent around his long axis. Applying force at the collar and upper lapel produces forward and backward movement or rotation.

Technique	Mechanics
4. With grips on each other's tunics, judoka circle each other, waiting for the opportunity to initiate a throw.	4. Preparation for a throw is a series of pull-push rotary actions in which the prime objective is to move the opponent's center of gravity into a position of minimal stability.
5. In a hip throw, which incorporates a lifting action, the attacker grasps the rear part of the collar with the right hand. The left hand grips the opponent's sleeve below the right arm. The attacker pulls the opponent forward with both hands (see figure 9.10a).	5. Pulling the opponent's body forward, the attacker lessens the opponent's stability by moving the line of his center of gravity closer to or beyond the edge of his supporting base. A high grip on the rear collar maximizes the force arm from the collar to the axis of the attacker's hips, over which the opponent will subsequently be rotated. The grip below the right arm will spin (i.e., rotate) the opponent around his long axis.
6. The moment the opponent is successfully drawn forward and destabilized, the attacker quickly initiates a pivot on his left foot by stepping across with his right while flexing the legs (see figure 9.10b).	6. By rotating his body to the left, the attacker prepares to use his hip as an axis of rotation over which the opponent will be pulled. Flexing his legs, the attacker not only increases his own stability but also prepares to totally destabilize the opponent by lifting him out of contact with the mat (and the earth) when he extends his legs.
7. The attacker's lower back is pressed against the opponent's thighs. The attacker's upper back is pressed against the opponent's abdomen. The opponent's upper body is pulled downward over the attacker's hip with both hands while the opponent's feet are lifted off the mat when the attacker straightens his legs (see figure 9.10c).	7. The attacker's upper hip becomes an axis of rotation over which the opponent rotates. A force is applied by the attacker's hands, which pull (and rotate) the opponent's upper body downward. Another force is applied by the extension of the attacker's legs, which drive the opponent's lower body upward. The opponent spins around the axis of the attacker's hips.
8. The opponent's lower body is forced upward, taking his feet out of contact with the mat. The opponent's upper body is pulled downward. The opponent rolls over the attacker's hip (see figure 9.10c).	8. The attacker's downward pull and leg extension eliminate the opponent's contact with the mat. Frictional forces between the opponent and the mat no longer exist. The opponent is now defenseless and will be thrown onto his back.
9. A breakfall is performed by the opponent on contact with the mat (see figure 9.10d).	9. Totally destabilized, the opponent prepares for the impact with the mat that will occur after the fall. The opponent correctly enlarges the time frame and area over which the impact is made with the mat.

Arresting Motion

Skill Highlights

1. Catching, landing, slowing down, and stopping are all forms of arresting motion. Arresting motion involves the forces present in a collision when two or more objects come together. Athletes need to use correct technique when arresting motion, whether it's their own body, an opponent, or the motion of an inanimate object such as a baseball.
2. To safely arrest their own motion, athletes apply a stopping force to their bodies over as large a distance and time frame as possible. In addition, the impact is spread over as big an area as possible. In this way the pressure applied at any one spot to the athlete's body is reduced. Pulling the arm back when catching, flexing the legs and rolling when landing, and using padding and crash pads gradually dissipate the forces that occur during an impact. An athlete can then avoid injury (which occurs when great force is stopped in an instant in a small area) and maintain control and stability when the impact occurs.

Spotlight on . . . Judo Breakfall

FIGURE 9.11 Judo breakfall.

Technique	Mechanics
1. Mastery of ukemi, or the art of falling, is necessary in judo, not only to facilitate movements that follow the fall but also to prevent injury when a judoka is thrown.	1. Falls in judo occur from various heights (e.g., from as high as an opponent's shoulders). They also occur at various velocities. The opponent can add his own muscular force to that of gravity. The momentum generated during the fall equals the mass of the falling judoka multiplied by his velocity.
2. When a judoka hits the mat, the athlete's velocity is reduced to zero.	2. The judoka's velocity (and momentum), which built up during the fall, is reduced to zero on contact with the mat. The force with which the judoka hits the mat is applied against the athlete's body by the reaction of the earth.
3. Ukemi teaches the judoka to strike the mat simultaneously with the arms and legs as the trunk makes contact. In this way the trunk is protected because it does not absorb the full force of the fall (see figure 9.11d).	3. By striking the mat with the arms and legs, the judoka uses the shoulders, thighs, and lower leg as shock absorbers. By carrying out this action, the judoka extends the time frame during which force is applied against the earth and, in reaction from the earth, against the athlete. An extended time frame over which force is applied causes a more gradual change in the judoka's momentum (and kinetic energy), and with this comes less chance of injury.
4. Judoka attempt to turn vertical movement in a fall into an oblique and subsequently horizontal movement by flexing and bending the body. They roll on the shoulders, back, and arms (see figure 9.11, a-d).	4. A situation where the judoka falls vertically onto the mat brings an athlete to a sudden stop. Considerable force is applied to the earth, and from the earth against the athlete, in a small time frame. Flexing the legs, squatting down, and rolling extend the time frame during which force is applied by the judoka against the mat and, in reaction, by the earth via the mat to the judoka.
5. By rolling and contacting the mat with the extended arms and legs, the judoka enlarges the surface area of the body that contacts the mat (see figure 9.11, b-e).	5. Enlarging the contact area of the body as the judoka's body contacts the mat decreases the pressure applied at any point on the athlete's body. Enlarging the contact area and extending the time frame that force is applied to the earth (and from the earth to the athlete) significantly reduce the possibility of injury as a result of being thrown.

ANSWERS TO REVIEW QUESTIONS

Chapter 2

1. Yes. Weight is a function of the attractive force of gravity. Standing on the surface of the earth, the closer an athlete is to the earth's core, the greater the pull of gravity and the more the athlete weighs. The earth is not perfectly round, and an athlete is closer to the center of the earth while standing at the North or South Pole than while standing at the equator. Height above sea level also plays a factor, so an athlete standing at the top of a mountain at the equator reduces the effect of gravity even more. At the equator not only will an athlete be farther from the earth's core, but because of the extra distance from the earth's axis as the earth spins around, the athlete's mass also pulls away more from the earth than when he stands either at the North or South Pole. A very sensitive scale could register the difference.
2. The earth and the shot pull on each other with the same force. The earth moves toward the shot relative to its mass. The earth's tremendous mass gives it a phenomenal amount of inertia, so it moves an immeasurably small distance toward the shot. The shot, having little mass and inertia, does most of the moving, so it accelerates toward the earth.
3. An athlete (e.g., a springboard diver) gains an additional velocity of 32 ft/sec (9.8 m/sec) with every second that he falls toward the water. Gravity's acceleration adds a velocity of 32 ft/sec for every second of fall (i.e., 32 ft/sec/sec, or 32 ft/sec^2). In the metric system this would be 9.8m/sec^2. If we wrote 32 ft/sec (with one distance unit and only one time unit), it would indicate speed but not acceleration.
4. When a skier extends her legs, she presses down against the earth, and the earth reacts by pushing upward with an equal and opposite force. This increases the friction between the skis and the snow. By flexing her legs, the skier momentarily reduces the pressure of her body against the earth. The earth, in response, reduces its reaction force. The reduction in these forces decreases the friction between the skis and the snow. Weighting (extension) and unweighting (flexion) assist in carving turns and in other skiing maneuvers.
5. Lean, lightweight athletes tend to have a better strength-to-weight ratio than do more massive athletes. Having less mass, squash and badminton players also have less inertia to fight against than do more massive athletes; so they can stop, start, and maneuver more easily.

6. The quarterback assesses the velocity of the receiver cutting across the field. He also assesses the distance from where he is throwing the ball to the place where the ball will be received. The quarterback then applies the right amount of force in a vertical and horizontal direction (taking into account environmental forces, such as the force of the wind and its direction). The ball must also be spun to better maintain its trajectory.
7. Newton's first law is his law of inertia. This law states that any object that has mass will have inertia, and the characteristics of inertia are resistance and persistence. If an object is motionless, it will want to remain motionless. If it is moving, it will want to continue moving in a straight line and at a uniform velocity. Newton's third law is his law of action and reaction. This law states that all forces act in pairs. If "A" acts on "B" then "B" acts on "A" with the same magnitude and in the opposing direction.

Chapter 3

1. The answer is the collective mass of the goalkeeper plus the ball. They form a single large mass and accelerate backward into the goal.
2. The baseball is traveling at tremendous velocity and has phenomenal momentum. The forces exerted by the ball on your hand (when you don't pull your arm back) are reduced to zero in an instant. Reducing this great force over a minimal time frame with your bare hand is what causes so much pain. Drawing your arm back at the instant of contact lengthens the time frame that the momentum of the ball is reduced to zero, lessening the force (or pressure) that the ball exerts on your hand at any instant during the catching sequence. When you wear a glove, it helps extend the time frame over which the ball's momentum is reduced to zero. The glove also spreads the pressure exerted by the ball over a larger area than if you used your hand alone.
3. When two football players collide with each other, the total momentum that they bring into the collision is the same as the total momentum that exists after the collision. What one player loses in momentum the other player gains. Kinetic energy is dissipated in various ways. Some goes into the work that the two players perform on each other, and the rest of the kinetic energy is dissipated as heat and noise.
4. The athlete has finished traveling upward and is no longer moving, so his kinetic energy is zero. Because he's at the top of his flight path, his potential energy is maximal. The pole, having done its job and straightened out because the athlete is at the peak of the vault, has zero strain energy. The force of gravity still exists, even though for an instant the athlete is traveling neither upward nor downward.
5. The distance the sled slides is an expression of work performed by an object that has motion. The energy that an object possesses by virtue of its motion is called kinetic energy. If you double the velocity of a moving object, you square its kinetic energy. On the second run the sled has four times the kinetic energy and four times the ability to do work. Consequently, the sled will slide approximately four times farther. This distance is an approximation because kinetic energy is also used in generating noise and heat.
6. The mass of the athlete's body in a one-hand handstand is supported by a much smaller base than in a two-hand handstand. If the gymnast could balance on the tip of a single finger, rather than on one hand, the pressure on the finger would be even greater because the supporting base has been reduced even more.

Chapter 4

1. The distance between the insertion of the biceps on the forearm and the elbow joint (i.e., the axis of rotation) is virtually the same for athletes of the same age and maturity. What varies greatly is the length of athletes' forearms. A long forearm lengthens the size of the resistance arm, and thus it increases the resistive torque produced

by the dumbbell. Consequently, an athlete with long forearms must produce more muscular force than an athlete with short forearms.

2. Holding the bat at the hitting end shifts the mass of the bat closer to the body axis around which you are swinging the bat. Mass that is shifted closer to the axis of rotation reduces rotary inertia. Consequently, it is easier to accelerate, decelerate, and maneuver the bat when you hold it at the barrel end. This does not mean that you will hit a baseball farther when holding the bat in this fashion!
3. When you swing your arms upward at takeoff, the momentum of the arms is transferred to your body as a whole. The arm swing combines with your leg thrust downward against the earth. In reaction, the earth thrusts upward with greater force. Adding the arm swing to the leg thrust means the earth returns more thrust to the athlete.
4. An athlete's angular momentum is the product of three components: the athlete's rate of spin (i.e., angular velocity), the athlete's mass, and the distribution of the athlete's mass (i.e., how extended or compressed the athlete is relative to his axis of rotation). The flight of athletes in the events mentioned lasts only a very short time. The athletes cannot push or pull against the air to increase or decrease their angular momentum. As a result, the angular momentum they give themselves at takeoff stays the same (i.e., it is conserved) for the duration of the flight.
5. The athlete's hips lift upward. Timed correctly, this reaction helps the athlete clear the bar.
6. The cat twist.

Chapter 5

1. Linear stability is a measure of the ability of an object to withstand forces (and torques) that would disrupt its linear motion. The more mass an object has, the more inertia it possesses. With greater inertia, the object has greater ability to continue in a straight line at a uniform speed. Consequently, it has greater linear stability. Athletes have mass so these principles apply to athletes as well as inanimate objects.
2. The athlete's foot on the beam is an important axis of rotation, although other axes can occur where there are joints in the athlete's body. Using the athlete's foot on the beam as the only axis of rotation means that the following can occur: If the athlete tips to one side, the athlete's center of gravity is no longer above the supporting foot, which provides the sole base of support. Gravity provides a constant force that pulls the athlete toward the earth. The distance that the athlete's line of gravity shifts from a position above the axis of rotation increases the size of the force arm and with it, the magnitude of the destabilizing torque. The athlete must use vigorous leg and arm swings to counteract this destabilizing torque to regain a balanced position. These actions are not always successful.
3. Assume that the attacker pushes against the defender from the left and continues to push against the defender from left to right. If the defender enlarges his base in the same direction, then more torque is required to destabilize him than if his base was not enlarged. In this situation the defender will usually shift his center of gravity toward the left as well.
4. Maximal stability means widening the base, lowering the center of gravity, and shifting the line of gravity into a central position within the base. Such a position would mean that far too much time is taken shifting the athlete's line of gravity toward the perimeter of his supporting base when the athlete is required to move quickly in a particular direction. It is better if the athlete's base is reduced in size and the legs are partially flexed, ready to drive quickly in whatever direction is necessary.
5. The heavier the discus, and the farther its mass is distributed from its axis of rota-

tion, the better it can fight against external forces that would disturb its flight.

6. When a weightlifter weighing 200 lb raises twice his body weight to arm's length above his head he is elevating the combined center of gravity of himself and the barbell a considerable distance. Raising the mass of any object above the surface of the earth reduces its stability. The weightlifter's feet are close together. This position decreases the size of the weightlifter's supporting base. The slightest movement of the barbell is likely to destabilize the weightlifter.

Chapter 6

1. When a scuba diver breathes compressed air through a regulator, the diver is breathing air at the same pressure as the surrounding water. At a depth of 66 ft, the water pressure is triple that on the surface. This pressure must also exist in the lungs. If divers hold their breath at 66 ft and ascend to the surface, the compressed air in their lungs triples in volume. Serious injury or death can occur from this action.
2. Fat weighs less per unit of volume than bone and muscle. The sumo wrestler will occupy an immense space in the water and, in return, is likely to experience a buoyant force greater than his weight of 400 lb. The lean, muscular gymnast will likely experience a buoyant force less than 100 lb and therefore will sink. In this scenario we assume that the water in the swimming pool is fresh water and not salt water.
3. Fred was pulled along in the low-suction wake of the pace vehicle. In this position, streamlining was unnecessary.
4. Drag forces are reduced dramatically when cycling in a recumbent position and when a cyclist is enveloped in an aerodynamic shell. The top speed by a sprint cyclist is about 45 mph. Athletes cycling in a recumbent position, and totally covered with an aerodynamic shell, have reached speeds of 80 mph.
5. For many years, lift forces were thought to be the predominant propulsive force in swimming. It was felt that water flowing over the upper and lower surfaces of the hand and forearm generated a lift force similar to airflow over the wing of a plane. The in-out-up-down action of the swimmer's hands in a stroke like the freestyle were thought to be generating lift forces that propelled the athlete through the water. Although lift forces do exist, they are now considered secondary to drag propulsion. As the swimmer's hands and arms are thrust backward, the drag generated in opposition to this motion is the force that propels the athlete forward through the water. Long pulling and pushing motions produce considerable drag forces and these are responsible for thrusting the athlete forward through the water.
6. A spinning ball drags around with it a boundary layer of air. On one side the boundary layer collides with the air passing by. Airflow is decelerated at this point, and a high-pressure zone is set up. On the opposing side, the boundary layer is moving in the same direction as the air passing by, so there is no collision and the air collectively moves faster. As a result a low-pressure area is set up. The pressure differential, high on one side and low on the other, creates a lift force that causes the ball to move in the direction of the pressure differential (i.e., from high to low). This is called the Magnus effect.

Chapter 7

1. The objectives and the special characteristics of the skill ultimately determine what technique is used in the skill. If you do not fully understand a skill's objectives and characteristics, you may recommend a technique that satisfies one objective of the skill but does not satisfy other important objectives.
2. A predictable environment is a situation in which an athlete can perform a skill without expecting sudden changes in weather or field conditions and without having to face and counteract the maneuvers of an opponent. Two examples are

weightlifting and a synchronized swimming routine.

3. No two athletes are exactly alike. A young, immature athlete will not have the power, coordination, or endurance of an elite athlete. Modifications must be made, but the basics of a technique used by elite athletes can be taught to novices as the foundation of skilled performance.
4. The process of analysis is made easier when a skill is divided into phases. It lessens the possibility of confusion that can occur when you try to analyze all of the complex movements of a complete skill simultaneously.
5. A skill can be viewed as a building: The phases are the building's walls, and the key elements are the bricks in the walls. Breaking down a skill into phases and key elements helps in its analysis and in the correction of errors.
6. Each element in each phase of a skill serves a particular mechanical purpose. If you do not know the mechanical reasons for performing each key element in a particular way, you will be unable to discern why one action is better than another, therefore allowing any manner of performance.

Chapter 8

1. Begin by observing the whole performance and assessing its results. Then shift your attention to the preliminary movements and mental set. Progress to the windup (or recovery) and to the force-producing phase and follow-through. This sequence is worthwhile because errors in earlier phases influence the phases that follow.
2. The recovery in a swimming stroke sets the athlete up for a repeat of the force-producing phase. The follow-through in a throwing skill does not set the athlete up for a repeat of the skill but rather works to dissipate the momentum and energy of the force-producing action.
3. Usually there is an appearance of tension, stiffness, and poor coordination. The athlete appears tight, and the performance appears awkward and jerky. In throwing and hitting skills, novices commonly apply force with the muscles that move smaller body segments before the larger muscle groups have accelerated the athlete's body as a whole, and immediately thereafter, they move the larger, heavier body segments. In throwing skills this incorrect sequence gives the appearance of the athlete throwing with the arm alone and forgetting to use the rest of the body.
4. Too much or too little muscular force over too long or too short a time frame produces less than optimal performance. Superior coaches talk of timing and coordination when they coach a skill. They may say, "Easy at the start" and "Fast at the finish" or "Float through this section" and "Explode at the finish." Such commands teach athletes to apply the right amount of force for the proper duration.
5. A major error is any action that seriously detracts from an athlete's ability to apply force optimally. Poor balance, poor timing, and incorrect application of force in amount, timing, distance, and direction are all major errors in the performance of a skill.
6. You can refer to texts and also to information provided on Internet sites on your chosen sport that offer successful teaching methods and techniques for the correction of errors. You can attend coaching seminars and workshops, where you can gain information from experienced coaches and specialists in the field. You can make use of information gained from research using the most up-to-date video and computer technology.

ANSWERS TO PRACTICAL ACTIVITIES

Chapter 2

1. Any object that is hung from a point and allowed to move freely will hang so that its center of gravity is directly below the point from which it is suspended. The cardboard will perform in the same manner as a gymnast hanging from a high bar with one hand. The gymnast's center of gravity will be directly below the gymnast's hand gripping the bar. A weight hanging from a string and suspended from the same point as the cardboard will act as a plumb line and indicate the line of gravity. So the center of gravity of the cardboard must be somewhere along the line indicated by the string. Suspending the cardboard and the weight together from each of the six punched holes will provide a series of lines that intersect. The center of gravity of the cardboard is at the point of intersection. The mass of the cardboard will be equally balanced on either side of the center of gravity along a line that passes through the center of gravity. When spun in the air, the cardboard will always spin so that its mass is balanced around its center of gravity. The same principle applies to springboard divers, gymnasts, high jumpers, and long jumpers. Rotating in the air, these athletes will spin around their center of gravity.
2. The two characteristics of inertia are resistance and persistence. When the large cart is accelerated quickly to the left, the inertia of the smaller cart tries to keep the smaller cart in the same spot that it was prior to accelerating the larger cart. This is an example of inertia's first characteristic of resistance. When you accelerate both carts together to the left and then suddenly stop the larger cart, the smaller one wants to keep going to the left, and this action portrays inertia's second characteristic of persistence. Athletes fight against the inertia of their mass when they accelerate and, once on the move, the inertia of their mass wants them to continue moving in a straight line. Athletes who are very massive must have tremendous muscular power to accelerate, to change direction, and then to stop.
3. Newton's third law says that all forces act in pairs and that a force acting in one direction causes an equal and opposite force to act in the opposing direction. Provided that the skateboard rolls easily, a heavy medicine ball thrown to the left will cause the thrower to move to the right. The distance that the thrower moves to the right will be relative to his inertia combined with the inertia of the skateboard. The friction of the wheels of

the skateboard will reduce the movement to the right. The medicine ball will move to the left according to its own inertia. In sport, it is really important that we have a stable base when we apply force. If we apply force to the left (e.g., throwing a shot or tackling an opponent), we will stagger to the right if we do not have a stable base. Any movement by our bodies requires that we have sufficient friction and an adequate supporting base to apply force against the earth. The reaction of the earth then thrusts us in the desired direction. Novice ice skaters frequently fail to position their skates correctly so that they can thrust themselves along the ice. Experienced skaters thrust simultaneously outward and backward with their skates. This action uses the length of the blade to provide both sufficient friction and a stable base to thrust forward along the ice.

4. When an athlete stands on the surface of the earth, the athlete's mass is drawn downward to the core of the earth. The surface of the earth responds to the athlete's mass pushing downward by pushing upward. The upward push is called a ground reaction force. More mass means more push downward and this causes more push upward. When you stand on a scale and swing your arms vigorously upward, the reaction (Newton's third law) to the arm movement is the same as suddenly and momentarily increasing your body mass. The scale will indicate an increase in your weight. The reverse happens when you swing your arms vigorously downward from an extended position above your head. Similar motions are performed by flexing the knees in downhill skiing. This action helps to "weight" and "unweight" the skis (i.e., increase and decrease the friction of the skis with the snow) during the performance of a turn.

Chapter 3

1. Newton's second law written as a formula is force = mass × acceleration. If you increase the mass (i.e., increase the weight) of the object hanging from the string, then the acceleration of the block will increase. Gravitational attraction pulls the weight toward the floor. Conversely, if you keep the force unchanged (i.e., you don't increase the size of the weight) and increase the mass of the block, then the acceleration is less, or possibly the applied force is insufficient to overcome the static friction of the block resting on the table. In sport, an athlete, who by training, increases his ability to apply muscular force and keeps his body weight unchanged will be able to accelerate at a greater rate. However, if an athlete increases his mass (i.e., his body weight) without increasing his muscular force, then his ability to accelerate will be decreased.
2. The simple scenario that occurs in this practical activity occurs every time that a tower diver dives down to the water or a pole vaulter is lifted up in the air on her pole and then falls down toward the landing pads. Any time that an athlete or an object rises above the surface of the earth and then falls down again replicates the scenario in the example.
   a. Mechanical work is performed on the block when it is lifted up from the floor and placed on the edge of the table.
   b. The gravitational potential energy of the block is greatest as it rests on the edge of the table.
   c. The kinetic energy of the block is zero when it rests on the floor and when it rests on the edge of the table.
   d. The kinetic energy of the block is maximum at the instant it hits the floor and has the greatest velocity for the distance that it has fallen.
   e. The kinetic energy of the block is dissipated by applying force to the floor and to itself, and also in generating noise and heat during its collision with the floor. Any damage to the floor and to the block will result from the block's momentum and kinetic energy.

You can substitute the word *athlete* for *block, diving board* for *table, water* for *floor.*

3. Speed skating will show similar stride length and cadence (stride frequency) as sprinters and likewise for rowers when these athletes accelerate from the start.
   a. When the gun goes off, the athletes' stride length will be short and the cadence (stride frequency) will be fast with many quick strides taken over a short distance. The reason for these actions is that the athletes are overcoming inertia and are accelerating.
   b. Once up to speed the athletes lengthen their stride and lower the cadence (stride frequency). The athletes' mass is now on the move and their inertia is now of assistance rather than a hindrance as it was at the start.

   Impulse equals force multiplied by the time that force is applied. At the start, when the athletes are overcoming inertia, it is more efficient for the athletes to apply considerable muscular force over multiples of short time intervals (meaning that they use fast, powerful, short strides). Once up to speed, there is no need to battle inertia and stride length can be increased and the cadence (stride frequency) can be reduced.
4. An egg that lands on a hard surface will easily crack. The bed sheet held by your assistants provides a cushioning effect and the egg is slowly and gently brought to a halt. In this situation, a small force is applied over a large time frame to bring the egg safely to a stop. The same principle applies to athletes when they drop from a height to the ground or when they catch an object like a ball moving at speed. The impulse of stopping safely and without injury always emphasizes the application of a small force applied over a large time frame rather than a large force applied over a small time frame. The latter situation will cause pain and injury.
5. Starting an object moving and overcoming static friction requires more force than that required to keep the object moving. The reading on the spring scale will indicate a higher value the instant that static friction is broken than when the object is moving across the table. The same relationship will exist with different surfaces, although the values measured on the spring scale will be higher with surfaces that have a greater frictional component (such as a rubber surface). The spring scale will also indicate the force necessary to overcome resting inertia and once under way, the scale will also show the lower value of inertia on the move. These principles of static and sliding friction apply to sport. Rolling friction (i.e., of wheels) is significantly smaller than sliding and static friction.

Chapter 4

1. To make any object rotate (e.g., a soccer ball or an athlete), force must be applied at a distance from the axis of rotation. In this way the turning effect of torque is produced (i.e., force multiplied by perpendicular distance from the axis of rotation that force is applied). When a gymnast performs a front somersault in the air during a floor exercise routine, the gymnast applies force with her feet against the floor while leaning slightly forward. The body lean positions her center of gravity ahead of the line of application of force with her feet against the floor. This provides the turning effect of torque, which ultimately spins her in the air.

   In closing the door by pushing at the handle, a small force applied by the index finger is multiplied by the large distance from the handle to the hinges of the door, and the torque that is generated is sufficient to close the door without causing pain to the index finger. Closing the door by pushing with the index finger close to the hinges means that there is very little distance involved in generating the torque necessary to close the door. The force required from the index finger is considerable and the pressure on the index finger is tremendous.
2. The spinning top competition involves applying as much torque as possible to an object with optimum levels of rotary

inertia. The following advice should be taken into consideration in the design of your spinning top.

a. The turning effect of torque is going to be applied by your thumb and index finger. The upper shaft of your top should have sufficient diameter and friction for you to produce the greatest possible turning effect on the top.
b. It would be best to make your top out of a fairly heavy metal, but not have it be so massive and heavy that you are unable to spin it.
c. The top should have as much rotary inertia as possible so that once it is spinning, it wants to continue to spin. For this purpose, as much of its mass as possible should be positioned in its outer perimeter.
d. The top should be designed so that it is as close as possible to the surface that it will spin on. This increases its stability.
e. The top should have a sharp point at its base to spin on and it should be spun on a hard, smooth surface such as steel or glass. This reduces friction to minimal levels while it spins.

In sport, the application of torque to an object that has great rotary inertia occurs often. Springboard divers, gymnasts, and ski aerialists frequently want to perform a skill that contains several somersaults. It is essential that the athletes apply torque to their bodies while they are fully extended and have maximum rotary inertia. In this way they give themselves considerable angular momentum. When the athletes pull themselves inward to a tight tuck from an extended body position, their angular velocity (rate of spin) is dramatically increased, making multiple somersaults possible.

3. A golf driver has a considerable amount of its mass positioned in the head of the club. It's the same with a bat, which has most of its mass positioned in the barrel end. When the more massive end of these clubs is closer to an athlete's body, there is a reduction in the rotary inertia of the clubs and they are easy to maneuver relative to the axis provided by the athlete's body. With the clubs held at the upper extremity of the handle, the rotary inertia of the clubs is increased because more mass is now positioned farther away from the athlete's body. Moving the clubs back and forth in front of the body or following a figure-eight path will be quite difficult when gripping the upper extremity of the handle. It will be much easier when the athlete grips the striking end of the clubs. The formula for rotary inertia is mr^2, indicating that small changes in the position of the mass of the clubs dramatically influences their rotary inertia. To rank the order of the clubs relative to their difficulty in moving through the required path will depend on the mass of the clubs, their length, and the distribution of their mass. In sport, less torque is necessary to rotate an object that has its rotary inertia reduced. Likewise, a spinning object or athlete that reduces rotary inertia will increase in angular velocity. This is seen in figure skating, gymnastics, diving, trampoline, and trapeze.
4. Your shoulder provides the axis of rotation. During a full circle—up, over, and down—your arm and the bucket provide a centripetal force that constantly holds the water in the bucket to a circular path. The water in the bucket wants to travel in a straight line following Newton's first law of inertia. If the base fell out of the bucket during the swing, the water would fly away at a tangent to the circle that it is being forced to follow. If you release your grip on the bucket, the bucket and the water would fly away at a tangent to the circle around which it is moving. Any rotary action in sport must have a centripetal force. These are seen in giants on the high bar and in hammer and discus throwing. An athlete's walking and running actions are made up of rotary actions, all of which require torque to cause the rotation and centripetal force (supplied by ligaments, bones, and tendons) to make the limbs follow a circular rotary motion.

5. This activity demonstrates Newton's third law applied to rotation and takes into consideration the rotary inertia of the body parts involved. The action is the movement of the left arm through 90 degrees. The reaction is the movement of your body in the opposing direction. Both the action of your left arm and the reaction of your body depend on their individual mass and how their mass is distributed. Your arm is not particularly massive but what mass it has is spread out (i.e., distributed) along the length of your arm from the shoulder to the hand. Muscular contraction makes it move 90 degrees. The remainder of your body, which is considerably more massive than your left arm, is concentrated close to the seat of the swivel chair. A vigorous arm swing of 90 degrees counterclockwise produces approximately 30 degrees of response in a clockwise direction. This indicates that your body in its particular position on the swivel chair has three times the rotary inertia of your left arm. Factors that alter these results might be long massive arms or, conversely, a massive body and short, thin arms. The response of one part of the body moving in an opposing clockwise direction to an action performed counterclockwise (or vice versa) is seen in high jump, long jump, diving, gymnastics, volleyball spikes, hurdling, and trampoline.
6. The following information will help you make your diagram and position arrows indicating lines of force. Assume that both Jan and Lance are the same body size, and ride similar bikes at the same velocity around a similar curve. The formula for centripetal force—the inward acting force that will help them around the curve—is mv^2/r (m = mass, v = velocity, and r = radius of the curve). Both Jan and Lance are traveling at the same velocity around similar curves. But Jan's mass is greater than that of Lance. So Jan's mass needs more centripetal force to get around the curve. This can only be provided by leaning more into the curve. For the original comparison, we must make Jan taller, so raise his center of gravity more in comparison to Lance. This requires even more lean. And to complete the picture, if Jan entered the curve at a higher speed than Lance (which he probably did), then even more lean is required. On the slick surfaces that existed, there was not enough friction to accommodate an athlete of Jan Ullrich's size and weight traveling through the curve at the velocity that he was moving. What should he have done? Temporarily, slow down and lean less.

Chapter 5

1. This experiment shows how the larger the area of the supporting base, the more stable the block becomes. The block will be most stable when it rests on a base with an area of 4 × 3 in., and with the 4-in. side parallel to the long axis of the plank. It will be least stable when it rests on its 2-in. side with this side parallel to the long axis of the plank and with the 4-in. side as its height. This experiment shows how stability of an object or an athlete is improved with (a) a larger base and (b) a reduction in height.
2. You will find that the most stable position is when you lie flat on the floor. If your attacker is trying to roll or lift you from the left to the right, then the following actions should be carried out:
   a. You should lie either face down or face up on the floor. This lowers your center of gravity as much as possible.
   b. You should position your center of gravity as close to the attacker as possible. Your left arm is pulled in close to your body. Your left leg is extended and moved toward the right, away from the attacker.
   b. Your right arm and leg should be extended and as far away from the attacker as possible. The extended right arm and leg stops the attacker from rolling you from left to right (i.e., applying torque from left to right). The position of your body mass and the extended right arm and leg maximize

the resistance and act as the resistance arm in a lever situation. If, on the other hand, the attacker tried to roll you from right to left, then you must extend your left arm and leg toward the left. These actions are fundamental to the sport of wrestling.

3. Males tend to have a higher center of gravity than females with the males' center of gravity above the line of the hips. When males stand with their heels and back against a wall and then flex forward into a toe touch, the wall forces their center of gravity to shift forward. This causes most males to tumble forward. The majority of females have a lower center of gravity and are able to flex into a toe touch with their heels and back against a wall. When males perform a toe touch away from a wall, they shift some body mass backward to counterbalance the movement of mass forward. This keeps their line of gravity above their supporting base.
4. Running around alternating curves on a flat surface at increasing speed is extremely difficult. You must lean into the curves, first one way and then the other. Increasing speed demands a continuous increase in the degree of lean. A point will occur in which the friction between your shoes and the flat surface becomes insufficient, and you will slide outward. You lean into the curve in order to push outward against the earth. The earth in reaction provides the inward push of centripetal force. If you increase your speed, you dramatically increase the requirement for the inward push of centripetal force. Your inertia will want to carry you outward at a tangent to each curve. Your diagram should show a series of stick figures with increasing lean and progressively larger arrows should indicate the increased force applied outward and the increased reaction of the earth pushing inward. Your line of gravity will always act perpendicularly and pull your center of gravity vertically downward. So you will have an arrow of the same size and length indicating the force of gravity.

Chapter 6

1. When you put hot water into the plastic container and swish it around, you cause the air in the container to expand and be driven out. Quickly screwing on the cap does not allow any air to get back into the container. As the expanded air inside the container cools down, its volume is reduced and there is less pressure inside the container than outside. When the plastic juice container is crushed, it demonstrates the tremendous pressure of the atmosphere (14.7 lb per square inch). If you go to a depth of 33 ft in the ocean, you will have two times 14.7 lb psi or 29.4 lb psi pressing on your body. Your lungs and other air spaces cannot withstand this pressure unless you breathe air that is at the same pressure. Scuba equipment provides air at the correct pressure so that breathing is comfortable.
2. When it rotates, the cylinder drags around a layer of air (called a boundary layer) which comes in contact with the surface of the cylinder. As the cylinder is pushed forward away from the snapping action of your thumbs, the boundary layer air on the underside of the cylinder rotates into the air flowing by the cylinder. The boundary layer air on the upper side rotates in the same direction as the air flowing by. The air on the upper surface of the cylinder is accelerated and the air on the lower surface is decelerated. According to Bernoulli's principle, a fluid (in this case, the air) that is accelerated reduces its pressure. A fluid that is decelerated increases its pressure. In the case of the cylinder, there is an imbalance of pressure with more pressure being exerted below the cylinder than above. Consequently, the cylinder will rise in the air. The cylinder replicates the movement of golf balls, baseballs, tennis balls, and table tennis balls that are given backspin. Sufficient backspin causes the cylinder and balls to fight against gravity and temporarily rise in the air.
3. What occurs in this buoyancy experiment will differ from one person to the

next. If, when your lungs are full, you sink, then you are displacing (with the volume of your body) an amount of water that weighs less than you do. If you float, then the water you displace weighs more than you do. If you float even after exhaling—that is without increasing the volume of your body by inhaling—you weigh less than the water that you displace. The amount of water that you displace is the water that you "take the place of" when you get into the water of the swimming pool.

4. The following will occur during your experiments:
   a. Swimming with your head out of the water—Raising the head causes the legs to drop in the water and as a result, drag forces increase.
   b. Swimming with your hands as fists—your hands and the active surfaces of your forearms are largely responsible for your propulsion through the water. Swimming with your hands as fists reduces the "paddle" surface area of your hands. Your movement through the water will be much slower.
   c. Swimming with no leg action—This tends to have the same effect as raising the head. Leg kicks help keep the legs at the surface of the water and help the swimmer maintain a horizontal position. When there is no leg kick, the legs tend to drop downward in the water increasing form drag.
   d. Swimming using one pull-buoy—For many athletes, a single pull-buoy is sufficient to hold the legs up at the surface of the water. Form drag is minimized when the swimmer's body lies horizontal in the water.
   e. Swimming with two pull-buoys—For athletes with "heavy legs" and a poor leg kick, two pull-buoys can be used to hold the legs at the surface and minimize profile drag. Whether using a single or double pull-buoy, the speed of movement through the water with the legs held close to the surface dramatically indicates how form drag is reduced the more horizontal the body position in the water is.

Chapter 7

1. The answers below give examples of the performance objectives of many sports and sport skills.
   a. shot put, javelin, long jump, and triple jump (propelling or projecting an object or the athlete's body for the maximum horizontal distance and within a prescribed area or zone)
   b. high jump, pole vault (propelling the athlete's body for maximum vertical distance and successfully crossing a bar)
   c. archery, darts, volleyball serve, tennis serve, baseball pitch (propelling or projecting an object for maximum accuracy)
   d. diving, figure skating, gymnastics (maneuvering the human body with the intent of achieving a prescribed manner of performance)
   e. weightlifting (elevating a resistance and maintaining it above the head for a certain time)
   f. wrestling, judo (maneuvering a human resistance)
   g. surfing, hang gliding, scuba diving (successful interaction with the natural environment)
   h. sprinting, middle- and long-distance running, cross-country skiing, swimming (moving the body over a prescribed distance and measured by a stop-watch)
2. The results of this activity depend on the body type and experience of the five athletes chosen. When performing the same skill, elite athletes use the same basic movement patterns, so the fundamentals of their technique will be similar. Variations occur in the degree with which they emphasize certain aspects of the skill. Personal idiosyncrasies will be visible in minor aspects of the skill's performance. The following sport skills are good for

comparing the fundamental movement patterns of five elite athletes.

- sprint starts and sprinting
- swimming freestyle
- high jump
- golf drive
- tennis serve
- a multiple twisting skill (e.g., a Lutz) in figure skating

3. The skills listed in the answer for activity 2 are excellent for comparing and contrasting the performances of elite athletes and novices. The following factors may be apparent in the performances by novices.
   - The novice is unstable during the application of force, and force is often applied in the wrong direction.
   - The novice's base of support is too small, too large, or not extended in the correct direction.
   - Not all the novice's muscles that could make a contribution to the skill are used.
   - The novice has not learned to relax the antagonist muscle groups when it is necessary to do so.
   - The application of force by the muscles is in an incorrect sequence.
   - The time used for the application of force by the muscles is incorrect.
   - The novice simply does not yet have the power to maneuver his or her body segments in the correct sequence or in the correct amount of time.
4. Elite athletes performing these skills turn their bodies into whips. They use a whip-cracking action to generate high velocity to the racket, the club, or the discus and javelin. All of these high-velocity actions (which also include punting and baseball pitching) require the following:
   - The athlete's body is initially accelerated as a unit.
   - The club, racket, or throwing implement is positioned to the rear of the body. In golf, this is called a windup.
   - The athlete steps in the direction of throw or strike with the throwing or hitting arm trailing to the rear of the body. In javelin throwing, the athlete performs a run-up and steps forward into a backward tilted throwing stance, which is the beginning of the throwing action.
   - The larger, more massive, and longer segments of the body (i.e., the legs), with their greatest lever length, mass, and rotational inertia, provide stability for force application. With the exception of the chest, the athlete's body segments get progressively smaller in mass, rotary inertia, and length, and simulate the reduction in segment size on a whip.
   - All segments are joined by axes—the joints of the body. Crossing each joint of the body are the agonist muscles, which are stretched and then contract vigorously. This stretch reflex action adds to the velocity of each progressively smaller body segment. Through training, antagonist muscles are taught to relax and provide little opposition to the contraction of the agonist muscle groups.
   - The body, followed immediately by the base of the larger body segments, is suddenly decelerated. The action of arresting the motion of the base of each segment in sequence is repeated throughout the whole body to the final segment (i.e., the hand holding the club, racket, or javelin), which will have the smallest lever length, the least mass, and the least rotary inertia.
   - The sudden deceleration of the base of each successive segment passes the velocity of its upper extremity on to the next segment in line by way of the joining axis.
   - In a golf drive, badminton smash, javelin throw, and tennis serve, the sequence is body as a whole, then legs, hips, chest, upper arm, forearm, and, finally, the club, bat, or throwing hand.

- The same principles are applied in kicking and punting for distance and velocity.
- In total, the athlete is summating (i.e., adding or building up) each successive torque, each successive angular acceleration, and each successive velocity from one body segment to the next.

5. Cycling has been chosen as a sample answer. Over the past 10 to 15 years, there have been dramatic changes in the design of bicycles and bicycle clothing. For example:
   - Bicycle frames and other components are made of lightweight materials such as graphite. This material is replacing steel and aluminum.
   - Bicycle frame tubing is designed to have aerodynamic teardrop shapes rather than drag-producing circular shapes.
   - Bicycle frames are "raked" (i.e., sloped downward) toward the front wheel. This puts the rider into an aerodynamic position.
   - Wheels are lightweight and use blade spokes or composite teardrop-shaped spokes. Disc wheels are also used, which have no spokes and generate minimum drag.
   - Super-high-pressure tires are used, which generate minimal rolling friction.
   - Helmets are designed to have aerodynamic shapes. Cycling clothing is slick, allowing air to pass by easily. Laceless shoes and fingerless gloves reduce the drag forces generated by normal shoes and gloves.
   - Cycle shoes now clip onto the pedals, allowing the cyclist to apply force throughout the pedal rotation.
   - Bike frames are stiffer and reduced in length. Together with smooth gear changes, this provides for minimal loss of power from the pedals to the back wheel.

Chapter 8

1. The following are answers to the *Freestyle wrestling competition* practical activity:
   a. The attacker usually puts himself at risk during an attack on his opponent. His center of gravity and line of gravity will be close to the forward perimeter of his base. Usually the attacker sees a weakness in the stability of his opponent (i.e., height of center of gravity, position of line of center of gravity, or size of supporting base) and directs his attack against this weakness.
   b. The attacker uses his own body to apply force and applies this force at some distance from an axis of rotation. The axis of rotation can be provided by his own body or by the opponent's body.
   c. The defender lowers his center of gravity, spreads his legs or arms to widen his base, and moves to lessen the torque applied by the attacker.
   d. A successful attack usually means that the torque applied by the attacker overwhelmed that produced by the defender.
   e. A successful defense usually means that the torque applied by the attacker was insufficient to destabilize the defender.
2. The following are answers to the *A mount on the beam in women's gymnastics* practical activity:
   a. Mounts on the beam are usually one- or two-footed—in the majority of cases, two-footed. The gymnast is more stable parallel to the length of the beam in a two-footed stance than at right angles to the beam.
   b. Mounts on the beam usually have one foot positioned slightly ahead of the other. This enlarges the base of support and improves stability parallel to the length of the beam. Occasionally, the feet are splayed outward for greater control.
   c. Gymnasts most commonly lower their center of gravity by flexing their legs, and they keep their line of gravity above their base of support. Keeping the line of gravity above the base of

support is easier parallel to the length of the beam than at right angles to the beam.

d. Gymnasts extend their arms and rotate them clockwise if they feel that they are going to fall off the beam in a clockwise direction. They rotate them counterclockwise if they feel they are going to fall off the beam in a counterclockwise direction. The same action applies to the action of the head. The clockwise action of the arms causes a counterclockwise reaction of the body. A counterclockwise action of the arms causes a clockwise reaction from the body.

e. Gymnasts want to make sure that they are stable when they land from a mount on the beam. Only then do they move to their next skill. If they are destabilized and immediately move to the next skill, the error in the landing from the mount is likely to be magnified in the next skill.

3. The following are answers to the *A fastball pitch in baseball* practical activity:

a. The athlete extends the throwing arm backward and away from the direction of pitch. The athlete also steps forward to the batter. The shift of the throwing arm backward and the step forward maximizes the accelerative path of the baseball.

b. By stepping forward toward the hitter, the pitcher extends the accelerative pathway of the ball and turns the hips 90 degrees away from the hitter. This allows the hips to be rotated vigorously during the pitch and stretches the large muscles of the abdomen and chest. By stepping out toward the hitter, the pitcher is angling his body "backward" (i.e., away from the hitter) and this helps increase the impulse (force × time) that power will be applied to the ball.

c. The sequence of limb and body movements is as follows: (1) The body as a whole moves toward the hitter, (2) the throwing arm is simultaneously extended backward, (3) the leg muscles contract and thrust the legs forward toward the hitter, (4) the hips are rotated toward the hitter, (5) the hips pull the chest around, (6) the chest pulls the throwing arm, and (7) the throwing arm pulls the hand and ball forward to its release position.

d. The linear and angular velocity of the large body segments, which have great length and mass, is passed on and multiplied to those that have less mass and length. The thrower turns his body into a whip, with the hand and the ball simulating the tip of the whip.

e. The rotary motion of the hips, chest, and forearm is combined with the linear motion of the body as a whole as it moves forward toward the hitter. The pitcher's body weight is shifted from the rear foot toward the forward foot in a linear manner.

4. The following are answers to the *A volleyball spike* practical activity:

a. The approach to the takeoff is used to move the athlete's body forward from a position where the athlete's center of gravity is to the rear of the feet to one where the athlete's center of gravity is above the feet or slightly ahead. This range of movement provides more time during which force is applied to get the athlete up in the air.

b. During the penultimate and last strides of the approach, the athlete's feet are positioned ahead of the center of gravity. The legs are partially flexed. Extended arms are swung back to the rear of the body ready to be swung forward and upward during the jump portion of the spike.

c. The arm swing forward and upward from a position to the rear of the body provides momentum transfer. The angular momentum of the arm swing is added to the downward thrust of the legs during the upward jump toward the ball.

d. The high jump approach is much longer than the volleyball approach. The high jump takeoff is from one foot and momentum transfer comes from both arms and the swing-up leg. Most commonly, a volleyball spike uses a jump from both legs at the same time.

e. Almost all of the jumping force generated in a volleyball spike is used to drive the player upward. The player does not want to travel forward for fear of hitting the net. In the high jump, the athlete has to clear the bar, so there is a strong horizontal component as well as a vertical and rotational component.

f. When the striking arm is swung backward prior to the spike, there is a reaction from the legs—they move in a rotary fashion (relative to their rotary inertia) in an equal and opposite direction. The powerful forward strike of the hitting arm causes the athlete's legs to move forward in reaction, and simultaneously, the athlete's seat moves backward.

g. The hitting arm is positioned to the rear of the shoulder. The muscles of the abdomen and the muscles of the chest, which have been stretched by the position of the hitting arm, contract violently, pulling the arm forward. The shoulders rotate and drag the hitting arm forward. The hitting arm is flexed at the elbow, with the elbow leading the striking hand. The whole sequence is whip-like, with the hand acting like the tip of a whip.

h. A top spin on the ball is caused by the striking hand hitting forward and down on the upper perimeter of the ball. The top spin is caused by the force of the striking hand hitting at the perimeter of the ball. The distance of the perimeter from the center of gravity of the ball (the axis of rotation) gives the ball a top spin. A floater is a strike on the ball in which the force of the hitting hand is directed through the center of gravity of the ball. No spin occurs and the ball has the haphazard flight of a lightweight ball given no spin.

5. The following are answers to the *Long jump* practical activity:

a. The run-up (approach) in the long jump has the same objective as the run-up in the high jump and the approach in the volleyball spike. The velocity of the run-up carries the athlete through the body position(s) assumed during the takeoff.

b. In the penultimate and the last strides of the run-up, the athlete's center of gravity is lowered and the jumping foot reaches forward for the takeoff board.

c. In the high jump, there is a much greater backward lean as the takeoff foot is positioned prior to driving up in the air. In high jump, most of the athlete's thrust is in a vertical direction. In the long jump, the takeoff is from one foot. Distance is of primary importance so the backward lean of the athlete is much less when the takeoff foot is placed on the takeoff board. In a volleyball spike, the athlete commonly uses a two-foot takeoff with both arms being swung upward simultaneously. This also occurs in the high jump but not in the long jump. The approach for a volleyball spike is frequently no more than two or three strides. Both the long jumper and the high jumper have to travel forward. The high jumper has to cross a bar and the long jumper has to achieve the greatest possible linear distance. In the volleyball strike, minimal forward travel in the air is preferred.

d. Lowering the center of gravity puts the body into a position where it can be driven forward and upward. A backward lean when the last stride is taken provides time over the jumping foot during which muscular force drives the athlete forward and upward. This is the mechanical principle of impulse.

e. Momentum transfer occurs during the takeoff of the long jump with the upward swing of the leading leg and the opposing arm. In the high jump, both arms contribute to momentum transfer more than in the long jump. The upward swing of the free leg also provides momentum transfer. In volleyball, both arms contribute to momentum transfer. Both legs drive from the gymnasium floor (or the sand, in a beach volleyball tournament).

f. Because of the tremendous velocity of the approach run in the long jump and the emphasis on linear distance, the takeoff angle of the athlete's body is approximately 22 degrees from the vertical. The takeoff angle in the high jump, which emphasizes vertical impulse, is between 30–40 degrees from the vertical. The takeoff angle of the athlete's body in a volleyball spike varies from a perpendicular position (0 degrees) to 30 degrees from the vertical. This angle depends on the position of the ball relative to the hitter at the instant of spiking, and also on the speed of approach toward the ball and the net.

g. If all of the athlete's force was used to drive the athlete horizontally through the air, then there would be no force thrusting the athlete up. With 22 degrees being the angle of takeoff, approximately 78 percent of the athlete's thrust is in a horizontal direction and 22 percent is used driving the athlete up in the air.

6. The following are answers to the *Sprint starts in a 100-meter race* practical activity:

a. The athlete is in a position of minimal stability.

b. The athlete's supporting base in the set position will cover an area enclosed by the athlete's hands and feet. The athlete's center of gravity will be above or almost above the hands when in the set position. The athlete will feel considerable pressure on the hands (and fingers) in this position. The athlete's line of gravity will be above or just to the rear of the hands. This "set" position makes the athlete highly unstable.

c. The leading leg is thrust forward and is strongly flexed at the knee. The arms are also flexed at the elbows at approximately 90 degrees. The flexion in the leg and the arms reduces their rotary inertia and makes it easier for the athlete's muscles to bring them forward. The arm action balances that of the legs.

d. An athlete with a quick reaction time will have the leg driving at the blocks extended or almost extended immediately after the gun goes off.

e. The center of gravity of the athlete is ahead of the leading leg and the foot will have made contact with the ground. The athlete's body is angled forward.

f. The accelerative strides of the athlete are short and progressively lengthen. The short strides are intended to overcome the inertia of the athlete's body, to gain momentum, and to accelerate.

g. The cadence (stride frequency) of the accelerative strides is fast.

h. The length and cadence of the strides in the first 5 to 10 m of the race is short and fast. In the middle portion of the race, the stride length is longer and the cadence is reduced because of the increase in length of the strides.

i. The center of gravity of an elite athlete will have a low parabolic rise and fall throughout the whole race. Too much rise and fall means that the athlete is expending excessive power and energy driving upward with each stride.

GLOSSARY

In this glossary we've tried to simplify explanations and terminology as much as possible. To help you understand these mechanical terms, examples are provided that refer to the movement of athletes or sport implements, such as javelins and volleyballs.

acceleration—The rate of change of velocity. An athlete can accelerate, decelerate, or have zero acceleration. In the latter case, the athlete can be either motionless or moving at a uniform rate.

agonist—Those muscles whose contractions cause or help action to occur; means *mover.*

airfoil—The cross-section of a wing. Airfoils come in different shapes. Many have a more curved upper surface than undersurface. Others are symmetrical. They are designed to produce lift.

angle of attack—The angle between the long axis of an object (e.g., a javelin or an airfoil) and the direction of fluid (e.g., air) flowing past. On a discus, the angle of attack is the angle by which the leading edge is raised or lowered relative to the airflow passing by the implement.

angular momentum—For an athlete, angular momentum is determined by the athlete's mass, the distribution of the athlete's mass (how stretched out or compressed the athlete's body mass is relative to the athlete's axis of rotation), and the rate at which the athlete is rotating. In mechanical terms the angular momentum of any object is determined by the object's mass × the distribution of its mass × its angular velocity.

angular motion—Motion that is circular or rotary. Somersaulting, twisting, rolling, swinging, rocking, spiraling, and pirouetting are all forms of angular motion.

angular velocity—The rate of spin of an athlete or object. This takes into consideration the number of revolutions, the time frame, and the direction of rotation. An example of angular velocity is 360 degrees in one minute in a clockwise direction.

antagonist—A muscle whose contraction opposes those muscles that cause movement (i.e., agonists). In sport, and antagonist is an opponent.

apex—The highest point in a trajectory. In a high jump or a dive the apex will be the top-most point of the athlete's flight path.

Archimedes' principle—Named after Archimedes, a Greek mathematician. Archimedes' principle states that "the buoyant force acting on an object is equal to the weight of the fluid that the object displaces." If an athlete weighs more than the weight of the water that the athlete takes the place of, then the athlete will sink. Conversely, if an athlete weighs less than the weight of the water that the athlete takes the place of, the athlete will float.

axis of rotation—An imaginary line that passes through the center of rotation of an object or an athlete. When an athlete is rotating in the air, the athlete's axis of rotation passes through the athlete's center of gravity. This is also the case for inanimate objects such as volleyballs, baseballs, and javelins. When in contact with the ground, an athlete's axis of rotation and center of gravity are often in different places.

balance—The ability of an athlete to control his movements for a particular purpose.

base of support—The area formed by the outermost points of contact between an object or an athlete and the supporting surfaces. The base of support does not necessarily have to be beneath an object.

Bernoulli's principle—Named after Daniel Bernoulli, a Swiss mathematician. Bernoulli's principle states that "pressure exerted by a fluid is inversely proportional to its velocity." This means that fluids (e.g., air and water) exert less pressure the faster they move. Conversely, fluids exert more pressure the slower they move.

biomechanics—A science that studies the effects of energy and forces on the motion of living organisms.

boundary layer—The layer of fluid that contacts the surface of an object or an athlete when moving in a fluid (e.g., air or water). All objects moving through the air will have a boundary layer. Swimmers can have a boundary layer of air passing over those parts of their bodies moving through the air as well as a boundary layer of water for those parts of their bodies moving through the water.

buoyancy—The tendency of a fluid to exert a lifting effect on an object or an athlete that is wholly or partially submerged in the fluid. Buoyancy exists in water, air, and other gases.

buoyant force—An upward force (i.e., opposing gravity) that is exerted on an immersed object by the fluid surrounding it. See also Archimedes' principle.

center of buoyancy—A point at which a buoyant force acts on an immersed object. The center of buoyancy for an athlete is usually higher on the body than the athlete's center of gravity.

center of gravity—A point at which the mass and weight of an object or an athlete are balanced in all directions. It is also that point where gravitational forces are centralized. The center of gravity for males is usually higher than for females.

centrifugal force—When objects or athletes are made to rotate, their inertia constantly tries to make them travel in a straight line. Inertia's pull competes against the inward pull of centripetal force. A common name for inertia's pull in this situation is centrifugal force. Centrifugal force is really inertia "in disguise."

centripetal force—Any rotating object will have a force that acts towards its axis of rotation. This inward force is called a centripetal force.

closed skills—Skills that are performed in a predictable, controlled environment. Synchronized swimming is an example of a closed skill.

conservation of angular momentum—A rotating object or athlete will continue to rotate with constant angular momentum unless an external torque is applied that increases or decreases this angular momentum. During the flight from the tower to the water, a diver will have virtually the same amount of angular momentum from takeoff to contact with the water.

conservation of linear momentum—In any collision between two objects or two athletes, the total amount of momentum before the collision is the same after the collision. Momentum is conserved.

deceleration—A decrease in velocity by an athlete or an object.

density—Weight per unit volume (or mass per unit volume). The amount of substance "contained" in a particular space—the greater the amount, the greater the density. Muscles and bones are more dense than fat.

dissipation of kinetic energy—In any collision between two objects or two athletes, the kinetic energy that is brought into the collision is dissipated in the following manner: Some kinetic energy is used in performing work by the objects and the athletes on each other during the collision. However, kinetic energy is also lost as heat and noise. See also the law of the conservation of energy.

drag—A force produced by the relative motion of an object or an athlete through a fluid (e.g., water or air).

eccentric thrust—A force that does not pass through the center of gravity of an object and that tends to cause rotation. Gymnasts, divers, and figure skaters use eccentric thrust to initiate rotation.

elasticity—The ability of an object to recover its original shape after being deformed. Clay is nonelastic. Golf balls, bows used in archery, and modern composite pole-vault poles are all highly elastic.

energy—In mechanics, energy is the ability to do mechanical work. There are several types of energy such as heat, chemical, electrical, and mechanical. The more energy an object or an athlete has, the greater the force with which it can shift or deform another object or athlete.

first-class lever—A lever in which the axis is positioned between the force and the resistance. A first class lever is able to magnify force, balance force with resistance, and magnify speed and distance. First class levers cause a direction change, with the force moving in one direction and the resistance moving in the opposing direction.

flexibility—The range of motion in an athlete's joints. Flexibility is also used to describe the "flex" of an object like a pole-vault pole.

follow-through—Motions in a skill performance that are performed after the force producing actions are completed. A follow-through progressively and safely dissipates the momentum and energy generated by an athlete in a skill performance.

force—Any influence that tends to change the state of motion of an object or its dimensions. Force applied by an athlete does not necessarily have to produce movement.

force arm—The perpendicular distance existing between the line of force application and the axis of rotation.

force-producing movements—These are the actions used by an athlete to generate force during the performance of a skill.

force vector—A symbol indicating the magnitude and direction of an applied force. Force vectors are frequently represented graphically by an arrow. The head of the arrow indicates the direction of the applied force and the length of the arrow represents the magnitude of the applied force.

form drag (also called profile drag, pressure drag, and shape drag)—When an athlete (or an object) travels through a fluid, high pressure occurs in front where the athlete contacts the fluid head on. Low pressure occurs immediately to the rear of the athlete. The greater the difference between high and low pressure, the greater the form drag.

friction—A force that acts in opposition to the movement of one surface on another. There are various types of friction (e.g., static, sliding, and rolling friction).

frontal axis—An axis that runs from the front to the rear. In the human body, the frontal axis passes by way of the center of gravity from the abdomen to the back of the athlete. At the center of gravity, the frontal axis forms right angles where it meets the long axis (from head to feet) and the transverse axis (from hip to hip).

fulcrum—An axis or hinge about which a lever rotates.

gravitational acceleration—The rate of acceleration of an object or an athlete toward the earth. Commonly stated as 32 ft/sec^2 or 9.8 m/sec^2.

gravitational potential energy—The energy that an object possesses by virtue of being within the earth's gravitational field and above the earth's surface. The more mass and the greater the height above the earth's surface, the more gravitational potential energy the object or athlete will possess. A diver will have more gravitational potential energy when standing on the 10 m tower than when standing on the 3 m board. However, it is possible for a massive diver standing on the 3 m board to have more gravitational potential energy than a lean diver standing on the 10 m tower.

gravity—The force of attraction that moves or tends to move objects with mass toward the center of a celestial body such as the earth or the moon. All objects that have matter have mass and exert a gravitational attraction—the more mass, the greater the attraction.

ground reaction force—The equal and opposite force exerted in reaction to a force exerted against the earth. An athlete pressing against the earth with a certain force will cause the earth to respond both equally and in the opposite direction with a force of the same magnitude.

horsepower—A term introduced by inventor James Watt and used as a measurement of power in the imperial (English) system of measurement. One horsepower is the ability of an object or an athlete to move 550 lb through a distance of 1 ft in a time frame of 1 sec. One horsepower = 745.7 watts.

hydrostatic pressure—The pressure (i.e., force/area) exerted by a fluid (e.g., air or water) to support its own weight or the weight of an object or an athlete immersed in the fluid. Atmospheric pressure is 14.7 lb per square inch at sea level. This pressure decreases with altitude. Sea water, which is virtually incompressible, increases in pressure by 14.7 lb per square inch for every 33 ft in depth. To this must be added 14.7 lb psi for the weight of the atmosphere resting on the surface of the ocean.

impact—A collision between two or more objects.

impulse—Force multiplied by the time during which the force acts. Athletes vary the amount of muscular force and the time frame over which this force is applied.

inertia—The tendency of an object or an athlete either to stay at rest or to move continuously in a straight line at a uniform velocity. Inertia is directly related to mass. A more massive athlete or a more massive object has greater inertia than one with less mass. See also Newton's first law.

key elements—Distinct physical actions that join together to make up a phase of a sport skill.

kinetic energy—The ability of an object or an athlete to perform work by virtue of being in motion.

Doubling the mass of an object increases its kinetic energy twofold. Doubling the velocity of an object increases its kinetic energy fourfold.

kinetic link (kinetic chain) principle—Simulating a flail-like action or the cracking of a whip. The kinetic link principle is used to accelerate lightweight objects to tremendous velocities in such skills as javelin throwing and baseball pitching. It requires a progressive increase in velocity from one limb segment to the next, starting at the most massive and ending at the least massive.

laminar flow—A flow pattern in fluids that is characterized by smooth parallel lines (i.e., like laminations in wood). Laminar flow occurs when fluids pass objects and athletes at very low velocities.

law of the conservation of energy—A law stating that the amount of energy in the universe is constant and cannot be created or destroyed, only changed in form. When an athlete slows down and gives up (kinetic) energy, the athlete's energy is transformed into other forms such as heat, noise, and the movement and distortion of whatever the athlete contacts. The total amount of energy remains unchanged.

lever—A simple machine consisting of a rigid object rotating around an axis or on an axis. In an athlete's body, bones, joints, and muscles work together as lever systems.

lift—The force acting on an object in a fluid that is perpendicular to the fluid flow. Lift is not always upward. It can be made to occur in any direction.

linear motion—Motion of an athlete or an object in a straight line. If all parts of the athlete or the object move in the same direction at the same speed, this motion is called translation.

linear stability—The resistance of a moving object or athlete to being stopped or having its direction of motion altered. Linear stability is directly related to mass/inertia.

line of gravity—A vertical line drawn from an object or an athlete's center of gravity to the earth's surface. A line of gravity is often called a "plumb line."

longitudinal axis—An imaginary line that runs the length of an object. Also called the long axis. When gymnasts twist, they rotate around their longitudinal axis.

Magnus effect—Named after Gustav Magnus, a German scientist. The Magnus effect is the movement of the trajectory of a spinning object such as a baseball or soccer ball toward the direction of spin. If the front, or leading surface, of a baseball is spinning to the left, the Magnus effect causes the trajectory of the baseball to "bend" to the left.

Magnus force—A lift force created by the spin of an object such as a baseball, soccer ball, or golf ball. See also Magnus effect.

mass—The amount of matter, or substance, in an object. A massive athlete has a lot of body mass. Mass is also a measure of an object's or an athlete's inertia. A massive athlete will have more inertia than an athlete who is less massive.

matter—Anything that has weight on the surface of the earth and that occupies space. Matter has mass and inertia.

mechanics—A branch of physics that deals with the effects of energy and forces on the motion of physical objects.

meter—A unit of length that is based on the metric system. One meter equals 3.281 ft. The 1,500-m race run in the Olympics is 120 yd short of a mile.

metric system—A decimal measuring system with the meter, liter, and gram as the units of length, capacity, and weight (or mass). Used throughout most of the world, although still not adopted for general use in the United States.

momentum—Quantity of motion. The mass of an object multiplied by its velocity. Increase the mass or the velocity of an athlete and you increase the athlete's momentum.

negative acceleration—A decrease in velocity of a moving object or athlete.

neutralizer—The name given to muscles whose role is to eliminate unwanted actions that might be caused when other muscles contract.

newton—A measurement used in the metric system and named in honor of Isaac Newton's contributions to science. A newton is a force that when applied to a mass of 1 kg gives it an acceleration of 1 m/sec^2. One newton = .2248 lb.

Newton's first law (the law of inertia)—All athletes and objects have mass and therefore have inertia. Their inertia is expressed by a desire to remain at rest. If a force is applied against them to get them moving, their inertia gives them the desire to travel at the same velocity in a straight line. (Forces applied by gravity, friction, and air resistance naturally change this situation.)

Newton's second law (the law of acceleration)—The acceleration of objects or athletes is proportional to the force that acts on them and is inversely proportional to their mass. A more massive athlete accelerates less than an athlete with less mass when the same force is applied to both.

Newton's third law (the law of action and reaction)—When an object or an athlete exerts a force on a second object (or athlete), the latter exerts a reaction force on the first that is both equal and opposite in direction. The law of action and reaction applies irrespective of whether the items applying force to each other are athletes or inanimate objects.

nonrepetitive skills (also called discrete skills)—Skills that have a definite beginning and finish and that do not repeat in a cyclic fashion. Examples of nonrepetitive skills include a hip throw in judo and a slapshot in hockey.

open skills—Skills that are performed in an unpredictable environment. Examples are skills performed against opponents (e.g., wrestling, soccer) and skills performed in changeable environmental conditions (e.g., skiing, yachting, surfing).

phases—A group of movements that stand on their own and which are joined in the performance of a skill. In the long jump, the following are phases: (a) mental preparation and preparatory motions prior to the approach; (b) the approach; (c) the takeoff; (d) movements performed in flight; and (e) movements performed during the landing.

positive acceleration—An increase in velocity of a moving object or athlete.

power—The rate at which mechanical work is done. (Mechanical work = force multiplied by the distance that the resistance is moved.) Power is work divided by the time taken to perform the work.

pressure—Force per unit area. Pressure is the ratio of force to the area over which force is applied. The same force applied over a larger area exerts less pressure per unit area.

projectile—Athletes or objects that are propelled in such a way that they follow a pathway (or flight path) resulting from the force of propulsion.

propulsive drag—This is a form of drag that acts in the same direction that an object or an athlete is traveling. In swimming, athletes are able to move their hands and feet in such a way that the drag force generated by the movement of their hands and feet assists in propelling the athlete through the water.

rebound—An action that can result from a sudden impact or collision. The type of rebound depends on the nature of the impact and the elasticity and motion of the colliding objects.

relative motion—A term applied to the relative motion of one or more objects or athletes as they pass one another.

repetitive skills (also called continuous skills)—Repetitive skills have a continuous, repetitious, and cyclic nature such as swimming, cycling, and cross-country skiing.

resistance—An oppositional force that tends to retard or oppose motion.

resistance arm—The perpendicular distance existing between the point of application of force applied by a resistance and the axis of rotation.

resultant—A single vector that results from the combination of several vectors. Several forces acting together on an object can produce a single equivalent force that results from their unified action.

resultant force vector—The direction and amount of a force resulting from several forces acting together on an object.

rolling friction—The friction that occurs when a round object such as a wheel rolls against a contacting surface. Rolling friction occurs in gears, axles, transmissions, chains, and engines.

rotary inertia (also called rotary resistance and moment of inertia)—The tendency of objects or athletes to initially resist rotation and then want to continue rotating once the turning effect of torque is applied and as long as a centripetal force keeps them following a circular pathway. Rotary inertia varies according to mass and the distance that mass is distributed relative to the axis of rotation. A somersaulting athlete has greater rotary inertia in an extended body position than when pulled into a tight tuck.

rotary stability—The resistance of an object or an athlete against rotation or being tipped over; if rotating, it is the resistance against forces and torque that might disturb its rotation. Rotary stability is directly related to rotary inertia.

scalar measurement—A measurement on a chosen scale of measures; a quantity that gives magnitude but not direction. For example, 20 mph is a scalar measurement indicating only speed. Velocity indicates both speed and direction.

second-class lever—A lever in which the resistance is positioned between the axis and the force. A second class lever favors the magnification of force rather than speed and distance. Both force and resistance move in the same direction.

skill—A movement pattern that is designed to satisfy the demands of a sport or a specific activity.

skill objectives—The goals or outcomes that occur as a result of the performance of a skill. The rules of a sport determine the skill objectives.

sliding friction (also called kinetic friction)—The friction generated between two surfaces that are sliding (i.e., moving) past each other.

speed—An athlete's or an object's movement per unit of time without any consideration given to direction. For example, 10 mph is speed with no direction indicated.

spotting—Assistance provided to an athlete during difficult and/or dangerous phases of a skill. Used frequently in gymnastics.

stability—Resistance to the disturbance of balance. Athletes are able to increase or decrease their stability according to the requirements of the skills they perform.

static friction—The friction that exists between the contracting surfaces of two resting objects. Static friction provides the resistive force opposing the initiation of motion between the two objects. Static friction exists between a non-moving athlete and the supporting surface.

strain energy—A form of potential energy that is stored in an object when the object is distorted or deformed. An archer's bow and a pole-vaulter's pole store strain energy.

strength—The ability of a muscle (or muscles) to exert force without taking into consideration the time that force is applied.

style—Personal variations that are made by an athlete to a skill, technique, or pattern of movements.

surface drag (also called skin drag, viscous drag, and skin friction)—Drag that occurs where a fluid (e.g., air or water) contacts the surface of an object or an athlete that is moving through the fluid.

technique—A commonly used method by which a skill is performed. In high jump, the Fosbury Flop has made the straddle technique obsolete.

third-class lever—A third class lever has the force positioned between the axis and the resistance. Third class levers favor the magnification of speed and distance. Both force and resistance move in the same direction. Third class levers are the most common lever systems in the human body.

torque—A rotary, turning, or twisting effect produced by a force acting at a distance from the axis of rotation. The initiation of rotation always requires the application of torque.

trajectory—The flight path of an object or an athlete.

translation—In mechanics, translation occurs when all parts of an object move the same distance in the same time frame. This situation is highly unusual because it would mean that there would be absolutely no rotary motion.

transverse axis—An axis that runs from side to side. In the human body, the transverse axis passes through the center of gravity from one hip to the other. At the center of gravity, the transverse axis will form right angles where it meets the long axis (from head to feet) and the frontal axis (from front to back—abdomen to back).

turbulent flow—A disturbed, nonlaminar flow pattern in a fluid.

uniform acceleration—Acceleration that is regular. Uniform acceleration occurs when velocity is increased at a regular rate.

uniform deceleration—A regular or uniform loss of speed or velocity per unit of time. An object or athlete that decelerates by a speed or velocity of 5 ft/sec for every second that passes would have a uniform deceleration of -5ft/sec^2.

velocity—The speed of an athlete or an object in a given direction. Velocity can change when direction changes even though speed may remain uniform.

viscosity—A term describing the stickiness of a fluid and the extent by which a fluid will resist a tendency to flow.

volume—The amount of space occupied by an object or an athlete.

vortex—Also called vortices. Vortexes are whirling, swirling masses of fluid. They occur in liquids (e.g., water) and in gasses (e.g., air).

wake—The track left by the rear or trailing edge of an object as it moves through a fluid. A wake is left by ships, swimmers, and, in air, by airplanes.

wave drag—The drag created by the action of waves at the interface between two fluids. In sport, wave drag occurs where air and water meet.

weight—The force exerted by the earth's attraction on an object or an athlete.

windup—Preliminary motions performed by an athlete that increase the distance and the time period through which force is applied. A "wind-up" is a term commonly used in the description of striking or throwing skills.

work—An expression of mechanical energy. Work is force multiplied by the distance through which an object or an athlete moves. A diver climbing the steps of the tower performs work because the diver's mass is raised a certain distance. No mechanical work is performed in isometric exercises because an athlete applies force against an object but causes no movement to occur.

BIBLIOGRAPHY

Abbott, A., & Wilson, D. (1995). *Human Powered Vehicles.* Champaign, IL: Human Kinetics.

Adrian, M.J., & Cooper, J.M. (1995). *Biomechanics of Human Movement.* Indianapolis: Benchmark.

Armenti, A. (1992). *The Physics of Sports.* New York: Springer Verlag.

Bartlett R. (1997). *Introduction to Sport Biomechanics.* New York: Routledge.

Blanding, S.L., & Monteleone, J.M. (1992). *What Makes a Boomerang Come Back: The Science of Sports.* Stamford, CT: Longmeadow.

Brancazio, P.J. (1984). *Sport Science: Physical Laws and Optimum Performance.* New York: Simon & Schuster.

Dyson, G. (1986). *Mechanics of Athletics* (8th ed.). Kent, England: Hodder and Stoughton.

Epstein, L.C. (1988). *Thinking Physics: Practical Lessons in Critical Thinking.* San Francisco: Insight.

Gonick, L., & Huffmann, A. (1991). *The Cartoon Guide to Physics.* New York: Harper.

Griffing, D.F. (1984). *The Dynamics of Sports: Why That's the Way the Ball Bounces.* Oxford, OH: Dialog.

Hall, S.J. (2003). *Basic Biomechanics* (4th ed.). Boston: McGraw Hill.

Hamill, J., & Knutzen, K. M. (1995). *Biomechanical Basis of Human Movement.* Baltimore: Williams & Wilkins.

Hay, J.G. (1993). *The Biomechanics of Sports Techniques* (4th ed.). Englewood Cliffs, NJ: Prentice Hall.

Hong, Y. (Editor). (2002). *International Research in Sports Biomechanics.* New York: Routledge.

Kent, M. (1996). *The Oxford Dictionary of Sport Science and Sport Medicine.* New York: Oxford University Press.

Knudson D., & Morrison C. (1997). *Qualitative Analysis of Human Movement.* Champaign, IL: Human Kinetics.

Kreighbaum, E., & Barthels, K.M. (1994). *Biomechanics. A Qualitative Approach for Studying Human Movement* (4th ed.). New York: Macmillan.

Luttgens, K., & Hamilton, N. (1997). *Kinesiology: Scientific Basis of Human Motion.* Madison, WI: Brown and Benchmark.

McGinnis, P.M. (1999). *Biomechanics of Sport and Exercise.* Champaign, IL: Human Kinetics.

Nelson, R. (1994). *Sport Mechanics.* Champaign, IL: Human Kinetics.

Nigg, B., MacIntoch, B., & Mester, J. (2000). *Biomechanics and Biology of Movement.* Champaign, IL: Human Kinetics.

Schultz, R. (1992). *Looking Inside Sports Aerodynamics.* Santa Fe, NM: John Muir.

Sports Illustrated. (1993). *1993 Sports Almanac.* Boston: Little, Brown.

Sprunt, K. (1992). *The Mechanics of Sport.* Leeds, UK: NCF.

Walker, J. (1977). *The Flying Circus of Physics—With Answers.* New York: Wiley.

Wallechinsky, D. (1991). *The Complete Book of the Olympics: 1992 Edition.* Boston: Little, Brown.

Watkins, J. (1992). *An Introduction to Mechanics of Human Movement.* New York: Kluwer Academic Publishers.

Zatsiorsky, V. (1997). *Kinematics of Human Motion.* Champaign, IL: Human Kinetics.

INDEX

Note: The italicized *f* following page numbers refers to figures.

ABOUT THE AUTHOR

Gerry Carr, professor of physical education at the University of Victoria in British Columbia, Canada, also teaches sport mechanics to coaches at the national Coaching Institute at the University of Victoria. In his university courses, Dr. Carr found that students responded better when he de-emphasized calculations and taught relationships and concepts rather than formulas. That approach forms the basis of this book.

Dr. Carr is a member of many professional organizations, including the Canadian Coaching Association and the Canadian Society for the History of Sport and Physical Education. He publishes extensively in the area of sport history and sport and politics. Dr. Carr has also written numerous books and articles on teaching and safety in gymnastics and on the fundamentals of track and field.

As a student, Dr. Carr competed for Britain in the Olympics as a discus thrower and earned a scholarship to UCLA, where he represented the Bruins in the throwing events.

Dr. Carr earned his PhD from the University of Stellenbosch, South Africa, in 1974.